Digital Dictionary

Series Editor
Fabrice Papy

Digital Dictionary

Edited by

Marie Cauli
Laurence Favier
Jean-Yves Jeannas

WILEY

First published 2022 in Great Britain and the United States by ISTE Ltd and John Wiley & Sons, Inc.

ISTE Ltd
27-37 St George's Road
London SW19 4EU
UK

www.iste.co.uk

John Wiley & Sons, Inc.
111 River Street
Hoboken, NJ 07030
USA

www.wiley.com

Library of Congress Control Number: 2022932447

British Library Cataloguing-in-Publication Data
A CIP record for this book is available from the British Library
ISBN 978-1-78630-788-0

Contents

Introduction

Marie CAULI[1], Laurence FAVIER[2] and Jean-Yves JEANNAS[3]

[1] Université d'Artois, Arras, France
[2] GERiiCO, Université de Lille, France
[3] AFUL, Université de Lille, France

Why a digital dictionary?

The project of a *Digital Dictionary* is a challenge, as the term "digital" has become omnipresent in our discussions and increasingly indispensable in order to describe the penetration of information technologies in our lives. For example, in French, to the adjective "*numérique*", meaning digital, has been added a noun – "*le numérique*" – which raises questions and reflections: "Dictionaries remain somewhat perplexed by the digital, and their definitions often refer only to the etymological and technical aspect – a sector associated with calculation, with numbers – and above all to devices opposed to the analog" explains Doueihi (2013). However, everyone agrees that "the digital designates something else" (Doueihi 2013). This "something else" refers to a cultural transformation whose importance continues to grow: "A digital man is not simply a man who uses digital tools, but a different man, who functions differently, who has a different relationship with what surrounds him: space, time, memory, knowledge...", explained Vitali-Rosati and Sinatra (2014). The *Digital Dictionary* proposes precisely to illustrate these "differences", starting with just a selection of them, before incorporating all those that are missing and may be added as later versions are released.

Definition of digital

Digital, as a noun, has entered our language to designate a mode of recording sounds, images or videos as an alternative to analog. The compact disk replacing the vinyl record is an example. The progressive digitization of all types of information (text, sound, audiovisual) has gradually established what some authors have called the "global screen" (Lipovetsky and Serroy 2007). The diversity of devices required to produce and read analog information has been replaced by the multiplicity of screens and the "all-screen". However, this overall screen is not yet enough to build "the digital": the Internet is its other constituent. The generalization of access to the Internet at university, at work and then in homes in rich countries has transformed human practices, which now take place in a hybrid world, both physical and digital. For Abramatic (1999), the Internet is the "first means of communication to have been conceived in the digital age". Its specificity is that it is "the first means of communication that combines telecommunications and computers from its conception" (Abramatic 1999). By integrating audiovisual signals, time-dependent signals that had not been taken into account at the origin of the Internet, it becomes the infrastructure of the digital world as a unified phenomenon. Media convergence is the result: printing, radio, television and cinema are no longer different technologies, the universality of digital language is asserted.

Digital as culture

This technical infrastructure, while it makes "digital" possible, does not sum it up. In English, the use of the term *digital* is rarely as a noun. When Brügger (2018) focuses on digitality, it is to show the essential duality to which it refers between a material dimension and signals, digital bits (binary digits), which can convey different layers of texts (those that are directly machine-readable and those that are human-readable). These two aspects are important, according to him, since they establish the framework for how users may interact with the digital medium (the technical device or artifact) and the digital text ("knowledge of both aspects of digitality is important, since each in its ways establishes a framework for how users may interact with the digital medium and the digital text"). What "*le numérique*", in French, introduces is the idea of a new culture rooted in machine-mediated reading, writing and social practices. This is not so much an industrial revolution, as Cardon (2019) notes, as a breakthrough comparable to that of the printing press, for "it is above all a breakthrough in the way our societies produce, share and use knowledge". Just as the printing press was much more than a new mode of production and reproduction of books by becoming the vector of religious (the Reformation), political and economic transformations, digital technology is creating a new world, a new civilization. According to some authors, and we do not yet know where it can lead us.

Digital words

Giving a view of and thinking about such a protean phenomenon by means of lexical "views", classically forming the entries of a dictionary, makes it possible to unravel the tangle of English terms, sometimes Frenchized, that designate half-technical, half-social realities, and which we have difficulty knowing exactly what they mean, even though they are so familiar to us. Neither quite a dictionary, nor really an encyclopedic dictionary, the *Digital Dictionary* highlights the words and things of an unstabilized culture which is being created before our eyes. It would like to contribute in its own way to what some have called "the intelligibility of the digital", because "the digital is a fact that we live with without always being able to understand what it is, or the meaning that should be given to it" (Bachimont and Verlaet 2020; online presentation of the journal *Intelligibilité du numérique*).

The power of usage

This dictionary, which is partial due to the number of subjects it covers and eclectic due to the diversity of the fields addressed, illustrates the extent of the interaction between humans and information technologies, as intended by the promoters of the Internet. Abramatic (1999) explained:

> The design choices of the Internet have pushed intelligence to the extremities of the network (i.e. in the computers that serve as interfaces with users or those that store and deliver information). This feature allows the Internet to take advantage of advances in computer hardware and software in "real time". It is this capacity to evolve that makes it possible to envisage using the Internet for uses that were not taken into account when it was conceived.

Technical choices have made the Internet an application-driven communication medium. The Web (a hypertext publication system operating on the Internet network), alongside electronic messaging and file transfer (FTP), remain the major applications of the Internet from which multiple uses are created, backed up by online services which are the source of wealth for digital companies (notably the famous GAFAM, Google, Apple, Facebook, Amazon and Microsoft), but which also give amateurs a new power. In this respect, the technical choices of the network have made usage a driving force in the technical development of the "information society" alongside technologies and standards (Abramatic 1999). The development of the Web toward Web 2.0, by offering unequalled possibilities of user-generated content, has added to this a questioning of the distinction between amateurs and professionals, multiplying the uses and categories of users involved in the development of the digital world. It is therefore to the Web that we owe, for the moment, the most spectacular development of our digital practices: "The Web, more than the simple presence of computers, has determined a major change in our practices and our relationship to the world. As a result of the omnipresence of the Web in our lives, the digital is everywhere" (Vitali-Rosati and Sinatra 2014).

Facing the "totalizing" digital

Digital technology is everywhere and no one can escape it, for better or for worse. Digital exclusion, as pointed out in several official reports in France (the Court of Auditors' report in 2020 (Cour des Comptes 2020), the Defender of Rights' report in 2019 (Défenseur des droits 2019)), is the corollary of widespread dematerialization, including public services that refer people to the Internet so that they no longer receive users at the counter nor authorize phone calls unless it consists of a connection with a robot. This exclusion can have multiple causes: lack of access to the network (the digital divide), disability, and illiteracy. Added to this is not only the omnipresence but also the fragility of digital devices and tele-procedures (online procedures) which have become compulsory for carrying out the most varied tasks: registering for university, making declarations to the authorities, buying, paying and sometimes even voting... The exploitation of personal data and traces of online activity left unintentionally feed personalized advertising. Violation of computer systems, disabling them and data theft are among the consistent dangers of our digital daily lives. In short, the "totalizing" nature of the digital world and the difficulties inherent in the new world it creates require the pooling of knowledge and its transmission to the large number of people by mediators of all kinds.

In search of inclusion

However, in addition to these difficulties, this digital world is also one that redistributes the capacity for action among users by overcoming the opposition between professionals and amateurs, between experts and non-specialists. Thus, the rise of self-education, on the one hand, and the ease with which applications can be designed, on the other, increase the number of non-passive uses of digital technology, which are not limited to the publication of information in the context of citizen journalism or collaborative encyclopedism, or to contributions to scientific projects in what is also known as "citizen" science. Low-code or even no-code development, which allows applications to be developed with minimal programming effort, enables professionals with little or no computer skills to build digital tools targeted at their activity. The same is true for the construction of "artificial intelligence" applied to a multitude of fields. However, at the same time as a wide and "active" use of digital tools is being deployed by making them accessible to the greatest number of people, the need for expertise is increasing and a specialized digital technology is being invented and perfected to transform medicine and science in general ("e-science", digital humanities), but also hospitals, agriculture and factories (Factory 4.0), a great variety of aspects that this first version of this dictionary is just beginning to cover. Far from being the exclusive technology of a service society, digital technology affects all sectors of activity, while at the same time adding the interconnection of objects to the interconnection of people (Internet of Things). On the one hand, digital technology makes the use of increasingly sophisticated tools commonplace; on the other hand, it reinforces the need for expertise and lifelong learning. It induces a never-ending race to update versions, systems, knowledge, etc. Consequently, its ubiquitous nature, or what we call its "pervasiveness" (from the Latin *pervadere*, meaning "to insinuate, to spread, to invade"), does not only create injustices: it excludes or includes. The *Digital Dictionary* is a tool for inclusion.

A research object for all sciences

While digital technology is both a set of techniques and a culture, is it the subject of a science? In French, the plural is required: digital sciences. Gérard Giraudon defines them as follows: "The digital sciences are fundamentally linked to microelectronics, mathematics, computer science, human-machine interfaces, signal processing (sounds, images, etc.) and their communication (protocol, networks, etc.), as well as to the design of more or less autonomous communicating systems (robotics, personalized assistants, etc.)"[1]. In English, we distinguish the term *informatics* from that of *computer science*, the former having a broader meaning than the latter (see the definition in the *Cambridge Dictionary*[2]). In short, while digital technology does indeed run across all the sciences, it is not itself the object of a single science, unlike computer science. In addition to the digital sciences, there are several initiatives that seek to build a multidisciplinary field that can bring together specialists in information technology and the human sciences. This is the ambition of web science, conceived in 2006 by Tim Berners-Lee. It is backed by the *Web Science Trust*, a non-profit organization whose goal is to support the global

1 INRIA, *Encyclopædia Universalis* [Online]. Available at: https://www.universalis.fr/encyclopedie/inria/ [Accessed March 14, 2021].

2 Available at: dictionary.cambridge.org.

development of web science, which was originally launched in 2006 as a joint effort between MIT and the University of Southampton. There is also a global network of "Internet and Society" research centers (the NOC), including, in France, the CNRS' *Centre Internet et société*, composed of its own "Internet and Society" research unit attached to the *Institut national des sciences humaines et sociales* (INSHS), created in 2019, and the "Internet, AI and Society" research group, created in 2020, bringing together researchers from different disciplinary backgrounds. These examples illustrate the attempts to construct the digital as a multidimensional object.

An alphabetical guide

The *Digital Dictionary* aims to identify the multiple facets in which the multidimensional object that is the digital world is embodied, thanks to the contributions of experts, academics and experienced practitioners. We had to conduct this project in an interdisciplinary manner. This is the significance we give to this work, which brings together contributions from a range of disciplines. It offers an overview of a boundary object and invites a collective intellectual approach. Indeed, there is a large number of works on the digital world, but they are not yet sufficiently interconnected. Thus, this dictionary brings together knowledge that is divided into fields and a specialized readership that still communicate very little with each other. It presents itself as the beginning of a fruitful synergy between the different contributors and their research objects. In this case, if it carries the risk of eclecticism or insufficiency, it is able to identify the areas to be taken up in future versions. This dictionary contains more than 80 entries. The choice to limit the number reminds us of what is the basis of a dictionary: to propose a certain number of notions that can be read separately or in relation to others. Some 50 authors invite us to discover new concepts and insights, be they technological or societal, that they have extracted from their fields of research. The entries differ in style and approach, but reveal the interdisciplinary dimensions of the digital issue. With this alphabetical guide, everyone – trainers, political leaders, associations, students and users – will find a foundation of basic knowledge to answer their curiosity and questions, but also to shed light on their practice, and even to influence political decisions for those who have responsibilities. At a time when the Covid-19 pandemic has further accelerated the digitization of our activities, this dictionary also wishes to offer avenues of reflection that could help us to prepare for a digital citizenship at the same time as a digital citizen, which cannot be reserved for the expertise of a few. In addition to being equipped and connected (which remains an unresolved issue for everyone), everyone must be made aware of and trained in the consequences of these tools and the issues involved. In order to be and remain in tune with the times, to regain control of one's autonomy and freedom, continuous learning and an enlightened attitude have become essential.

Thinking digital, thinking in the digital age

Promoting the development of skills for everyone means looking at access to knowledge and its dissemination in the school system and the public sphere. As far as research is concerned, science is undergoing a global overhaul in terms of the density of production, the contribution from more contributors and the speed of dissemination. It is renewing research practices and intensifying interdisciplinary collaborations in view of the interweaving of technological, educational, social and ethical issues. However, scientific sources are under-exploited and rarely reach the public sphere. They underline the importance of the organization of the knowledge system and the role of researchers in disseminating cross-disciplinary thinking as widely as possible. As far as the educational system is concerned, the teaching of IT and digital culture is encountering a difficult adaptation process within an institution that is constantly questioning what it should transmit and who should transmit it, and remains blocked in its evolution by a form of collective indecision.

In addition to these factors, there are others. These include the use of English, which, in addition to its terminology, leaves its mark on foreign sociocultural functioning and ways of thinking; the simplifying logic of the computer, which shapes our ways of thinking; the linguistic palette, which is insufficient to capture an unprecedented reality, as well as unanticipated categories that limit the formulation of questions; unequal access to equipment and its use; etc. Other problems arise from the difficulty of concretely representing the digital world, which is made up of software based on algorithms, Internet routers, satellite connections, cellular telephones and sensors. All these connected objects, between radiation and matter, mask material infrastructures and geostrategic issues. Moreover, this partial perception is part of a context where magical thinking is developing, flattering the *continuum* of full and immediate access to all our demands and the prospect of a future full of promises. All of this is fed by a daring science-fiction imagination, mobilized around the human–machine relationship and their possible fusion. Science fiction has become the preferred mode for imagining what might happen and for exploring, in a fantasized mode, the trajectories that these changes might follow.

This lack of knowledge spares neither the ordinary person nor the decision-makers involved in the development and implementation of policies, whether they are members of the administration or elected officials, leading them to make decisions based solely on budgetary considerations, or worse, under the aegis of digital communicators rather than experts in the subject.

However, we are at a turning point, both in the history of the digital world, but also in the history of humanity. These issues must not remain in the hands of a few. They must enter the public sphere, and citizens must have sufficient general digital literacy to get the fullest possible idea of the power of the changes and grasp all the issues at stake.

Paradigm shift

Indeed, the absolute interdisciplinarity of digital technology leads to a real paradigm shift in form and content. These transformations are distending boundaries, encouraging decompartmentalization, shifting hierarchies and modifying spatial and temporal representations. Today, information and images are available and instantaneous throughout the world without an intermediary or object (work). They trigger a kind of stupefaction in the face of "infobesity". Space is shrinking and bringing societies closer and closer together, but paradoxically creating distance and exclusion in living together. All areas and all human activities are concerned. New meanings are given to property, intelligence, information (or disinformation), trust, friendship, etc.

The economy

Digital technology is profoundly changing the way we produce, trade and consume, and is reshuffling the cards of entrepreneurship and business models. In tourism, for example, the Internet has become indispensable: no hotel can do without TripAdvisor or Booking.com. Similarly, consumers, who are not well trained or informed about alternatives and issues, can hardly escape the services of Google, Facebook or Apple to get around, communicate, entertain themselves, search for information or order products. This new model, which lowers the marginal cost of goods and services, operates on three main motivations: data, near-zero-cost copying and multitude. It allows for increasing volumes of data, facilitating the management, transmission and processing of increasingly big, rapid and powerful information due to the recent generations of algorithms. This exponential appropriation of information allows a very precise measurement of phenomena and their consequences. It is used in a wide variety of application fields. However, the "digital revolution" is not just about technology, it is inseparable from the liberalization of the telecommunications sector on a global scale, in which the United States, through a few players, has taken a hegemonic position. Moreover, the digital economy refers to the sector shaped by the computer, Internet, audiovisual industries, etc., but also to the induced effects in terms of multiplication of goods and services, dematerialization, profitability and productivity (efficient use of existing infrastructures, such as hotel rooms, professional or personal cars, etc.). The main driver of this metamorphosis, which is disrupting the traditional economy, is the Internet, which is opening up a service economy to a global clientele through large platforms. By integrating a service or a useful skill into the product (the mattress offers quality of sleep), by diversifying the range, it generates a system that assumes that the more applications and users there are, the lower the costs. But while the user accesses a volume of information, often by paying for the services, the operators take over the users' personal data. These new forms of contribution or conditionality make it possible to ultra-individualize the supply–demand relationship. They multiply the products, offers and services that are similar to a luxury that has become "natural". However, at the same time, they capture and direct consumers with profiling techniques. Moreover, the digital economy is one of the leading recruiters in France. It develops new forms of organization that are more horizontal, mobilizing teamwork, multidisciplinarity, shaking up the hierarchy without, however, superseding it, making generations cohabit without establishing parity, activating "collaboration" with partners and subcontractors worldwide. It diversifies and facilitates means of payment, and also develops cryptocurrencies that offer the hypothesis of an alternative to the official state currency (such as bitcoin). It is developing new ways of storing and exchanging value on the Internet without a centralized intermediary, presented as an infallible process for securing and archiving transactions (such as the blockchain).

Work

The digital economy also extends to the workplace. In many cases, developments in professional activities are only presented in terms of benefits. Thus, technical progress has so far affected material production tasks with mechanization and automation, but more and more "intellectual" operations can be carried out by information processing systems (telecoms, hotel business, water distribution, etc.). Other professions are likely to be reconfigured with artificial intelligence (training, law, medical and social services, etc.). Strategic jobs (organization of complexity, IT jobs, jobs in direct contact with users) will increase. Crowdwork (the execution of micro-tasks online and in telework, paid by the unit) is developing, as well as traditional work activities (delivery, cleaning, transport, etc.) passing through digital platforms and often ensuring the majority of the income. This crystallizes the difficulties faced by workers, who are obliged to declare themselves as micro-entrepreneurs, while the employer claims to be a "neutral intermediary", responsible only for putting people in touch with each other and escaping the obligations of labor laws or taxation. Thus, digital technology acts as a catalyst for and amplifies the organizational changes that are already underway: organization, standardization, centralization, chaining and splitting up of tasks, and the rise of networked coordination. For workers, it also increases the risk of permanent presence and connectivity, with a duty to respond immediately. The growth of online and mobile work is multiplying the issues of concern: increased surveillance of workers, more frequent job changes, management of work by algorithms, ergonomic risks caused by human–machine interfaces, psychosocial consequences of the resulting new work organizations, as well as the questioning of labor laws and social protection of workers. Finally, particular attention must be paid to the under-representation of women in this sector in order to avoid a social regression in view of the growth and need for jobs.

Technical changes and scope of application

In everyday life, the trend is to handle more and more equipment, computers, smartphones, tablets and other connected objects. They invite more and more simplification and user-friendliness. Intuitive and more ergonomic, the touch screen is the most successful example. The race for applications with the capacity to provide a granularity of use that is as close to the consumer as possible leads us to imagine a future where connected objects could anticipate our needs before they are even formulated. The idea that they are "smart" is creeping in and is made possible by the miniaturization of devices (privacy recorders, mobile phones, televisions, cars, bicycle sharing systems, alarms, surveillance, bank accounts, etc.). Worse still, the system interfaces themselves involve conditioning the user, even before they have access to the content of the tools.

Moreover, all the sectors in which we are led to evolve or that we approach in our daily or professional life are impacted. In the field of health, powerful algorithmic models are being developed (diagnosis, monitoring of an epidemic), and the same is true in precision agriculture (irrigation, weather), in the field of policing (law enforcement management) or in the military. Their impact raises questions about the profound changes they are causing. The increasing introduction of automatic systems in machines and the systems' levels of autonomy raise questions about the decision-making process and, consequently, the regulatory bodies associated with it. For example, the evolution of combat techniques with machines that are faster than humans and never tire raises the question of delegated decision-making when carrying out certain tasks, the overall control of the maneuver or the possibility for the human being to take over. Art is also a field of application for artificial intelligence. Its ability to analyze and reproduce all the properties of a work raises questions about the boundary between imitation and new works, whose esthetic originality should be legally recognized. New exhibition or collection access devices are reinventing museum visits and tourist activities through virtual or augmented reality, or through mobile applications, experimenting with new cultural concepts. In the same way, the interactions between digital tools and the mental functioning of users (novices or experts) or between intellectual approaches and artificial processes, can lead to profound changes in the behavior of experts in terms of overconfidence, complacency, loss of adaptability and expertise. Another dimension of these tools concerns extensions, psychic and corporal implants that replace or augment deficient human functions. Thus, brain–computer interfaces (BCIs) open up prospects for new forms of interaction that are likely to increase human performance, in particular, the ability to control material assistants or the communication of disabled people. However, unlike cyborg representations, BCIs do not decode intentions or read minds, and several technical, human and ethical challenges remain to be overcome in order to improve their use. Similarly, the robot, being outside of time and outside of conventional social space, is likely to meet with empathy and even adhesion to its requests. By reproducing the natural human state, the robot is not required to follow social conventions. Presented as neutral, it can serve as a reinforcement for disabled people, doubly penalized by social conventions and the gaze of others. It makes us forget that it is indeed a human being who controls the machine.

The uses of digital technology and its effects on the user

Digital technology is transforming human behavior in terms of access to information, consumption and leisure. The change is taking place through uses. These develop through a digital transition process and vary according to gender, status and age, but also the potential offered by the environment. Digital technology is also not self-evident for everyone. Apart from the promises of opening up, substitution and even re-enchantment, there are many risks of vulnerability, fragility and even exclusion. Insofar as it is now a major component of our ecosystem, it poses real challenges in terms of inclusion, which are not without ambiguity between positive effects and risks of cognitive and social division.

Moreover, while the classical concept of identity is based on external characteristics, namely name, date of birth, place of residence, signature and immutable biometric elements, such as eye color and fingerprints, identity on the Internet is first and foremost the result of the digital traces we leave behind: communication traces, location information, proof of consumption, but also forced choices. It also results from the way we present ourselves and the quest for visibility that is accentuated in the interconnected world. This creates a need for visibility that can be satisfied on social networks or on blogs, supported by the "likes" of many "friends". All these elements have an impact on private life, on a career and on relationships, and they can change the relationship with reality and tend to make intimacy and confidentiality disappear. Its amplifying effect increases the vulnerability of the most fragile. Finally, the values of freedom, democracy, justice and trust have been undermined by digital services. Ownership and control of personal information must be guaranteed by law. Similarly, the collection of data should not be possible without the informed consent of users (who often accept the terms of use without reading them, because the text is incomprehensible or deliberately too long), and the means of obtaining this consent should be fair. Anonymity must be preserved. Legislation pays special attention to maintaining and strengthening these, thereby defining the notion of "fairness of an online service". In France, the GDPR (European General Data Protection Regulation) entering into force was a new response in order to slow down the amount of unsolicited information being sent. It inaugurated the set of new legal challenges that the CNIL[3] identified and for which it is a reference.

The future and its risks

The future of the digital age requires genuine reflection, both on the scientific and technical level and on the social and political level. The various implications of digital technology, which are considered to be matters for experts, must be assessed in their proper light, especially as their effects are accelerating and amplifying and as other "revolutions" are underway (artificial intelligence, nanotechnologies, etc.). This is why, faced with the individual, social and political consequences of the development of technologies and their continuous renewal, and faced with the issues surrounding the transformative effects of digital technology, it is essential to acquire the basics of what are now called "digital sciences", to prepare for continuous professional development, to have a global vision of the possible applications of information technology at all levels of society and to regain control of a field that is beyond our control, in order to preserve its benefits without suffering its adverse effects.

The memory of knowledge and the transmission of culture

Digital technology is not just a communication space. It has also become the essential repository of the knowledge produced. Not only are publications of all kinds going digital, but so are the libraries that capitalize on them. Far from being

---

3 The *Commission Nationale de l'Informatique et des Libertés* is an independent French administrative authority. Its objective is to ensure the protection of personal data. It plays a warning role but also advises, controls and sanctions, providing information to the public.

only storage spaces, they are real as well as virtual places where one can find what is no longer available on the cultural market (books, photos, films, music and all cultural products), while discovering new ones. Access systems, whether they be search engines on the Web, content selection mechanisms on television channels or online library catalogs, condition the digital visibility of past and present culture. They are decisive for the transmission of culture and knowledge, today and tomorrow. Access to scientific publications on the Web is, for example, an essential issue for overcoming the health crisis of Covid-19 that we have been experiencing since March 2020. Similarly, the possibility of finding archives and unpublished documents is essential for the training of future generations. The digitization of libraries, as well as the systems of knowledge organization created for the digital world through the various systems of document classification, are areas where commercial logic and cultural logic must not overlap. Digital visibility and, beyond it, referencing and indexing, but also interoperability (the possibility of making different systems communicate), implement mechanisms that reconfigure the traditional logics of cultural transmission and access to information. The way in which digital systems classify and restore information is not free of bias, which can only be highlighted by a multidisciplinary study of these systems from a technical and ethical point of view. Nothing is or will be accessible today or tomorrow if it is not visible in one way or another on the Web. This shows the extent to which the conception of digital technology conditions the construction and memory of human knowledge.

Challenge to society

The first question concerns security. In a society where more and more elements are computerized, security flaws are numerous. In the absence of a response from the public authorities, or its inadequacy, self-education and associative support that is built on the fringes of institutions must provide the means to avoid systems being taken over by various powers. The industries themselves can easily fall prey to cyberattacks (nuclear power plants, hospitals), especially since they do not have the skills or the means to deal with them. These attacks can be dramatic, but they should not obscure the number of lucrative scams that many individuals fall victim to and need to protect themselves from. The arrival of voice search and visual search is gradually becoming more and more commonplace, as the range of smartphones is updated. These new uses will turn into regular uses, then into expectations and, finally, into requirements, foreshadowing a loss of expertise in certain areas. At the same time, however, they are opening up possibilities for surveillance and control. All these small changes and new uses, taken together, mask a hidden power that institutes a form of surveillance culture that could become a way of life. They represent a danger to our freedom. Without being alarmist, a real reflection on the loss or theft of data, as well as on the control of procedures carried out by digital technology, must be carried out in order to regulate them on a political, legal and ethical level. While some companies have become more powerful than states, who will protect the information? How can we prevent its exploitation?

The increase and intensity of risks are also identified around the hegemonic, concentrated and competitive phenomena of powerful economic players (the GAFAMs and companies with the same model). They have established themselves as the world's leading stock market capitalizations and have been able to reorganize the economic field by imposing new models and new types of competition that stifle all forms of genuine opposition. More important than a state, these actors shift the places of power. The same is true of the resources that enable the manufacture of technological materials, which are held by a few countries: China, the United States, Russia and, to a lesser extent, Canada and South America. In addition to the monopoly of material resources, there is the crucial issue of the future and stability of the infrastructures that are held mainly by the United States, which are in the process of being counterbalanced by the rise of China and Russia. This raises major political questions, made more complex by the competition from new, faster entrants. Thus, the virtual space is crossed by crucial geopolitical stakes, bearing the seeds of profound redistributions of geopolitical maps, which constitute factors of tension and are the subject of unprecedented legal battles. Moreover, the strategy of global control has extended, on the one hand, to many states that have set up access, blocking and surveillance procedures, practicing "state censorship" and, on the other hand, to democratic countries that have set up processes that compromise freedom. It calls for new forms of economic regulation and governance that are likely to loosen the grip that providers have on data, making data sharing mandatory with the consent of users or taking charge of new services in common goods. It questions the effectiveness of the rules governing the power of these companies, and, in particular, the current ineffectiveness of sanctions.

Challenge for the state

Looking back in history also makes us aware of the change in the relationship with the state and even the notion of the state, which are being redefined. The state is gradually being transformed into a platform state, a provider of applications, and we can only observe the withdrawal of the public administration from its traditional obligations. This is reflected in the dematerialization of the administrative relationship, and its counterpart in the relationship and support of the citizen. Thus, in the public sector as in the private sector, human relations are being eliminated in favor of a Web that offers standardized assistance, often leaving the user alone, helpless or disenchanted. Finally, social protection makes us aware of the need to rethink the social contract. Originally based on the idea of a citizen that is considered as unique, this principle is now challenged by the obligation of transparency and the desire to predict. For if we know that some people run more health risks than others, will it be possible to maintain equal treatment for all? Digital technology also raises questions about the radical changes in the perspective of the legal world, which should protect citizens. It disrupts the traditional way of interpreting the sources of law, and, in particular, laws, in the face of a predictive justice based on statistics, which puts aside the semantic understanding of texts and the individualized judgment of particular cases.

Emergence of new models

No one can deny the benefits of digital technologies, but new models need to be devised. The main benefit is the development of knowledge and its availability. Everyone can build up expertise without a degree. If digital technology makes people captive, it also allows creativity to be exercised. If it produces abandonment and disinvestment, submission and even addiction, digital technology brings together individuals and not societies. It gives rise to new forms of mobilization and citizen or political organization. The many actors involved, working in networks, articulating themselves around communities, have built up a treasure trove of experience, skills and references from which everyone can draw. They contribute to the transformation of territories and implement a culture of sharing and cooperation that spreads quietly and can be put at the service of democracy. However, the idea of a new age that is assumed by self-organized social dynamics must be refuted and must give way to a new model that combines bottom-up and top-down methods. Moreover, the digital economy fosters innovation. Citizens can now interact directly on a peer-to-peer basis, whether in market exchanges or in fundraising. They can collaborate, co-construct and so on. These fertilizations of collective intelligence represent unprecedented opportunities (collaborative consumption, Wikipedia co-production and knowledge dissemination, communities). In this sense, by fostering social links, this digital model feeds a cooperative mode and reactivates the notion of common goods. Moreover, we often talk about civic and political apathy, public debate undermined by false information, violence or radicalization online, and yet new figures of digital citizenship are emerging. These are expressed in different ways: the sharing of unprecedented information that thwarts censorship, a form of resistance to the generalized connection, and voluntary non-use. In the face of control and censorship, in the face of the exported economic model that has been imposed, new alternative means are developing and instantaneous information channels are changing the means of conducting mass actions. If not, it is still possible to refuse invitations to follow sites or reject useless products and services, to make good use of passwords, to have control over the tools one uses. Free software and licenses have been produced on a voluntary basis, allowing one to run the program as one wishes, to modify the source code to make the program do what one wants, to redistribute them to whomever one wishes. These "free" systems, benefiting from specific licenses, make it possible to avoid control by the developer or the sponsor. Digital technology therefore also provides the means to avoid being completely locked into the considerable monopolies it has created.

Education, citizenship and digital literacy

This is why digital literacy training for all is no longer an option but an obligation. It must encourage us to decipher our technological environment more accurately in the face of the disorientation into which technological progress is leading us, especially as it is ambiguous. As a means of surveillance, it also makes the freedom to express oneself and to create possible for millions of people who were previously silenced. Because digital technology has become a condition for the individual's development, and because it is at the heart of our country's economic, social and cultural future, everyone must equip themselves with a digital kit in order to act consciously in the digital environment, to understand the rudiments of computing and the Internet, to imagine the physical functioning of the computer, to grasp an English-speaking lexicon that escapes them doubly because of the distance of the language and the numerous neologisms, to discern the precautions to be taken to avoid the dispossession of personal data, but also to seize the opportunities. This approach is essential to avoid a simplistic or reductive vision based on clichés and generating defensive reactions, in short, to equip ourselves with the tools available in order to tame the techniques and mobilize them as vectors of collective development. To do this, the digital world, which is both omnipresent and addictive, must be understood in a way that is different from the everyday, familiar objects and services. This presupposes a form of culture that is capable of resisting the dominant strategies and influencing technological decision-making. In this context, schools play an important role in the dissemination of digital culture, even if this objective goes beyond this institution alone. For example, with massive online courses such as MOOCs[4], teaching has been broadened, opening up new ways of learning. The teacher–student relationship can give way to interaction and even intercreation. The teaching of disciplines and their evaluations have been transformed. Digital tools can become a remedy for school failure. However, this necessary digital culture implies a clear understanding of what it covers. Even if the reference points are

---

4 MOOCs: Massive Open Online Courses.

becoming blurred and the levels of expertise are heterogeneous, everyone is concerned: while some older people may be at a loss when it comes to digital technology, others have become very active on these networks and are aware of the issues. As for the digital natives, born with these tools, they handle them intuitively but do not necessarily understand the meaning of transformations and do not necessarily grasp the stakes. Indeed, while we are fascinated by the digital skills of some young people, this does not mean that they have a social and political understanding of it. The use of a technology has consequences for their vision and representation of the world. Moreover, although access and equipment are practically assured, even if some areas are still poorly served, many people have difficulty using the tools efficiently, solving the remaining computer problems (bugs, updates, etc.), entering the online procedures or making a reasoned choice about dubious advertising offers. This energy that is spent on clarifying the obscure or even unintelligible offers of the major operators, or on solving multiple technical problems, results in a general discouragement that leads to blindness with regard to the harmful consequences that must be protected against. This requires an initial, ongoing and permanent educational strategy in the face of the perpetual renewal of digital technologies. It is the first step in creating a collective awareness on this subject and taking advantage of the benefits it can bring. The culture of the free, the commons and collective intelligence should be an integral part of it. However, a victory against the threats that digital technology poses to our societies, for the benefit of a citizen and emancipator digital technology, implies a digital culture with several dimensions: scientific and technical, political, ethical and legal. This is the mission that the authors of this book have set themselves to raise awareness.

References

Abramatic, J.-F. (1999). Rapport de la mission développement technique de l'Internet. Report [Online]. Available at: http://mission-dti.inria.fr/.

Bachimont, B. and Verlaet, L. (2020). Présentation de la revue *Intelligibilité du numérique* [Online]. Available at: http://intelligibilite-numerique.numerev.com/.

Brügger, N. (2018). *The Archived Web: Doing History in the Digital Age*. The MIT Press, Cambridge, MA.

Cardon, D. (2019). *Culture numérique*. Les Presses de Sciences Po, Paris.

Cour des Comptes (2020). Rapport public annuel. Annual report [Online]. Available at: https://www.ccomptes.fr/fr/publications/le-rapport-public-annuel-2020.

Défenseur des droits (2019). Rapport annuel d'activité. Annual report [Online]. Available at: https://www.defenseurdesdroits.fr/fr/rapports-annuels/2020/06/rapport-annuel-dactivite-2019.

Doueihi, M. (2013). *Qu'est-ce que le numérique ?* PUF, Paris.

Giraudon, G. (n.d.). INRIA. *Encyclopædia Universalis* [Online]. Available at: https://www. universalis.fr/encyclopedie/inria/; https://www.universalis.fr/encyclopedie/inria/2-des-sciences-du-numerique-aux-sciences-numeriques/ [Accessed 24 August 2021].

Lipovetsky, G. and Serroy, J. (2007). *L'Écran global : culture-médias et cinéma à l'âge hypermoderne*. Le Seuil, Paris.

Vitali-Rosati, M. and Sinatra, M.E. (eds) (2014). *Pratiques de l'édition numérique*. Presses de l'université de Montréal, Montreal.

A

Accessibility

Nathalie Pinède
MICA, Université Bordeaux Montaigne, Pessac, France

Accessibility is emerging as a major concept with regard to the disability-digital articulation. Accessibility can be understood as a component of "access" (Fougeyrollas *et al.* 2014), even if we often encounter effects of superposition and confusion between the terms "access" and "accessibility". Nevertheless, while there is no consensual definition of accessibility, certain constitutive dimensions allow us to distinguish it from the concept of access, which is more positioned at a political and theoretical level. The geographical aspect, correlated with the notion of distance and the inscription in a space, constitutes a first dimension. But with accessibility, it is also a question of perception by people, particularly in terms of ease of access, which paves the way to a differentiated apprehension depending on the individual. Finally, accessibility is significantly operational in nature and linked to practices, particularly in the form of norms and standards. Thus, accessibility shifts the generic issue of access to a more subjective vision, at the center of which individuals, in all their diversity, must find their place. This is the meaning of the definition of accessibility proposed by the French Interministerial Delegation for Disabled People (*Délégation interministérielle des personnes handicapées*) in 2006, an open definition based on both personal and environmental factors:

> Accessibility enables the autonomy and participation of people with disabilities by reducing or eliminating mismatches between their abilities, needs and wishes on the one hand, and the different physical, organizational and cultural components of their environment on the other. (DIPH 2006)

From a digital perspective, accessibility therefore consists of "providing equal access to physical and digital environments, by offering safe, healthy places and resources adapted to the diversity of people likely to use them" (Folcher and Lompré 2012, pp. 89–90). These lines reflect the shift in responsibility that has marked the change in representations of disability: it is no longer the person who must adapt to their environment, but the environment that must provide the conditions for quality access for all people, regardless of their differences and beyond an approach restricted to disability/disabilities. As such, accessibility is indeed one of the facets of an inclusive society, as a necessary, though insufficient, condition.

The expression "digital accessibility" is not without ambivalence. It can be understood in two ways: accessibility *through* digital technology or accessibility *of* digital technology. The first approach, accessibility *through* digital technology, considers digital technology and information and communication technologies (ICT) as opportunities to access resources or services that provide added value in a situation of impediment or limitation (so-called "enabling technologies"). The second approach, digital accessibility, involves acting so that interfaces, tools and contents can be consulted and manipulated by people with disabilities (visual, auditory, motor, cognitive, for example) and therefore in acting on the obstacles that can make ICTs "disabling" technologies. In this context, digital accessibility mainly concerns websites, digital services of any kind, smartphone applications, digital documents, in connection with software, standards and various media. This includes issues of physical accessibility (perceptible dimension of the website or digital document for all), but also accessibility to content and knowledge. Accessibility therefore implies both access through *the* senses (sensory level), but also *through* meaning (production of meanings).

Of course, these two sides of digital accessibility are neither contradictory nor exclusive of each other. They are symmetrical and representative of a dual vision of accessibility that can be interpreted either from the point of view of opportunities for access and participation via digital technology, or from the point of view of the risks of impediment derived from use (Pinède 2018). On the latter point, digital accessibility has been a right for people with disabilities for the past 20 years or so, on pain of reinforcing exclusions. In this way, regulatory, normative and regulatory frameworks exist at different levels and they are all part of an inclusive political and social will, particularly in connection with the development of e-government and e-services, and the accessibility of web content.

At the international level, the Web Content Accessibility Guidelines (WCAG[1]), proposed by the World Wide Web Consortium (W3C), are an authoritative

---

1 Available at: https://www.w3.org/WAI/intro/wcag.php.

international reference framework setting out the technical methods for distributing Web content according to standards that meet accessibility criteria.

In France, in 2009, the implementing decree of May 14, 2009 of article 47 of the law of February 2005 defined the General Accessibility Guidelines for Public Administrations (*Référentiel d'amélioration générale de l'accessibilité* [RGAA]), based mainly on the WCAG, which in 2019 became the General Accessibility Guidelines (RGAA version 4[2]). The RGAA is both a predominantly technical standard and a methodology for checking compliance with international rules for public communication services. In 2016, the promulgation of Law No. 2016-1321 "for a digital republic" added in Chapter 3, "Access to digital for the vulnerable public", new obligations for administrations and an extension of accessibility obligations for other categories of companies. At the European level, since 2010, these accessibility principles have also been included in digital strategic plans. Directive 2019/882 of 17 April 2019, on accessibility requirements for products and services, is thus a significant step forward, both in terms of the scope of the services concerned and the emphasis placed on the principles of universal design.

This is another decisive aspect of accessibility approaches: the methodological dimension and the use of design methods, such as user-centered design (UCD), universal design (Choi 2005), associated with a consideration of the user experience (UX Design). The strength of these different approaches is to translate this concern for integrating the diversity of user figures and needs into the heart of the implementation process of technical devices or the design of goods and services. Moreover, the field of digital accessibility is one where innovation is very present in order to develop solutions in terms of systems and interfaces adapted to the variety of individual situations and contexts of use. Thus, many technical aids to communication are proposed to alleviate the difficulties of consulting content and browsing on computers or cell phones for people with disabilities, by offering compensation for the various types of sensory disability: interfaces with voice synthesis, virtual keyboards, voice amplifiers, access to the written or oral content of a message, assistance in understanding a written or oral message, etc.

These elements of framing thus show the presence of a consequent arsenal to accompany the accessibility of digital environments. However, several paradoxes can be highlighted. First of all, the reality of the application of these measures in the field remains very contrasted and struggles to impose itself, despite the regulatory obligations. On the other hand, the entry through WCAG and RGAA type standards is a necessary but insufficient one. Beyond access standards, it is also a question of uses and singular experiences of devices and contents in a multidimensional

2 Available at: https://www.numerique.gouv.fr/publications/rgaa-accessibilite/.

environment. Digital accessibility aims to "speak" to as many people as possible, which cannot be achieved without certain forms of translation, or even reduction, of the informational and esthetic richness of the interfaces. Finally, the search for inclusion for all through the prism of accessibility is not free of a paradox, that of an acceptance and legitimization of digital ineluctability, within which recourse to technology to access services and content supported by technology is often imposed.

While accessibility obviously involves regulatory, technical and normative aspects, it also calls the social representations at work into question, for example, in the relationship that system designers and content mediators have with sociotechnical devices. In this respect, information and awareness-raising initiatives are crucial to changing certain prejudices. It is therefore crucial to think about digital accessibility beyond interactions with tools, in an extended and revisited sense from the point of view of the technical, cognitive and symbolic mediations at the heart of digital uses by and for people with disabilities.

References

Choi, S. (2005). Universal design: A practical tool for a diverse future. *International Journal of the Diversity*, 6, 116–124.

Délégation interministérielle aux personnes handicapées (DIPH) (2006). Définition de l'accessibilité : une démarche interministérielle. Document, Ministère de la Santé et des Solidarités, Paris.

Folcher, V. and Lompré, N. (2012). Accessibilité pour et dans l'usage : concevoir des situations d'activité adaptées à tous et à chacun. *Le Travail humain*, 75(1), 89–120.

Fougeyrollas, P., Boucher, N., Fiset, D., Grenier, Y., Noreau, L., Phillibert, M. (2015). Handicap, environnement, participation sociale et droits humains : du concept d'accès à sa mesure. *Développement humain, handicap et changement social, RIPPH*, 5–28.

Pinède, N. (2018). Penser le numérique au prisme des situations de handicap : enjeux et paradoxes de l'accessibilité. *tic&société*, 12(2), 9–43.

Agricultural Robotics

Philippe Le Guern
CRAL-EHESS, Université Rennes 2, Angers, France

Robotization of the labor market and job destruction

During the debates between the candidates of the last presidential election in France, robots suddenly appeared on the public scene, prompting a reflection on the taxation of the surplus value of robotic labor, in order to finance the social protection system of humans. What was at stake? Nothing more and nothing less than the possible substitutability of robots and artificial intelligence for a large number of activities that were previously performed by humans, both active and salaried. Several studies have pointed to the possible disappearance of a significant number of jobs: according to France Stratégie, 3.4 million jobs have been threatened within 10 years in France. For Frey and Osborne, from Oxford University, 47% of American jobs were potentially automatable, whether they were manual or managerial occupations. However, other studies put this catastrophic viewpoint into perspective or contradict it, stressing in particular – from a Schumpeterian perspective – that automation is a factor of competitiveness and therefore of job creation: for the OECD, only 9% of jobs in the United States were facing the possibility of being automated. According to the International Federation of Robots (IFR), the countries with the highest density of robots have the lowest unemployment rates, such as Germany and South Korea.

Alternately considered as a new actor of economic and societal development or, conversely, as a decisive factor of future casualization, the robot questions the paradoxes of advanced capitalism, at the same time as the theory of the end of work, or at least of wage labor. Moreover, the question is not new, since analyses have often associated the rise of capitalism with the development of mechanization: economic thought in particular – Malthus, Ricardo, Sismondi, Marx, etc. – has drawn attention to the links between mechanization and the contraction of employment, the effects associated with the acceleration of the volume and time of production of goods, the deceptive nature of the use of machines to reduce the drudgery of work, the increase in capital without the concomitant distribution of the gains obtained, etc. The questions raised by the industrial revolution and machinism have been taken up and updated by contemporary authors who are committed to examining the effects of technical innovation on social dynamics, employment and the workforce: Jeremy Rifkin, Bernard Stiegler, Bruno Teboul, Paul Jorion or Daniel Cohen, to name but a few, have questioned the validity of the theory of creative destruction in a world governed by robotics and artificial intelligence, the redistribution of machine-generated wealth, the limits of a market incapable of absorbing all the goods produced, mass unemployment, the possibility of political and ethical regulation of the use of machines.

Thus, the link between robots and the world of work raises many questions, particularly ethical, legal and managerial ones: do robots modify the hierarchy of skills and the recognition of qualifications in a given professional field? Do they contribute to the de-skilling or exclusion of certain employees from the labor market, or even to the disappearance of certain professions? Does automation fulfill the promise of a life free of labor constraints or does it, on the contrary, contribute to a deregulation of social norms? What is happening to the notions of wages, calculation of remuneration and investment in the productive forces?

To answer such questions, several field surveys have been conducted: on the place of robots in art (Grimaud and Paré 2011; Pluta 2013; Becker 2015), in the agricultural world (Vereecken 2018; Le Guern 2020), military (Sparrow 2016), medicine (Wannenmacher 2019), precision industry (Perres and Kechichian 2012), etc. They open up a series of questions on safety in cobotics (Sghaier and Charpentier 2012), intellectual property (Larrieu 2013) or the ethical and legal aspects of the use of robots in space (Nevejans 2019), military (Randresta 2013; Daups 2017) and sexual (Danaher and McArthur 2018; Tondu 2020) contexts, to take only these examples.

Agricultural robotics and its environment

In this respect, the agricultural world could be a particularly relevant field of investigation for studying the effects of robots and, more broadly, digital innovations with regard to work situations, insofar as agricultural machinery is constantly being strengthened because of information technology and the use of satellite data. According to Axema, the union of manufacturers in the sector, in Europe, France is the country where investment in agricultural machinery is the highest: from milking cows to picking fruit, agricultural robotics is one of the leading sectors in professional service robotics. A sign of the importance given to this sector is the presence on the market of Japanese, German or American firms established in France (Kubota, Claas, etc.), as well as numerous projects initiated by organizations such as IRSTEA (*Institut national de recherche en sciences et technologies pour l'environnement et l'agriculture*), for example, and competitiveness clusters or agro-equipment manufacturers. All of these innovations are in fact part of a context of accelerated development of uses related to digital technologies in the agricultural sector. As early as 2013, the first significant surveys on the adoption of the Internet and the use of smartphones and computers indicated that 79% of farmers were connected, mainly to check the weather and production prices. GPS, drones, sensors and Big Data are redefining what is known as precision agriculture, where tasks such as recognizing the surface of plots, modulating inputs, guiding farm machinery, weeding, monitoring weather conditions, etc., are controlled by guidance systems coupled with sensors. The increasing number of companies developing new decision support tools for the agro-industrial sector – Weenat, Géosys, Satplan, Airinov, Wanaka, RedBird, Agroptimize, etc. – testifies to the shift from agriculture based on human expertise, acquired through the transmission of inherited knowledge and empiricity, to techno-scientific agriculture: to give an example, we can cite the use of capable drones, thanks to spectrometry and remote sensing, to map each square meter of plot with an accuracy of 5 cm/pixel to give the status of parameters such as nitrogen level, water stress, leaf density, etc., with such tools facilitating the estimation of yield and the optimal time to harvest. At the same time, this movement to produce innovation has been accompanied by a set of measures taken by public policies to support and regulate its uses: in August 2017, the Minister of Agriculture chaired a Higher Council for the Orientation and Coordination of the Agricultural and Food Economy, dedicated to the digital transition. Encouraging innovation, the ministry also organized the *DigitAg Challenge*, for the development of apps to serve farmers, and commissioned a common data portal, AgGate, to collect mass data from the sector in a protected cloud so that it is not pre-empted for the sole benefit of dominant groups (Monsanto, John Deere, Google, etc.).

Typology of agricultural robots and determinants of equipment choices

Ploughing robots, milking robots and harvesting robots are the main automated devices used on farms today. These are supposed to fulfill three main types of functions:

– logistical assistance (harvesting assistance, etc.);

– data acquisition assistance (with on-board sensors on dairy cows, or mapping tools for areas to be treated or irrigated, etc.);

– intervention assistance (i.e. performing precision or strenuous tasks, etc.).

In this respect, it is interesting to observe the "promises" made by equipment manufacturers to users of agricultural robots: in line with the analyses devoted to "technological solutionism" (Morozov 2014; Joly 2015), we can see how robots are supposed to facilitate the daily tasks of farmers by limiting work fatigue, improving productivity, making up for a shortage of manpower, and even helping to take environmental and food health issues into account.

By observing farmers who use milking robots, I was able to study the motivations that guide the choice of such equipment and their impact on the working and living conditions of farmers. Although the majority of respondents emphasized the vocational dimension of the farming profession, the pleasure of being in contact with nature and their strong feeling of autonomy in the running of their farm, the vast majority also emphasized the weight of the constraints they face: long working hours and reduced free time; insufficient remuneration; the weight of administrative and management costs; fluctuating milk prices; daily stand-by duty linked to milking (1,300 hours/person/year on average). Equipping a milking robot, even if it represents a substantial investment – for an average of 120,000 euros, not counting the improvement of the buildings – may therefore appear to be the solution to better manage the farm, in an unstable context (seasonal work peaks, climate change, complex and changing economic and legal environment, etc.) marked by a strong increase in productivity, but without any real increase in the workforce. Equipping ourselves with a robot therefore corresponds less to a desire for the intensification of activity than to the search for a better distribution of tasks, between flexibility of time and new leisure time. Other arguments are also taken into account, relating in no particular order to health concerns, the size of the herd to be managed (which can include more than 100 cows), the feeling of a reduced social life, or the desire to modernize and therefore enhance the image of the farmer, from "bumpkin" to "geek".

"It's not the same job anymore"

It remains to be seen whether the promises offered by the milking robot are actually fulfilled. The farmers I interviewed said that the very nature of their activity had changed by switching to the robot, in short: "It's not the same job anymore". In fact, the automation of many tasks taken over by the robot changed the typical day of the farmer, the twice-daily milking schedule no longer structuring their activity since it was the cows themselves that chose the moments to be milked. Various quantitative surveys confirm this point: according to a study conducted by the University of Leuven in 2004, the reduction in weekly working time after the installation of the robot was 22 hours (Wauters and Mathijs 2004). However, rather than saving time, we should probably talk about reallocating time resources to new tasks. Because of the data provided by the sensors associated with the robot, it is the objective of anticipation that is now put forward, combined with a fine reading of numerous variables (robot attendance for each cow, drop in production, condition of the animal, etc.) that are provided in real time. As one vet said to me, "it's more like continuous monitoring now".

While the robot, coupled with sensors and algorithms, offers fine and immediate knowledge of the status of each animal, we can, on the other hand, wonder about the effects of the knowledge of the herd that is now broadcast by the computer: because what is at stake is not only a form of substitution of empirical knowledge for machine expertise, but also the organization of a growing dependence on mass data and algorithms that accentuate the hold of techno-capitalism in agriculture 3.0. If we look at innovation in this sector, we can see that the issue is no longer simply the ability of robots to centralize milking data and make it available to the farmer, but rather the exploitation of Big Data to organize an economy of services whose level of sophistication is likely to make the farmer totally dependent on the system. Some manufacturers are now developing tools that not only recommend, but also organize the action to be taken (triggering veterinary care, for example) cow by cow from the data collected by the robot, while other applications adapt the quantity or type of food supplied to the cow according to the price of milk or feed. It is therefore not too difficult to imagine how, from upstream to downstream of the livestock production chain, integrated solutions could be offered to the farmer, in a sort of growing techno-economic autarky. I would add that being assisted by a robot and artificial intelligence creates a form of information dependence whose effects are questionable. Thus, many farmers are worried about the "beeping syndrome" induced by untimely automated alerts. Although farmers feel that they have gained time flexibility by freeing themselves from physical constraints, the stress induced by SMS alerts or technical breakdowns can represent a form of psychological strain that leads to the abandonment of automated milking in some cases.

To conclude: techno-capitalism and alienation

What lessons can be drawn from the partnership between the farmer and the robot? In the agricultural world that I have observed, the robot appears above all as an answer to questions as central as the demand for free time and leisure, the possibility of passing on one's farm in the best possible condition, continuing one's activity in spite of impaired health, or coping with the growth of the herd. In this sense, it seems to contribute to a better quality of life for the farmer and to a valorizing reconfiguration of their social image. However, if the limits of this disalienation through automation are apparent, it is the dependence on an increasingly globalized system – of which data and the services that result from it are the cornerstone – that raises questions: supposedly facilitating work, these services are in fact part of an economic logic of services, leaving us to wonder whether they will eventually alienate the farmer.

References

Becker, J. (2015). *Humanoïdes. Expérimentations croisées entre arts et sciences*. Presses Universitaires de Paris Ouest, Nanterre.

Danaher, J. and MacArthur, N. (eds) (2017). *Robot Sex: Social and Ethical Implications*. The MIT Press, Cambridge, MA, and London.

Daups, T. (2017). Le droit des conflits armés oblige légalement à des robots militaires "Ethical by Humanity". *Revue Défense Nationale*, 157–166.

Grimaud, E. and Paré, Z. (2011). *Le jour où les robots mangeront des pommes : conversations avec un Geminoïd*. Éditions Petra, Paris.

Hostiou, N. (2016). Nouvelles organisations de la main-d'œuvre agricole et dans le travail des éleveurs. *Pour*, 3(3), 249–254.

Jarrige, F. (2016). *Technocritiques : du refus des machines à la contestation des technosciences*. La Découverte, Paris.

Larrieu, J. (2013). La propriété intellectuelle et les robots. *Journal International de Bioéthique*, 24, 125–133.

Le Guern, P. (2020). Robots, élevage et techno-capitalisme : une ethnographie du robot de traite. *Réseaux*, 2(2–3), 253–291.

Nevejans, N. (2019). Hommes ou robots dans l'espace. Approches éthique et juridique. *Journal international de bioéthique et d'éthique des sciences*, 30, 135–157.

Perres, C. and Kechichian, A. (2012). L'application robotisée de peintures dans l'industrie automobile : des solutions permettant des économies d'énergie grâce au concept de charge interne. *Annales des Mines – Réalités Industrielles*, 19–23.

Pluta I. (2013). Robots sur scène : (en)jeu du futur. *Jeu : Revue de théâtre*, 149, 145–148.

Randretsa, T. (2013). L'autonomisation des robots sur le champ de bataille. La guerre, le droit et l'éthique. *Revue internationale et stratégique*, 92, 18–27.

Rifkin, J. (1997). *La Fin du travail*. La Découverte, Paris.

Sghaier, A. and Charpentier, P. (2012). La problématique de l'utilisation des robots industriels en matière de sécurité. *Annales des Mines - Réalités Industrielles*, 24–31.

Sparrow, R. (2016). Robots and respect: Assessing the case against autonomous weapon systems. *Ethics & International Affairs*, 30(1), 391–400.

Tondu B. (2020). Les robots sexuels : objets auto-érotiques, fétiches ou nouvelle forme d'objets transitionnels pour adultes ? [Online]. Available at: https://hal.archives-ouvertes.fr/hal-02497286 [Accessed 16 December 2021].

Vereecken, N. (2018). Les abeilles-robots peuvent-elles sauver notre agriculture ? *Pour*, 2(2–3), 117–123.

Wannenmacher, D. (2019). Impacts et enjeux du robot chirurgical dans les blocs opératoires : passer des compétences techniques individuelles aux compétences non-techniques collectives. *Journal de gestion et d'économie de la santé*, 4, 316–334.

Anthropology

Marie Cauli
Université d'Artois, Arras, France

Digital technology: an anthropological revolution?

The transformations impacted by technological developments have generated a plethora of terms to account for the multiple changes linked to the rise of digital techniques, in particular that of "revolution", and more precisely of "anthropological revolution". But we still need to agree on these words. The diversity of names has made a fortune in the digital field more than in any other. Thus, taking the time to clarify, by mobilizing the teachings anthropology, understood here as the study of the human being in general, of its system of values and representations, productions and practices, can be of valuable assistance in understanding the contemporary technological context. All the more so since many classical concepts of anthropology, "augmented" with the digital qualifier, seem to have migrated toward new approaches.

Qualifying and putting into perspective

The major challenge, which is not a simple one, is to try to identify what exactly is being reversed and turned upside down, to find out whether we are witnessing a real disruption, not only technological, but also anthropological and cultural, or whether we are witnessing a transformation of the uses of technical means, as humanity has experienced on several occasions throughout history. In other words, what is the share of permanence and change in the diffusion and appropriation of discontinuity and disruption provoked by the technological wave? Are we dealing with a change of scale or a change of "nature"?

Political revolution

From the outset, there has been much to be said about the meaning of the word revolution. The definition given by the dictionaries themselves is not very clear. However, the word implies the idea of a major change occurring suddenly to move an order of things from one state to another. It emphasizes a historical event that is part of a long process and is eventually triggered. Thus, in political terms, the term revolution evokes the prospect of radical social change and presupposes the intentionality of the actors who bring it about. In this sense, this notion may echo in part the proponents of American counterculture, which presided over the development of computer technology, the watchword being to disrupt the social order by making the potential of networks available to everyone. Now, 40 years later, if we take up Marxist thinking, far from having eliminated class relations and reconfigured the distribution of wealth, a new globalized empire, based on data rather than on productive forces, has imposed itself, developing a "surveillance capitalism" and leaving a power that exceeds that of the states in the hands of a few private actors. However, this movement had the seeds of liberation in it, whether it was the return of decision-making to the grassroots, the place given to collective intelligence, the possibility for the greatest number of people to be producers, etc. However, the political revolution has not had the desired scope. While it accompanied revolts, transforming the content of social struggles, the expected social change was produced by rather unexpected external events, such as Covid-19, which forced us to modify some of our behaviors, probably in an irreversible way.

Technical revolution

As far as the technical revolution is concerned, history shows us that human activities can be transformed by technological accelerations. The term "revolution" has been used to describe the emergence of the printing press or the steam engine. Nevertheless, for these discoveries, as for those of digital technology, it is difficult to determine when and how such a phenomenon occurred and when it would have ended. In fact, like all previous technical developments, the spread of digital technology is the result of long efforts and a continuous process, itself a transition to other ongoing transformations, whose effects are both a stage and a culmination. It is a process marked by inventions and inventors who have made it possible to improve performance and reach decisive milestones because of the convergence of several scientific disciplines and the alignment of instruments and computer programs. Using the term "revolution", we can describe the rapid temporality, to which the recomposition of the whole of the contemporary technical system is added, which has not only integrated the previous levels in order to surpass them, but has also acquired a convergence, a coherence which makes it a systemic and cumulative process. The unprecedented power of these developments is to be linked to the total effects diffusing into all spheres of human existence and activities, in the manner of the total social fact claimed by Marcel Mauss. It is not a new machine, but a "total effect" machine. We keep referring to it because it seems to be the only way to expose the shock effect of its consequences and what it portends for the future. This capacity to be total, conferred by computerization, networks and, now, the interoperability of all connected objects, sets up a new world outside the ground which, since the 1970s, has allowed the older generation to computerize us, then to bring us into the "global village", then into a world of services where we now "naturally" acquire the necessary automatisms. What we experience on a daily basis as micro-changes have generated an end to end, global transformation of the system.

A "total social fact"

Its transformative effects should not be underestimated, especially as they destabilize the matrices in which human symbolic thought is developed. We are witnessing the transformation of economic models, of the contemporary state, of temporal and spatial changes, of a different relationship with time and space. Digital technology is shaping our territories, homes, mobility and personal equipment. Now, and in an unprecedented way, everyone is equipped, and the technical objects that we handle on a daily basis structure the field of our perceptions, practices and representations, and disrupt our values. These objects, from which we are never separated, constitute our existence and condition us on the basis of computer-simulated realities. They act on all the properties of the person – corporal, relational or cognitive – which determine the conditions of their identity. They significantly modify social practices, relations between humans and therefore representations, values and culture in the sense in which anthropology defines them. They transform our experience of the world and our existential sphere, questioning the most secure principles.

Desymbolization

Moreover, digital technology, in conjunction with science, is pushing back the limits of the living world even further. Their alliance has led to advances in major aspects, such as the control of reproduction and heredity, and is now focusing efforts on controlling the nervous system. By reducing the complexity of living phenomena, thanks to logical modeling processes, to automatic operations, they tend to expel the fallible, unpredictable, spontaneous and intrinsic finality of humanity. This is why technological innovations are credited with altering the boundaries between man and machine, freeing humans from their current biological limits, and even ending death. It is in this context that the digital world feeds many fantasies with regard to increasingly spectacular innovations (robots, surgical implants, artificial prostheses) concerning the possibility of improving human capacities, and therefore of pushing back the traditionally established anthropological limits. These beliefs and fears on which the transhumanist ideology is grafted today are not new; they are even constitutive of the history of humanity. But they can be put into perspective by mobilizing a few anthropological concepts, the most enlightening of which are those of "anthropological invariants" and "bumpers of thought" in the sense that Françoise Héritier gave them. These anthropological invariants at the intersection of the biological and the cultural, such as the unavailability of the human body, the uniqueness of Man, the programmed death of humans and the order of generations, constitute the foundations of social life. These incompressible, irreducible facts, which are beyond the control of the will and which humanity has been confronted with (for example, the alternation of day and night), feed a cognitive frame of reference and symbolic structures. Although they are subject to the transformations that social and cultural history imposes on them, they retain their structure. Thus, a good number of questions arise above all from a cultural disruption, from the difficulty of conceptualizing, of separating what is of the order of nature, culture or technology. For technology, by pushing back the limits, without reaching the ultimate point of explanation of the functioning of the, living, produces new "natural" facts and fuels new cultural problems. It is therefore the benchmark that are missing from each new technological development, but also the meaning that we are committed to determining, rather than the phenomenon itself. These displace our certainties and confront us with new ways of thinking that contradict our habits and destabilize what was considered certain. Some people therefore agree that this paradigm shift is an anthropological revolution.

A major bifurcation of our societies

It is, in any case, a major bifurcation of societies. We are therefore called upon to take this technical dimension seriously, which is advancing at a lightning speed and contrasts with the slow pace of change in our mentalities and behavior. In view of the number of machines, communicating objects and connected human beings, the world has been transformed into a vast site of everything connected. Digital technology has created a cultural disruption that reduces humans and living beings to their functions, and for which machines, housing and screens are the intermediaries. It replaces the world with a simplified equivalent where phenomena, freed from their cultural baggage, are reduced to logical processes manipulated by automatic operations and reoriented toward practical purposes. This process, oriented toward improvement, optimization, performance or prediction, advocates values of efficiency with a short time span, whose surging wave should be tempered.

Conclusion

The desire to account for the organization of the world has been the subject of elaborate theories since the dawn of humanity. From all of these questions, myths have proposed a pool of possible answers around biological constraints and conceptual representations. The choice of a new narrative in the face of a maladjusted system where new symbolizations are lacking goes beyond the laboratories. We do not really know how far this "revolution" will take us, even if everyone feels its effects. Anthropology does not dispense with the urgent and joint mobilization of all disciplines and researchers to examine the phenomenon, but it does teach us to stand back when faced with the observations of new phenomena specific to hypermodern societies, which update or reinforce pre-existing logics. In the light of history, it has also shown that individuals are never passive in the bundle of constraints and determinations that are exerted on them. Thus, modern humans can decide what place and what regime they want to give to technology. It is up to them to evaluate where they want to place the cursor of their existence and their relationship to others, to choose the degree of distance that suits them and what potentialities they want to exploit.

References

Babinet, G. (2015). *Big Data, penser l'homme et le monde autrement*. Le Passeur, Paris.

Beckouche, P. (2017). La révolution numérique est-elle un tournant anthropologique ? *Le Débat*, 193, 153–166.

Fouré, L. and Obadia, C. (2009). Entretien avec Françoise Héritier. *Le Philosophe*, 31, 9–25.

Letourneux, F. (2021). Le numérique, révolution anthropologique. *Sciences Humaines*, 1(1), 22.

Art and Robotics

Philippe Le Guern
CRAL-EHESS, Université Rennes 2, Angers, France

Robots and AI, operators of social deconstruction?

Artificial intelligence (AI) and robots are now ubiquitous in most everyday situations, from conversational agents (Velkovska and Relieu 2020) to vacuuming robots to recommendation algorithms, among other applications. The analysis of the content of publications posted on social networks thus involves AI, search engine index mass-produced data and making personalized predictive proposals: remotely controlled military robots are capable of carrying out missions and making strategic decisions relatively autonomously, surgical robots are becoming widespread in the context of precision therapeutic interventions, etc.

A first step was taken in the 1980s and 1990s, when the first generation of robots was transformed from stationary servomechanisms without external sensors or AI to programmable robots controlled by microprocessors and equipped with visual and tactile sensors. Able to recognize and synthesize speech, integrating AI and having navigation systems, a third generation extended the field of action of robots from the 1990s onwards, leading to a fourth generation, marked by two major exploratory fields: the design of animats, that is, artificial systems inspired by biology or botany that develop adaptive capacities in a dynamic, complex and unpredictable environment, and deep learning, which is inspired by the different neuronal layers of the human cortex (Boyer and Farzaneh 2019).

Beyond their technological and ergonomic diversity, these different robots raise a series of anthropological and ethical questions: since they perform tasks by substituting themselves for human action and since they have a degree, albeit relative, of autonomy, who can the responsibility for their actions be attributed to? Moreover, should we – especially in the case where the robot makes a "choice", and not a simple automated and repetitive task – favor the moral value or the utilitarian value of the resulting consequences? Furthermore, how is the robot, capable of increasing productivity and indifferent to social norms in the labor market, likely to change our conception of employment, wage-earning, training, etc.? Finally, by transforming the nature of interpersonal relations between humans and non-humans, does the robot – for instance, sex robots – not open up a new kind of social interaction space, likely to modify our affects, our emotional responses and the processes of subjectivation (Dumouchel and Damiano 2016; Becker 2020)? In other words, if robots are anthropologically revolutionary, it is in the sense that they are likely to act as operators of social deconstruction, leading us to consider norms and values in a new light: what is work, power, responsibility, love in a world where new machine agents deconstruct the division of labor, gendered assignment, etc.? But if they appear revolutionary, it is also in the sense that they question the limits of autonomy and therefore the independence that they could acquire in relation to humans and their control: if it is a given that robots are limited by the nature of the script with which their designers endow them, so that they can prove to be particularly efficient in executing specific tasks, but very limited in situations that are nevertheless considered simple where their lack of common sense is lacking, it is quite difficult to foresee how far AI will be able to go in the more or less near future. This is a question that a rich cultural imagination, with more or less humanoid artifacts, has never ceased to address: from Galatea – a statue brought to life by Aphrodite – to Golem, via the replicants of *Blade Runner*, Goldorak, Fritz Lang's *Metropolis*, *2001, A Space Odyssey* or even *The Wizard of Oz...* .

Art, a world impervious to AI?

If there is a field, a priori unexpected, where robots and AI exert their influence and bring new questions, it is that of the art worlds. To claim that this is "unexpected" is in fact due to the very nature of the art worlds and the conventions they are based on: as art historians or cultural sociologists – Bourdieu, Moulin, Becker, Heinich, Sapiro, etc. – have shown, artistic activities are considered to be the domain of subjectivity, of the exceptionality of talent, of selflessness, as opposed to routine tasks where the participants are interchangeable and where the search for profit seems to be a primary source of motivation. This idealized vision of art is, of course, debatable insofar as it is the product of a social construction corresponding to a moment in history and to the type of worldview associated with it, but it is clear how it is opposed a priori to any idea of automation of tasks or delegation of creativity to machines.

Yet art is not impervious to AI – as the exhibition *Artistes et robots*, inaugurated at the Grand Palais in 2018, attested – and the question is whether advances in robotics are not likely to shake the conventions that underpin our vision of the artist, the work, esthetics and the spectator experience. Of course, the upsurge of robots is not new: the "Meta-Matics" presented by Jean Tinguely at the legendary exhibition *Le Mouvement*, held at the Denise René Gallery in Paris in April 1955, prefigured the invention of artifacts designed to automate the act of painting. These first machines were, however, rudimentary: a pencil attached to the end of a metal rod inscribed circular patterns on a disc covered with a sheet of paper and subjected to a rotating movement. Tinguely's work may well have been a way of pointing out the fetishization of the commodity in painting. The fact remains that this is not a robot in the true sense of the word, but a machine-like device, which contrasts with the computational nature of Paul and e-David, robots designed, respectively, by Patrick Tresset and Oliver Deussen to produce original drawings and portraits. Kac (1997) situates the birth of computational art in the 1960s, citing three major works that he sees as emerging robotic art: Nam June Paik and Shuya Abe's *Robot K-456* in 1964, Tom Shannon's *Squat* in 1966 and Edward Ihnatowicz's *The Senster* in 1969-1970. The example of *K-456* is particularly edifying: the robot wanders through the streets of New York during the second *Annual Avant-Garde Festival* in 1964 and reproduces a speech by J.F. Kennedy while excreting beans. As such, the device questions the robot's autonomy and the nature of interactions with the public. In 1982, this same robot was used again in a retrospective performance in which it was hit by a car, a way of questioning what happens when the machine escapes human control. This way of making the robot the center of the artistic device as much as a focus for political and metaphysical questioning was made particularly evident in the early 1980s with the use of the robot in performance art produced by the Survival Research Laboratory (SRL), founded by Mark Pauline in particular, at the end of the 1970s.

Mimetic art or disruptive art?

In short, robotic art invests in two opposite polarities, each of which in its own way questions the nature of artistic creation: on the one hand, it is a matter of acting "in the manner of" by appropriating the formal rules characteristic of a given artistic movement or author. For example, the IArtist program consists of a type of AI that learns art history: the algorithm can automatically search, via Google Images or Flickr, for images of works belonging to a given style and precisely analyze their characteristics to invent new images. In 2016, in a different vein, a team of researchers associated with Microsoft succeeded in combining visual recognition techniques and deep learning to analyze the pictorial characteristics of Rembrandt's 346 paintings, from the way he applied layers of paint, to the canvas, to the recurring stylistic elements in his portraits. The result is an original 3D-printed portrait called *The Next Rembrandt*. In the field of music, the AI Song Contest, a kind of computerized equivalent of the Eurovision Song Contest, invites researchers to compose melodies that take up the codes of the contest, while the song "Daddy's Car", elaborated by an AI, resembles a work by the Beatles, as much from the point of view of the harmonies, as the rhythms or lyrics. On the other hand, there are creations that no longer reproduce the formal rules of a particular artist, but try to produce something new. But to what extent is AI capable of creating a new work and leaving the mere imitative plane, or, in other words, of generating new esthetic experiences? A work produced by an algorithm, designed by the Obvious collective, *Portrait of Edmond de Bellamy* sold for $432,500 in 2018 at Christie's: the algorithm learned to assimilate the rules of portraiture after "studying" 15,000 portraits painted from the 15th to 20th century. It is true that the signature at the bottom of the painting, designed by AI, seems to authenticate the originality of the work, but can we speak of a stylistic revolution or a decisive breakthrough in art history brought about by the machine? No, probably not. The debate has thus shifted from the question of esthetic originality to that of legal originality. This is how Yanisky-Ravid and Schlackman, American lawyers specializing in copyright, became interested in the case of the *Next Rembrandt*, considering that it was not a pastiche, but an original creation and, consequently, that the laws in force which are the basis of copyright were inadequate for thinking about art designed by AI. Yanisky-Ravid then concluded that AI systems could be viewed as independent contractors or employees, so that the ownership and responsibility for the works were attributable to the human users of these machines.

In conclusion, we can see how art is an exciting field of application for AI creators. Being able to analyze and reproduce all of the stylistic and technical properties of a work or an artist, for example, has reached a fascinating point of completion. But can we talk about "creativity" when it comes to machines? While the elaboration of works based on stereotypical patterns – such as mushy literature – is within the reach of AI, works expressing an original worldview seem to be beyond the reach of artificial creativity for the moment. Although an example of AI made it to the shortlist of a literary competition in Japan – out of 1,450 participants, eleven were robot authors – its lack of understanding of emotions in relation to situations and characters is an undeniable hindrance that does not yet allow it to compete with Gustave Flaubert or Philip Roth.

References

Becker, J. (2020). Concevoir des machines anthropomorphes : ethnographie des pratiques de conception en robotique sociale. *Réseaux*, 2(2–3), 223–251.

Boyer, A. and Farzaneh, F. (2019). Vers une éthique de la robotique: Towards an ethic of robotics. *Question(s) de Management*, 2(2), 67–84.

Dumouchel, P. and Damiano, L. (2016). *Vivre avec les robots : essai sur l'empathie artificielle*. Le Seuil, Paris.

Guesdon, M. and Le Guern, P. (2016). Une voix sans corps pour des corps sans voix. À propos des hologrammes en général et d'Hatsune Miku en particulier. In *Où va la musique ? Numérimorphose et nouvelles experiences d'écoute*, Le Guern, P. (ed.). Presses des Mines, Paris.

Yanisky-Ravid, S. (2017). Generating Rembrandt: Artificial Intelligence, Copyright, and Accountability in the 3A Era – The Human-Like Authors are Already Here – A New Model [Online]. Available at: https://digitalcommons.law.msu.edu/lr/vol2017/iss4/1.

Artificial Intelligence

Jean-Michel Loubes

Institut de Mathématiques, ANITI, Université de Toulouse, France

Artificial intelligence (AI) is a generic term for a wide variety of algorithms that seek to replicate human reasoning. Historically, the computer was programmed to accomplish a predetermined task chosen by the programmer. The algorithm is thus a succession of steps to perform, all previously determined by the person who designed the program. The algorithms resulting from AI try to simulate human intelligence to adapt to the problem encountered. They are two main types.

Formal or symbolic AI is based on the implementation of a large number of rules that the computer will follow. These rules, inspired by biology or the functioning of the brain, will determine the steps that the computer program will follow. This is historically the first direction that has been studied by computer researchers, who have built "expert systems". Nevertheless, the excessive complexity of the rules that are supposed to reproduce human reasoning makes it difficult to build efficient algorithms that can be used in practice. Thus, encoding the functioning of the human mind through a succession of logical rules to reproduce its behavior exactly is such a complicated problem that it has required a paradigm shift. Instead of trying to reproduce it, researchers have tried to simulate it.

This approach is mainly based on machine learning methods. An initially "unintelligent" system acquires knowledge by making connections (learning) between initially unconnected data. In the same way that a human being learns to reason by doing their own experiments or by being guided by someone who wants to teach them, systems based on machine learning need to learn in order to be able to develop human-like decision rules on their own. Thus, the principle of machine learning algorithms is based on the fact that they can develop, from a set of examples, called a learning sample, a decision rule that will apply to all new cases encountered. From a large amount of previously collected and stored data, mainly containing decisions that have already been made and the variables that explain these decisions, the machine will try not only to understand how these decisions were made, but also to identify the rules that govern these choices. In this learning phase, the machine will try to detect characteristic behaviors in the data, trends (patterns or features) in the observations and similarities between characteristics and decisions already made. More precisely, machine learning is divided into unsupervised and supervised learning. In the first case, the computer observes data and seeks to discover similarities between these observations. In this way, it seeks to discover reasons for segmenting individuals into homogeneous groups. Supervised learning consists of presenting the computer with pairs of observations and decisions. The machine will then build relationships that explain why decisions were made by finding rules on all of the examples it observes. In both cases, it is the computer alone, without human intervention, that will discover rules in the data to make sense of them using mathematical principles related to the notions of correlation and similarity, and not on causal reasoning. The most efficient algorithms use deep neural networks to build these rules. Their complexity is such that it is not possible for an observer to clearly understand why one decision was made rather than another. They reach such high levels of performance that, for a large number of tasks, they are not only able to match human experts, but can now surpass them, especially in image analysis.

These algorithms have become widespread throughout society. They guide us in our daily activities (consultation of Internet sites, recommendations on market sites, but also conditioning access to music and culture stored on digital media), and in our relations with banks and insurance companies. They allow us to be better treated (development of diagnostic aids, personalized medicine), to move around (autonomous vehicles) and, soon, all aspects of our lives could be analyzed by these algorithms. Thus, according to our characteristics, our typical profile can be defined in relation to other individuals already analyzed and the algorithm will issue a fixed rule according to our membership group or in relation to similarities (resemblances) with respect to individuals already studied.

The process of discovering typical behaviors is automated, without any post facto control. However, it is from these typical behaviors that models are created, decisions are made and future events are predicted.

We can perfectly see the danger inherent in such algorithms. In a traditional Cartesian reasoning, a theory makes it possible to elaborate a model, fruit of a human reflection. Then this model is confronted with reality through data collected during experiments planned to confront the data with the model. In this way, the theory can be clearly refuted or accepted on the basis of facts. The model can then be analyzed from an ethical or moral point of view, even discussed. But in learning-by-doing, the creation of the model comes from the study of the data, without any post hoc analysis. From the moment we decide to entrust the algorithm with decision-making power, it will shape reality to conform to its model. It "freezes" reality on the basis of what it has seen through the prism of the sample provided for learning and then reproduces the model ad infinitum. Naturally, the model no longer evolves and adjusts reality to its own prediction. As the behavior is learned, the rule of prediction can then be clearly expressed: there is no more room for randomness, only for "repeatability". Often, the confrontation of ideas allows each person to search for the "truth" by becoming aware of their mistakes, even if they knowingly make a wrong choice. AI is otherwise categorical: the algorithmic matrix aims to optimize decisions "justly or coldly". Of course, the morality or fairness of this judgment is not predefined, but depends, on the one hand, on the way the rules are learned (the objective criterion that has been chosen) and, on the other hand, on the way the learning sample has been constituted.

The choice of the mathematical rules used to create the model is crucial. An algorithmic decision is said to be explicable if it is possible to give an explicit account of it based on known data and characteristics of the situation. In other words, if it is possible to relate the values taken by certain variables (the characteristics) to their consequences on the prediction, for example of a score, and thus on the decision. An algorithmic decision is said to be interpretable if it is possible to identify the characteristics or variables that participate most in the decision, or even to quantify their importance. However, in the case of an opaque algorithm, it becomes impossible to simply relate values or characteristics to the outcome of the decision, especially in the case of a nonlinear model or one with interactions. A high value of a variable may lead to a decision in one direction or another depending on the value taken by another unidentifiable variable, or even a complex combination of other variables. An opaque model that cannot be easily explained, for example to a candidate for employment, and which would result in a form of disempowerment of the decision-maker allowing them to hide behind the algorithm.

Thus, the first source of equity failure in machine learning comes from a bad constitution of the learning sample, leading to non-representativeness. Indeed, the whole principle of Big Data is based on the availability of a large number of data allowing precise rules to be extracted. Normally, this would lead us to believe that the entire population generates data that can be used.

In fact, there are communities that are much less represented and minorities that are left out. If the algorithm is wrong about a small number of cases, the mathematical rule may lead to it not trying to improve the accuracy of this minority, because the error committed is less than the error linked to the total variability of the observations. This type of error therefore tends to favor the treatment of people from the majority at the expense of those from a minority. If the accuracy of a method differs for different categories of the population, fairness of treatment is no longer respected. Would we accept that, for HR recruitment, some files are carefully examined, while others are simply drawn at random? Will certain population groups, because they are socially disadvantaged or participate less in the economy than others, be left out of this digital revolution? It would no longer be individuals who would be neglected, but entire groups of the population that would be excluded from the analysis by AI.

The second potential source of discrimination depends on the fundamental role of the relationship between the variables to be predicted and the explanatory variables in the training sample. Indeed, the training sample may, voluntarily or not, present a social bias that may include discrimination against a part of the population. This bias will be transformed into a rule by the algorithm, which will then pass it on. Thus, the rule may depend on variables with a discriminating character and lead to a processing bias depending on these variables.

To remedy this problem, one might think that it would be sufficient to simply prohibit the use of certain characteristics of individuals. However, this argument is clearly invalid, since the principle of learning is the discovery of correlations between data with redundant information, whether or not they are present in the training sample. But decisions based on discriminating variables open the door to unfair decisions. AI optimizes a criterion that is supposed to relate features without being aware of the consequences, focusing only on its mathematical optimality to fit a model to a reality. If the reality bears the seeds of discrimination, AI will not only witness it, but will become its vector by generalizing it to all of the cases studied.

Conclusion

AI researchers have become aware of the dangers it represents. For some years now, research dealing with the explicability of algorithmic decisions and the ethics of AI has been developing, in collaboration with multi-disciplinary teams combining computer science, mathematics, law and human sciences. While it is difficult to define a mathematical concept that measures the fairness of AI, this research aims to be able to measure the biases in the algorithms and thus guarantee equal treatment by the machine. However, we must not forget that humans are not perfect and that their decisions are also biased in many ways. Thus, AI, created by humans and based on past decisions, will not be perfect either. AI must not be allowed to operate without control: decisions must be analyzed without placing blind trust in a digital god created by the human mind and therefore retaining all of its imperfections.

References

Besse, P., Castets-Renard, C., Garivier, A., Loubes, J.M. (2019). Can everyday AI be ethical? Machine learning algorithm fairness. *Statistiques et Société*, 6(3).

Le Cun, Y. (2019). *Quand la machine apprend : la révolution des neurones artificiels et de l'apprentissage profond*. Odile Jacob, Paris.

Nilsson, N.J. (2014). *Principles of Artificial Intelligence*. Morgan Kaufmann, Burlington.

O'Neil, C. (2016). *Weapons of Math Destruction: How Big Data Increases Inequality and Threatens Democracy*. Crown, New York.

Villani, C., Schoenauer, M., Bonnet, Y., Berthet, C., Cornut, A.-C., Levin, F., Rondepierre, B. (2018). Donner un sens à l'intelligence artificielle : pour une stratégie nationale et européenne. Report, Conseil national du numérique.

B

Between Digital Transformation and Cultural Evolution

Ilham Mekrami-Guggenheim
CYBERELLES, Paris, France

Digital transformation

Digital transformation is a term that first appeared in 2014 and has continued to fascinate experts ever since, to the point of becoming a trendy term describing contrasting realities. It has even given rise to professions such as CDO (Chief Digital Officer), leader of digital transformation within an organization.

Digital transformation refers to all the organizational, commercial and technological strategies and actions that a public or private organization must implement in order to cope with the changes brought about by the use and development of digital technology.

Digital transformation, a paradigm shift

In many respects, the science fiction of yesterday has become a reality with the advent of digital technology. Digital technology is not just a communication or marketing channel, it is an industrial revolution that has had a profound impact on the way we live, communicate and work with each other. *Digital technology is the Fourth Industrial Revolution that our humanity is experiencing* (Shwab 2017). The First Industrial Revolution took place in 1784, marking the advent of mechanization with the invention of the steam engine and the exploitation of coal, and bringing about the emergence of new industries. The Second Industrial Revolution, in 1870, introduced the concept of mass production with electricity and assembly lines. The Third Industrial Revolution gave rise to computer science in 1969, with the first programmable computer. The Fourth Industrial Revolution began at the end of the 20th century, with the advent of digital technology.

Three key factors will contribute to making this industrial revolution special: the widespread use of the Internet, social networks and smartphones. Technological disruptions are exponential and their adoption, even standardization, is confusing organizations, governments and companies.

Transforming, adapting and acclimatizing quickly is becoming an imperative in a world that is constantly moving, even changing.

This imperative for digital transformation within organizations gives rise to the concept of digital Darwinism (Goodwin 2018). The digital technology used by start-ups results in the emergence of new business models that destroy what was taken for granted. Organizations, like living species, must evolve, reinvent and transform themselves. Moreover, this change has to happen quickly and agilely.

The two faces of digital transformation

Digital transformation refers to those actions or survival reflexes that organizations put in place to adapt. We are witnessing a technological race with the emergence of a new consumer and a new employee.

The new consumer, better informed thanks to the Internet, now has an unprecedented echo chamber with social networks, and the power of ubiquity with the smartphone. Organizations have had to abandon top-down communications and establish a dialogue where they try to align their actions with their promises. For example, supermarkets have had to display more transparency in processed products as a result of the widespread use of applications that scan the barcode and show the true composition of products and their impact on health.

The new employee is also one of the components that organizations address when implementing digital transformation strategies, through recruitment and retention. The shortage of expert talent in digital fields imposes a new balance of power. Company comparators have emerged, allowing candidates to compare opinions on their future employer. *Collective intelligence is becoming a Holy Grail, and bringing generations and expertise together is becoming a key issue.* To transform, and therefore survive, organizations must get the old and new generations or professions to collaborate. The old professions have a vertical and specialized knowledge of the organization, while the new professions are those that will translate it into this new digital world.

Digital transformation strategies

Digital transformation strategies are generally organized around two main families or workstreams:

– technology: quickly adopts new disruptions, from the simplest to the most complex – from the development of a website adapted to all mobile devices (called responsive) to artificial intelligence;

– cultural: breaks down silos within the organization and fosters a culture that challenges existing processes. We have seen the emergence of the concept of digital acculturation, which goes beyond training because it is iterative and multiform. Collaboration is becoming one of the most powerful levers: collaboration with external partners, direct collaboration between the hierarchy and those in contact with the field and, finally, collaboration between generations and professions.

References

Goodwin, T. (2018). *Digital Darwinism: Survival of the Fittest in the Age of Business*. Kogan Page, London.

Google Trends (n.d.). Volume and Internet search history of the terms "transformation digitale" and "transformation numérique" [Online]. Available at: https://trends.google.com/trends/explore?date=all&geo=FR&q=%2Fg%2F1q6m6j0h0, Transformation%20digitale.

Shwab, K. (2017). *The Fourth Industrial Revolution*. Currency Books, New York.

Blockchain

Jean-Paul Delahaye
Université de Lille, France

Conceived at the end of 2008 by one or more people hiding under the pseudonym Satoshi Nakamoto, the idea of the blockchain only came into existence initially with bitcoin, the cryptocurrency whose capitalization is now (August 2021) worth more than 800 billion euros. It soon became clear that this idea could be applied in a multitude of ways. This makes it difficult to define a blockchain, since it is difficult to know how far the bitcoin blockchain model, which is perfectly precise, can be relaxed. We will try to formulate a general definition anyway.

A blockchain is a computer file copied identically in the memory of each node of a peer-to-peer network, that is, without a central coordinating node. This file is composed of a chain of pages (or blocks) linked linearly to each other by a system of cryptographic traces ensuring its immutability. It evolves only by adding pages, so everything that is written to it remains there indefinitely (we speak of a *register*).

The bitcoin blockchain holds information about all transactions between bitcoin accounts since its inception on January 3, 2009. This account information is used to calculate the balance of each account. It is the trust that bitcoin users have in the indestructibility of the blockchain and in this calculation of each account's balance that ensures that everyone can find the money they keep on the shared file.

For the blockchain to be robust and reliable, the nodes of the peer-to-peer network must monitor each other, controlling what is written to it and conducting checks that generate a carefully maintained general consensus. In addition to the nodes – which we will call "master nodes" – of the network that each hold a copy of the blockchain and monitor its progress, there are simple user nodes for all those who wish to have bitcoin accounts and trust the operation of the network without participating.

In the case of bitcoin, there are several million users and only about 10,000 master nodes each holding and managing a copy of the blockchain. To ensure that there will always be master nodes, the inventors of bitcoin devised a special incentive system in which many variants are possible. In the case of bitcoin, the incentive to be a master node comes from the protocol that periodically allocates a certain amount (decreasing over time) of bitcoins to one of the master nodes. Today, 6.25 bitcoins (about 240,000 euros at the price of August 18, 2021) are thus distributed every 10 minutes to one of the master nodes. These 6.25 bitcoins are created ex nihilo and all the bitcoins that exist were created by this process.

To be the winning node, one must be the first to solve a difficult numerical problem (partially inverting the SHA256 hash function), which is repeated for each new page. This requires some ability to compute the SHA256 function. The rule is simple: the more SHA256-computing capabilities one has, the higher the probability of winning. Users interested in winning have therefore accumulated SHA256 computation capabilities. To avoid having too low a chance of winning, bitcoin users who want to access these bitcoins distributed every 10 minutes form pools. In a pool, everyone calculates SHA256 values to win, but only one actually manages the blockchain. This is called mining the bitcoin blockchain. Those who participate are called miners. There are several tens of thousands of bitcoin miners for about 10,000 master nodes.

This particular mode of operation of the bitcoin blockchain has led to a competition between miners, creating a somewhat absurd situation: today, the bitcoin network calculates 120×10^{18} values of the SHA256 function per second, mostly using specialized chips (ASICs) that can do nothing else. This results in a large amount of electrical energy consumption, equivalent to the electrical output of at least six medium-sized nuclear reactors.

The incentive model for owning and monitoring a blockchain designed by Nakamoto is not absolutely inevitable. Indeed, *private blockchains* have been devised where, unlike the bitcoin blockchain (which is public), only a finite number (e.g. set at the creation of the network) of master nodes can write new pages. They are possibly also the only nodes allowed to access the blockchain for reading. For example, in the case of an association of 50 banks that want to quickly manage exchanges between them using a blockchain, the node that writes the new page every *n* seconds can be chosen at random, or can even be fixed by a periodic rule that is applied once and for all. There is then no SHA256 computation required, no reward to distribute, no unreasonable power expense. The blockchain is updated and monitored by the 50 banks, which have an interest in it because it simplifies their relations and creates trust between them, since the calculations of the account updates are made by all and monitored at every moment collectively.

Other blockchain models are possible, where only some can write and everyone can read. If, for example, schools and universities wanted to collectively make public all the degrees they issue, they could create a dedicated blockchain. Only the schools and universities would write (digitally sign their messages) on a blockchain, with each school or university managing a master node. Everyone would be able to read to find out absolutely secure information about the diplomas issued. Here again, there is no need for the costly incentive system provided for the operation of the bitcoin blockchain: the advantage for each school or university would be the ability to participate in the management and monitoring of this blockchain, which, moreover, would be very modest in volume and monitoring cost, compared to that of a cryptographic currency like bitcoin.

There are now more than 14,000 cryptocurrencies. Some, called stablecoin, are designed so that a common currency equivalent (for example, dollars) is set aside for each unit issued on the blockchain. Unlike bitcoin, the price of stablecoin is perfectly fixed.

Many central banks are considering issuing a cryptographic version of their currency, and some countries, such as China, have already started large-scale experiments. Thousands of blockchains sometimes exist simply for the purpose of sharing information (without associated currencies).

Smart contracts (the most important of which are those of the Ethereum blockchain) are programs deposited on blockchains whose operation is controlled by all the nodes of the network. This ensures the security for executing these programs, opening the door to a multitude of new applications, sometimes called "decentralized applications", because it is the entire network that executes the program, whose results are therefore like bitcoin transactions, validated by all the master nodes of the network. Decentralized finance (DeFi) is based on the possibilities created by blockchains to circulate value on a network (monetary units, securities, various assets) and on the capacity of smart contracts to carry out complex operations on these digital assets in complete security and very quickly.

References

Antonopoulos, A. (2017). *Mastering Bitcoin: Programming the Open Blockchain*. O'Reilly Media, Sebastopol.

Faure-Muntian, V., De Ganay, C., Le Gleut, R. (2018). Les enjeux technologiques des blockchains (chaînes de blocs). Rapport de l'Office parlementaire d'évaluation des choix scientifiques et technologiques [Online]. Available at: http://www.senat.fr/notice-rapport/2017/r17-584-notice.html [Accessed 26 February 2022].

Favier, J., Huguet, B., Bataille, A.T. (2018). *Bitcoin, Métamorphoses : de l'or des fous à l'or numérique ?* Dunod, Paris.

Brain–Computer Interfaces

Frédéric Dehais[1] and Fabien Lotte[2]
[1]*ISAE-SUPAERO, Université de Toulouse, France*
[2]*INRIA Bordeaux Sud-Ouest, Talence, France*

Origins

Brain–computer interfaces (BCIs) have their origins in Hans Berger's seminal work in electrophysiology on brain waves in 1920. However, it was not until the 1960s that direct communication between a brain and a computer was demonstrated. Edmond Dewan trained volunteers to control certain brainwaves in order to manage a light or send Morse code messages by "thinking" (Dewan 1967). If the power of these brainwaves, measured by an electroencephalograph (EEG), increases or decreases, then the computer understands that the subject wishes to turn the light switch on or off, or to generate or not generate a Morse code pulse. A year later, he collaborated with the artist Alvin Lucier, who created a musical performance by making live percussion sounds through his brain activity. At the same time, Joseph Kamiya pioneered neurofeedback, demonstrating that it was possible to regulate one's own brain activity with visual, auditory or tactile feedback and, in turn, improve cognitive performance. In 1973, Jacques Vidal introduced the term BCI, which continues to describe all types of direct communication between a machine and a user, passing only through brain activity, the latter being measured and analyzed by the system.

Recent developments in the field of neurophysiological sensors, artificial intelligence and embedded computing have led to renewed interest in BCI. These neurotechnologies are now expanding in the clinical field (assistive technologies, motor rehabilitation, etc.), in gaming to enhance the gamer's experience and are invading the general public through applications for well-being (meditation experience, sleep improvement). Having become less intrusive, they also offer interesting perspectives for neuroergonomics, through the study of the brain "at work" in the field (Dehais *et al.* 2020). They promote new forms of interaction and promise to improve human performance and its coupling with the machine. Basically, all invasive (i.e. implanted electrodes) and non-invasive brain imaging techniques, such as EEG or even near-infrared spectroscopy, can be used to implement BCI that can be either "active" or "passive" (Clerc *et al.* 2016a, 2016b).

Active BCIs: controlling and acting on your environment

Active BCIs allow a user to interact with an artifact (the cursor of a mouse, for example) by generating a "mental command" that will be interpreted by an algorithm. These BCIs were first developed to provide people with motor disabilities with the ability to control effectors, such as a wheelchair, or to communicate by displaying letters on a screen. At present, BCIs do not "decode" intentions or read an individual's mind. Generally, the user must train themselves to consciously produce "simple" and "clear" brain signals, for example by relaxing, concentrating or performing mental motor imagery (e.g. imagining a hand movement), in order to induce the right command. In parallel, the machine learns to recognize the said commands, during a calibration phase, from examples of brain signals from the BCI user. Once this human–machine co-learning is achieved, it is possible to interact with a system (a robot, an airplane) to direct it mentally. However, current systems are far from perfect and regularly misunderstand the recognized mental commands. Therefore, they cannot necessarily be used for critical control applications. Another so-called "reactive" approach is to flash small checkerboards at a certain frequency (e.g. 6 Hz) on a screen. If the subject focuses on the checkerboard at 6 Hz, then their brain response, measured with the EEG, will increase in that specific frequency and the BCI will be able to decode that attention is being paid to it (Allison *et al.* 2010). Thus, by placing four checkerboards with different frequencies, for example at the top, bottom, right and left of a screen, the user can focus their attention on them to make a robot move forward, backward, turn right or left. The advantage of this technique is that it considerably reduces, or even avoids, the need for a calibration and learning phase. However, the flickering of the stimuli induces eye fatigue in the user and a risk of photosensitive epileptic episodes. In addition, these devices require the user's full visual attention, which is no longer available for anything else, for example, to pay attention to the environment.

Passive BCIs: toward "implicit" interaction

In the case of passive BCIs, the goal is no longer to voluntarily control an effector but to use brain activity to enrich the human–machine interaction in an implicit way (Zander and Kothe 2011). The challenge is to infer the mental states of an operator, such as their level of mental load, fatigue, stress, in order to dynamically adapt the interaction and maintain an optimal engagement. The idea is then to adapt either the interface (e.g. changing the modality of an alarm to make it more salient), or the sharing of task and authority between the human and the systems, or the operator through different stimulation techniques (Dehais *et al.* 2020). Recent demonstrations of such passive BCIs have been made in real operational conditions to infer the state of human operators such as aircraft pilots. Finally, a last use of passive BCIs, particularly interesting for ergonomics, is the objective evaluation of interfaces and user experience. These BCIs can provide real-time indications to objectively quantify the mental effort, cognitive fatigue or emotional dimensions related to the use of an object or an interface. Many manufacturers in the transportation, video game and social network industries are investing in this approach to ensure increased comfort and safety.

BCIs: challenges

BCI is a promising area of research for many usability applications. However, several challenges still need to be addressed to improve their usability. First of all, BCI requires the wearing of solid electrodes whose discomfort limits their use over a long period of time. The development of new hardware (for example, the use of conductive fabrics) may provide a first step toward a solution, but the design of more ergonomic headsets must be considered. Second, most BCIs require a calibration phase where the user is required to produce specific commands at specific times so that the algorithm can learn to recognize them. This step can sometimes be long, tedious and must generally be performed for each new use. New machine learning algorithms have been developed to increase the accuracy of the classification and reduce the duration of the calibration phase. A complementary approach is that of open science (i.e. sharing data on the Internet) to create a large database of neurophysiological data and train the algorithms on more examples. Beyond this formal work, the improvement of the BCI-user relationship seems to be an avenue too often overlooked by designers. Previous works (Allison *et al.* 2010; Jeunet *et al.* 2016) have shown that many individual factors can explain success or failure in mastering a BCI. For example, during the calibration and training phase, the user has little information about their performance (especially when it is poor) and the feedback provided to them is usually not very informative and unimodal (i.e. often visual only). Training tasks are too often repetitive and directed, and do not adapt to the user. This work recommends building a user model that could predict the user's performance according to their profile (personality, skills, etc.) and use it to customize the design of the BCI and the training of the user to control it. Indeed, controlling a BCI is a skill that should be learned, just like learning to ride a bike. At present, however, researchers do not yet know how to make this learning process effective and useful. Finally, the questions of acceptance of these technologies in a professional setting and the related ethical issues should be carefully considered. Thus, ergonomics is a discipline that can not only contribute greatly to the improvement of BCIs, but can also benefit in return from these devices to better understand the relationship between performance, brain and technological environment.

References

Allison, B., Luth, T., Valbuena, D., Teymourian, A., Volosyak, I., Graser, A. (2010). BCI demographics: How many (and what kinds of) people can use an SSVEP BCI? *IEEE Transactions on Neural Systems and Rehabilitation Engineering*, 18(2), 107–116.

Clerc, M., Bougrain, L., Lotte, F. (2016a). *Brain–Computer Interfaces 1: Foundations and Methods*. ISTE Ltd, London, and John Wiley & Sons, New York.

Clerc, M., Bougrain, L., Lotte, F. (2016b). *Brain–Computer Interfaces 2: Technology and Applications*. ISTE Ltd, London, and John Wiley & Sons, New York.

Dehais, F., Lafont, A., Roy, R., Fairclough, S. (2020). A neuroergonomics approach to mental workload, engagement and human performance. *Frontiers in Neuroscience*, 14, 268.

Jeunet, C., Lotte, F., N'Kaoua, B. (2016). Human learning for brain–computer interfaces. In *Brain–Computer Interfaces 1: Foundations and Methods*. ISTE Ltd, London, and John Wiley & Sons, New York.

Zander, T.O. and Kothe, C. (2011). Towards passive brain-computer interfaces: Applying brain-computer interface technology to human-machine systems in general. *Journal of Neural Engineering*, 8(2), 025005.

C

Coding[1]

Nicolas Pettiaux
Collège Saint-Hubert, Auderghem, Belgium

To understand new concepts and words, there is nothing like using dictionaries and encyclopedias, and then referring to the uses. According to the famous French dictionary *Le Robert*, "coding is in computer science the transformation (of data) according to a code". It is short, but coding is also a word that is increasingly fashionable. It refers of course to the source code of software, that is to say the very human words that programmers have written to, once compiled and interpreted by computers, execute the actions we want them to do.

Let us continue with Wikipedia: "The noun encoding and the verb to encode have been attested in computer science since 1969, in the sense of capturing and translating into code simultaneously, and used as antonyms of decoding or to decode". The word coding therefore refers to encoding. The Wikipedia article goes on to provide important definitions and examples, but these are probably not the ones that are in common use today.

In the collaborative dictionary *Wiktionary*, it reads:

– the process of encoding or decoding;

– the process of writing computer software code.

Indeed, this definition facilitates the perception of the different forms of coding usually used by software (color, flashing, etc.).

It is this second meaning that is most used today. Probably in order to make programming more accessible to everyone, rather than leaving it in the hands of computer scientists, the word coding has usurped some of its clear meaning. We therefore refer to coding to say that users write computer programs and, with them, control information processing or robots, and in particular object robots. This is now within the reach of everyone, from elementary school onwards. For example, programming and the associated thought process of putting ideas into successive actions. Preparing a meal and implementing a recipe is already a kind of computer program. So, writing a recipe for cooking, writing the complete procedure for constructing a geometric figure or writing explanations to achieve a clear objective, precisely enough to leave no ambiguity for pupils, children, but also for colleagues in relation to what is expected of them, are forms of programming or coding in everyday contexts.

Today, computers play a major role in day-to-day life. Smartphones are simply very powerful computers that are in our hand. Coffee machines or dishwashers are controlled by computers, which are themselves controlled by computer programs. Our lives are therefore, in one way or another, likely to be controlled by such programs, whether we like it or not.

This calls for a re-examination of the necessary learning of a digital culture, which could be called digital literacy, in which the basics of coding (and therefore programming) should be taught to all children and, while we are at it, adults. Because it allows us to better understand the logic of all of these computer programs that control our lives. Not everyone is destined to become a biologist or a chemist, yet everyone is taught biology and chemistry at school. A citizen must know the basics of these subjects in order to be able to make decisive choices for themselves and for society, to be able to vote as an informed citizen. It is necessary for this to be extended to the digital world, that training in the analysis of the media narrative as well as in coding and programming be part of everyone's training today, perhaps as early as elementary school, for example, in the form of games and to support other training.

References

Wiktionary (n.d.). Coding [Online]. Available at: https://en.wiktionary.org/wiki/coding.

Communication[2]

Serge Tisseron
Université de Paris, France

Digital tools: the presence of the absent

The benefits of digital tools for socialization are obvious. The UNICEF report, "The State of the World's Children 2017: Children in a Digital World"[3], even concludes that smartphones reduce feelings of isolation, foster existing friendships and strengthen the quest to share and seek out others. However, remote communication is fundamentally different from face-to-face communication. This is why, while long-distance relationships lent themselves to lockdown, we must now question their sometimes problematic consequences and train professionals not only in their uses, but also in their limits and pitfalls.

The smartphone as a barrier to bonding

While digital tools can strengthen existing ties, they do little to create new ones. During lockdown, they essentially brought together people sharing the same lifestyle and concerns. As a result, they reinforced strong ties and the withdrawal of each person into a small circle of people who shared their lifestyle and beliefs, with the risk of forgetting those who were different from them and their loved ones.

Furthermore, it has been shown that a parent who uses their smartphone while talking or playing with their child responds with shorter sentences and poorer facial expressions. This reduces the educational support they provide. The child may react to this with a sense of abandonment that disrupts development of a secure attachment (Beamish *et al.* 2018).

Finally, by claiming to constantly inform us about the world, our smartphone invites us to not only be less attentive to smells and sensory stimuli around us, but also to our inner state and environment (Turkle 2011). And the situation is likely to be made worse by the development of talking machines (Tisseron 2020).

Absent bodies, sideways glances

In face-to-face communication, unconscious motor coordinations are established between bodies which result in a greater sense of co-presence and trust (Hall 1966).

---

2 This text is a reworked version of a more complete version of my initial text that appeared in Tisseron and Tordo (2021).

3 Available at: https://www.unicef.org/reports/state-worlds-children-2017.

The exchange of glances also plays an essential role in the construction of an empathic resonance.

On the other hand, with distance, the unconscious motor resonances between the interlocutors are absent and the bond of shared trust is less assured. Moreover, as the interlocutors are not looking at their webcam, but at their screen, each one gives the other the impression of looking elsewhere. This is why communication by webcam can lead to the worry of not feeling listened to, and even less understood. This can lead to the temptation to give a more dramatic expression to one's emotions, or even to exaggerate one's words in order to be sure to be understood (Tisseron 2020b).

The dramas of teleworking

In face-to-face meetings, participants often get together before a meeting to exchange informal comments in a jovial atmosphere. During the meeting, it is possible to have brief exchanges with our colleagues and to ask their opinion on a question or a remark before sharing it with the group. Finally, at the end of the meeting, everyone can get together with a few friends to "debrief" in the moment.

On the other hand, in a remote setting, the arrival of each of the participants is abrupt, with no possibility of interacting without the other people present, either before, during or after. This can lead to a feeling of insecurity. Those who feel sure of themselves are more tempted to attack their interlocutors, because the safeguard represented by the gaze of others does not exist. And, conversely, those who feel fragile cannot find in the gaze of a friend the silent complicity that would allow them to intervene, or even to defend themselves if they were attacked. Finally, when two interlocutors share a common sense of power, the relationship between them can evolve into a confrontation. The result is a greater risk of professional fatigue for all.

These situations can give fragile or vulnerable people the impression that they are being left to manage their tasks alone and they feel guilty for never doing enough. The misunderstandings and conflicts are sometimes so serious that it becomes difficult for some employees to envisage meeting again in person with "colleagues" by whom they have felt mistreated or even persecuted. This increases the risk of burnout.

This is why remote communication is all the more successful when it alternates with face-to-face communication, and when participants, especially those in positions of authority, compensate for the loss of sensoriality by showing more verbal empathy. This is done by avoiding long silences and by bouncing back more often on each other's words to avoid a painful feeling of solitude. As for meetings, the remote part should be reserved for the mutual provision of information, and the face-to-face part for making important decisions.

References

Beamish, N., Fischer, J., Rowe, H. (2018). Parents' use of mobile computing devices, caregiving and the social and emotional development of children: A systematic review of evidence. *Australasian Psychiatry*, 1–12. doi:10.1177/1039856218789764.

Hall, E.T. (1966). *La dimension cachée*. Le Seuil, Paris.

Tisseron, S. (2020a). *L'emprise insidieuse des machines parlantes : plus jamais seul !* Les Liens qui Libèrent, Paris.

Tisseron, S. (2020b). *Facilités et pièges de la communication à distance : les leçons du confinement. COVID-19 : vers un nouveau monde ? Une analyse de la pandémie à travers le regard des sciences sociales et humaines*. MA Éditions, Paris.

Tisseron, S. and Tordo, F. (2021). *Comprendre et soigner l'homme connecté : manuel de cyberpsychologie*. Dunod, Paris.

Turkle, S. (2011). *Seuls ensemble : de plus en plus de technologies, de moins en moins de relations humaines*. L'échappée, Paris.

Community

Edwige Pierot
Aix-Marseille Université, Marseille, France

Community, social classes and social stratification

Historically, the notion of "community" comes from the dissociation made by the sociologist Ferdinand Tönnies from that of "society", but also from research on "social stratification". The concept of "social stratification" is understood with reference to that of "social classes", which establishes different orders of correspondence between social phenomena. The notion was studied in the 19th and 20th centuries, and the relationships between social phenomena are explained from the analysis of inequalities: from an economic point of view by Karl Marx, by including all the components of life with Max Weber, and by interweaving the economic, symbolic and cultural levels with Pierre Bourdieu. While "social classes" are used to describe relations of inequality, these relations must not be reduced to inequalities, because they have multiple scales and cannot be superimposed. Since the second half of the 20th century, the notion of "social classes" has been weakened by economic, social and cultural transformations: the evolution of production relations, the erosion of class cultures combined with the development of mass culture, as well as the non-recognition of social and class movements have diffracted the notion of "social classes", which the notion of network has contributed to weakening. While the notion no longer operates to describe attitudes and behaviors, it is difficult to replace it with an analysis by "social groups", even if the study of the digital practices of the working classes asserts that the appropriation of culture cannot be disarticulated from the social group where it is produced. The dynamic perspective, which combines the disarticulation of class relations and the historical presence of communities, enables us to show that online territories remain marked by social stratification, while movements such as open data and open archives are sources of creation of knowledge qualified as lay forms. This sociological detour informs us about the protean nature of the communities present in technological environments marked by the quantitative and qualitative increase in skills in searching for online information, on the one hand, and the effects of increased access, marking the re-establishment of a symmetry between classes of experts and classes of laypeople, on the other hand.

Communities and the digital environment

The properties of the digital document, the technologies dedicated to its processing and the digital economy are among the components from which digital communities have been able to develop: the transition from the Document Web to the Social Web has initiated the generalization of digital social networks (especially in companies). These new forms of collectives can be linked to a multitude of sociotechnical devices and to the implementation of various types of activities: personal and recreational, economic, administrative, professional or educational.

The term "community" refers to these new spaces of social structuring articulated to networks: cooperation and collective action can be based on technological resources, while interactions are manifested in friendly domains and for activities, be they amateur, professional or political. Common interests, institutions and prior social structures are also essential for the formation of such communities. These communities of the digital environment are distinguished according to specificities from which a typology could be established. This was done by observing the nature of community relations, the level of its organization, the type of use of information and communication technologies (ICTs), the structure of exchanges, the profile of members or the modes of cooperation and belonging. Thus, the digital environment includes virtual communities, communities of practice, learning communities or epistemic communities.

Communities: activities, properties and dynamics

The observation of several types of activities carried out in communities has revealed that they produce information assets. These are goods or services that either carry information or need information to be used. Some assets contain information that allows them to be produced or used immediately (buying advice, product information), while others, such as radical innovations, require acculturation even if they do not contain much information (software, for example). While these two criteria should be distinguished, they frequently overlap. The informational assets thus produced in communities can be put at the service of democratic activities and participate in the functioning of epistemic communities.

These communities can be identified from their purpose, the development of new knowledge and by the epistemic properties they develop while articulated to the uses of ICT. Indeed, the articulation between the modalities (their organization and structuring) and the contents of the exchanges, on the one hand, and the knowledge produced, on the other hand, leads to the shift from a community of experience to an epistemic community. The knowledge produced and manipulated by lay epistemic communities sheds light on the dynamics of activities instrumented by ICT. Furthermore, another consequence of the formation and circulation of knowledge should be highlighted: as it is organized and made public on networks, it becomes an information commons. This information commons is understood as the set of information and tools from which it can be freely shared.

Communities and collective intelligence

Making use of information commons implies thinking about the governance of these assets. Thus, in order for resources and knowledge to participate in activities beyond those of the community that produced them, so that they can be accessible, available and reusable between members of different communities, it is, on the one hand, essential that they be anchored in the principles of linked open data. On the other hand, to envisage the deployment of a collective intelligence, and thus the accessibility, sharing and uses of these resources and knowledge in and between the epistemic communities present on the Web, requires the achievement of semantic interoperability between the knowledge organization systems used by the different communities; semantic interoperability being a condition for access to cultural interoperability. Thus, the communities present in the digital environment and the epistemic properties they contain and deploy make possible the development of the project led by UNESCO for "knowledge societies".

References

Flichy, P. (2010). Les nouvelles formes des collectifs. *Réseaux*, 6(164).

Lévy, P. (1997). *L'intelligence collective : pour une anthropologie du cyberespace.* La Découverte, Paris.

Papy, F. (2016). *Digital Libraries: Interoperability and Uses*. ISTE Ltd, London, and John Wiley & Sons, New York.

Pasquier, D. (2018). Classes populaires en ligne. *Réseaux*, 2(208–209), 9–23.

UNESCO (2005). *Vers les sociétés du savoir*. UNESCO Éditions, Paris [Online]. Available at: http://unesdoc.unesco.org/ark:/48223/pf0000141907 [Accessed 5 May 2020].

Computer

Laurent Bloch
   Institut de l'iconomie, Paris, France

Invention of the computer

John von Neumann's "First draft of a report on the EDVAC" (Neumann 1945), published on June 30, 1945, was to launch the construction of a machine called the EDVAC, which would have been the first true computer. The project was delayed, but the principles of technical organization described in the text, known as the von Neumann architecture, remain those of virtually all computers built to date (2020).

How did von Neumann, a renowned mathematician with a well-established scientific position in several fields, from set theory to probability calculus, become interested in automatic computation? Firstly, von Neumann worked simultaneously with Alan Turing at the Institute for Advanced Study (IAS) in Princeton from 1936 to 1938 and was quite familiar with his work. Later, Goldstine (1972) recounted in his book how, in 1943, while working as a "contingent scientist" in the U.S. Navy on the ENIAC (electronic numerical integrator and computer) project for ballistic calculations of naval guns, he spotted von Neumann on the platform of Aberdeen (Maryland) train station, dared to approach him and told him about his work. Von Neumann was immediately enthusiastic and joined the project.

The ENIAC project was to produce a large computing machine and was started at the University of Pennsylvania in 1943, under the direction of J. Presper Eckert and John W. Mauchly. When completed (by the end of 1945), ENIAC was the largest calculator of its time. However, ENIAC did not meet the commonly accepted definition of a computer: a programmable, automatic, universal machine. Performing a calculation with ENIAC required manual interventions to adapt the machine's configuration, going against the requirement of being automatic and programmable. In fact, programming was done mainly by means of switches and switchboards, as on mechanographical machines. It was in thinking of ways to improve this operation that von Neumann designed his architecture.

In many respects, the designers of ENIAC had come up with the wrong ideas. For example, they had chosen decimal arithmetic operations, which were very cumbersome to implement, whereas binary arithmetic was much more elegant and easier to implement with electronic circuits. Von Neumann had the right ideas: before von Neumann, programming meant turning knobs and plugging pins into switchboards; after von Neumann's work on this, it has meant writing a text; this revolution paved the way to computer science (Goyet 2017).

Later, Eckert and Mauchly accused von Neumann of plundering their ideas, but this thesis does not stand up to a simple reading of the "First draft of a report on the EDVAC". It is fair to say that there is a difference between the EDVAC and the ENIAC of the same order, as between Galileo's telescope and the telescope made earlier by an anonymous Dutchman: Galileo did indeed benefit from the example of his predecessor, but as Koyré (1939) pointed out in a famous study, his telescope was the realization of a scientific theory, whereas his predecessor's instrument was the result of an empirical approach. The von Neumann text is one of the foundations (with the Turing machine) of a new science. In fact, far from being an improvement of ENIAC, EDVAC took the opposite view on several fundamental points and brought entirely new ideas, totally unsuspected by ENIAC's designers.

The construction of the EDVAC was delayed, and the first von Neumann machine was British.

In fact, the first real computers were the Mark 1 at the University of Manchester, built under the direction of Max Newman, a working prototype in 1948, and the EDSAC, built at the University of Cambridge under the direction of Maurice Wilkes in 1949. The principles behind these two machines had undoubtedly been developed by John von Neumann in the United States, the underlying theory was that of the Englishman, Alan Turing, but the achievements were British, hence the temptation of a commemorative American usurpation.

This question of the primacy or not of ENIAC is far from being merely a detail. Depending on how you answer it:

– the computer was invented by Eckert and Mauchly, or by von Neumann;

– the first achievement was American, or British;

– computer science was born from the normal technical evolution of mechano-graphical machines, or from an epistemological breakthrough whose source is in fundamental research in logic;

– computer science is either a clever engineer's tinkering, or a major intellectual breakthrough and a new science.

References

Goldstine, H.H. (1972). *The Computer from Pascal to von Neumann*. Princeton University Press, Princeton.

Goyet, S. (2017). Les interfaces de programmation (API) Web : écriture informatique, industrie du texte et économie des passages [Online]. Available at: https://codesource. hypotheses.org/241 [Accessed 15 January 2022].

Koyré, A. (1939). Galilée et la révolution scientifique du XVIIᵉ siècle. In *Études d'histoire de la pensée scientifique*, Koyré, A. (ed.). Gallimard, Paris.

von Neumann, J. (1945). First draft of a report on the EDVAC. University of Pennsylvania, Philadelphia [Online]. Available at: http://www.virtualtravelog.net/entries/2003-08-TheFirstDraft.pdf [Accessed 15 January 2022].

Computer Science

Jean-Pierre Archambault
EPI, Villejuif, France

Year after year, computer science is increasingly present in the educational system in different forms, modalities and statuses that should not be confused, which is not always obvious in the many debates that have accompanied the development of uses and the place of digital technology and computer science in education for decades. These statuses are complementary. Computer science is a pedagogical tool: it more or less transforms the "essence" of disciplines; it is a working tool for teachers, students and the school community; and it is an element of general school culture in the form of a school discipline as such (see the section "Why teach computer science?" in this dictionary).

An educational tool

The computer is a multifaceted educational tool that can make a significant contribution to improving the quality of education. All this is well known and, for the most part, has been known for a long time. Nevertheless, let us take a look at it.

It lends itself to the creation of "real" communication situations that make sense for students with difficulties. It helps to motivate them. It has given a new lease of life to Freinet pedagogy, particularly with the Minitel in the 1980s, then with the Internet. However, it is not a miracle tool that will finally bring happiness to the world of education. It is imperative that students, at some point, experience the pleasure of learning in order to learn.

It encourages intellectual activity. Indeed, with the computer, we can observe a transposition of the classic behaviors that we observe in the field of the manufacture of material objects. Like a craftsperson who prolongs their efforts until their work is actually finished, a high-school student, who will be satisfied with having solved nine out of 10 questions of their mathematics problem (which is already not so bad!), will work until the program for solving the equation of the second degree that their teacher has asked them to write works, so that they can better understand the notions of unknown, coefficient and parameter.

An example with iconic software is the word processor. Writing is rewriting: a triviality indeed, but a difficult task for teachers when they want students to "revise their copy". Rewriting means rereading. But students may be reluctant to do this, as a few annotations from the teacher are not enough. They often get at best a few spelling and punctuation corrections. Indeed, with a pen and a sheet of paper, moving a word, a sentence, a paragraph, correcting a few mistakes, copying a new version from a draft that has quickly become unreadable because of the many changes... . All this quickly turns out to be tedious and prohibitive if there is not a strong motivation. However, it happens that the students have to convince themselves that they have not maintained the implicit dialogue with a reader (they have killed data), that they have insufficiently differentiated between what they wanted to say and what they have actually written, and that they have misperceived the registers of language. With a word processor, almost everything changes. If repetitions need to be spotted or spoken language needs to be highlighted, the teacher can ask for words to be italicized. Errors, erasures and additions are no longer unacceptable. It is easy to repeat. The cumbersome task of rewriting by hand is avoided. Poor handwriting no longer stands in the way of others reading, illegible writing due to fine motor skill problems is no longer an obstacle. In addition to other tools (dictionary, pen, grammar, etc.), the contribution of the computer is rich and unique.

The computer turns out to be a (necessary?) condition for the existence of intellectual operations, in the sense that it effectively allows them to be carried out, by making them infinitely easier, by removing the "lowly material" constraints. It is as if the tool's impact is all the greater when its effect is insignificant.

A piece of software that magnifies at will the shape of a curve at a given point helps the mathematics teacher to highlight the notion of local flatness contained in the deep structure of the derivation.

Digital technology provides easy access to a mass of documents through portals and search engines. The document occupies a central place in certain disciplines, such as history and geography. The Internet is a favorable context in that it facilitates the location, availability and effective work on various documents.

There is simulation, dynamic phenomena and geometry figures, with the aim of ensuring that students know how to create mental images and see in space. These and other examples show that it is indeed reasonable to think that computers can be used and that they are a potentially effective teaching tool.

The challenge is to use the tool in a reasoned and reasonable way, to do things differently and better. An "all-digital education" approach makes no sense. One of the conditions to be met is the training of teachers. They must have a general computer culture, know the concepts underlying hardware and software in order to know what to expect from them, both generally and specifically in their subject area, in order to guide and support their students, and to set up learning environments and situations based on the didactics of their subject. But the truism is that computers must be in working order at all times. The commonly accepted "norm" for companies is one computer manager for every 100 or 150 computers, depending on the situation. We are still a long way from this in schools. Having recalled these prerequisites, we can think, in all "pedagogical secularity", that on the basis of the teacher's competence, voluntariness and pedagogical freedom, the computer has its place in the teacher's teaching kit.

The "essence" of the disciplines

But it is not only the pedagogical uses of digital technology for teaching. Computer science changes the "essence" of disciplines, that is, their objects of study, methods and tools. This applies to the sciences, the humanities and more or less to all disciplines. For example, can we imagine geography today without geographic information systems (GIS), its mapping and statistical data processing software? Or linguistics without natural language processing? Or astronomy without its computer calculation pipelines? Or genomics without its DNA sequence analysis algorithms? Digitization, databases, hypertext and the Internet are taking over the human sciences. The "four-color" theorem (which is enough to make any map) has been proven by computer. Everywhere, to very different degrees, the teaching of school subjects must take this into account. There is no experimental science without simulation and CAE. So there are other good reasons for having computers in the classroom which are, in fact, requirements for the educational system.

And then there are the technical and vocational courses, where computers are legion and where computing has become commonplace over the last 30 years, taking into account the changes in companies and the economy. The word processor has replaced the typewriter, the database has replaced the cardboard file, CAD software has replaced the drawing board, the numerical control machine has replaced the filing vice, etc. The car industry now recruits more computer specialists than mechanics. We no longer carry out flight tests to manufacture aircraft, we simulate them. All sectors of activity are concerned: surgery, medicine, arts, architecture, law, etc. We need computer specialists, of course – there is a shortage of them – but all staff, whether they use computers on a day-to-day basis or not, must have the general culture that gives them the necessary perspective and efficiency. Beware of the difficulties of dialogue when digital illiteracy is rife. The presence of computer science in the professional component of the education system cannot therefore be discussed. It is an obligation because of the transformations in the economy, companies and administrations.

Finally, let us mention a very edifying example. Michel Vovelle, a historian of the French Revolution, has compiled a considerable amount of data from documents of the period (the *cahiers de doléances*, in particular), from former and present-day historians, and, with the help of the computer, has mapped the immense documentation accumulated. In this computerized study, one can hardly find any trace of the cliché that made the opposition between Paris and the provinces the driving force of the revolutionary dynamism: 1789 pierced the whole kingdom. On the other hand, it is confirmed that the confrontation with Catholicism was indeed constitutive of French political space. The regions draw a very clear national plurality, and the roots of modern political temperaments are to be sought at the heart of the founding or structuring event. The teaching of history is led to show this "intrusion" of the automated statistical tool.

Teaching, cognitive learning, but also human relations

The episode of the "pedagogical continuity" of March, April and May 2020 showed that it was necessary not to forget the fundamentals of the educational act which impose limits on the technique. Teaching obviously includes a cognitive dimension, the appropriation of knowledge, the Pythagoras theorem, conjugation, the agreement of the past participle, but the construction of knowledge by each student is highly favored by relationships with others, the dynamics of the class. Working together is effective and allows for mutual enrichment, cooperation rather than competition. This social dimension is also an apprenticeship of life in society.

One of the reasons given by students for returning to school is to reunite with their friends who they miss. Friends are important in the pleasure of going to school. And we all have in mind those examples of vocations resulting from the courses of the "favorite" teacher that the students had and who knew how to share with them their passion for their discipline. The emotional dimension plays an important role in education.

So there is cognitive intelligence, but also bodily, cultural and social intelligence. It is well known that men and women are "social animals".

"Digital relationships" are far from replacing face-to-face interactions. If remote learning has its obvious virtues for autonomous and already trained adults, or "older" students, it has its limits with age, especially for young children. Beware of the adults we would train with students who are too well defined during their schooling.

Personal and collective work tool

For teachers and students, computer science is a personal and collective work tool. This aspect has entered on a large scale into customs, in particular on a bureaucratic level to prepare a lesson or a presentation. Both teachers and students surf the Internet. They retrieve various documents made with a word processor, a spreadsheet, graphics software, etc. Their research is of course facilitated by the fact that they are able to use the Internet. Their research is of course facilitated by search engines. New forms of cooperative work, pooling of resources and circulation of information are emerging.

The major institutional administrative operations (e.g. preparation for the start of the school year, organization and implementation of examinations and competitions) and communication with families cannot do without computers.

Computer science as a school subject

The teaching of computer science for all students, in the form of a school subject in its own right, is gradually being introduced. It is a question of general culture: giving all students the general culture of our time. Please refer to the section "Why computer science education?" in this dictionary.

References

Archambault, J.-P. (2000). Anatoly Karpov et la pédagogie : rencontre et échange avec le champion du monde sur les échecs et leurs relations avec la pédagogie, l'informatique, Internet [Online]. Available at: https://www.epi.asso.fr/revue/articles/a1410b.htm#BPAGE.

Archambault, J.-P. (2013). Ressources, instruments, ouverture des formations nécessairement diversifiées [Online]. Available at: https://www.epi.asso.fr/revue/articles/a1305b.htm.

Archambault, J.-P. (2015). Rapport OCDE et enquête PISA : pourquoi l'informatique à l'école ? [Online]. Available at: https://www.epi.asso.fr/revue/articles/a1510f.htm.

Archambault, J.-P. (2020). L'école de demain s'appuiera largement sur de l'enseignement à distance, selon Jean-Michel Blanquer : attention danger ! [Online]. Available at: https://www.epi.asso.fr/revue/articles/a2006a.htm.

Computer Security

Gérard Berry
Collège de France, Aigaliers, France

Security, computer attacks, viruses, secure networks and electronic voting

Definition

Computer security, or cybersecurity is about maintaining, in the face of malicious attacks, the availability and integrity of computer equipment, the computer data they produce, store and process, the confidentiality of access to these data, as well as maintaining the behavior of cyber-physical systems, that is, hardware systems monitored or driven by specific software which themselves use predefined data or data captured in real time. The subject is becoming critical because of the explosion in the number and strength of financially motivated attacks: by 2021, their direct financial return was estimated to be several times that of drug trafficking, and it is only growing. Attacks can also have more strategic targets, such as those on power grids or other critical military and industrial systems, launched by states; this is known as cyberwarfare.

The populations concerned

There are three distinct populations to consider: attackers, victims and defenders. The image of lone geeks attacking the Pentagon is obsolete, if ever it existed. Most of today's attackers are either mafias or states, which often hire highly skilled specialist organizations to carry them out. The victims are varied: individuals, industries, public services, administrations, armies, etc. The defenders are the IT departments of the organizations and too rarely the users themselves. The defense industry is developing and state bodies are dedicated to it in France (*Agence nationale de la sécurité des systèmes d'information* (ANSSI), the National Agency for Information Systems Security).

Types of attacks

Cyberattacks are highly technical and can be targeted or generic. They are often built from attack kits, available on the encrypted Darknet. They remain difficult to detect and stop in time, and it is often impossible to identify the perpetrators and sponsors, let alone gather real evidence.

They usually start by penetrating a computer, phone or other computerized object, either to target it itself or to use it as a base to further penetrate a network. They can be initiated by cracking a weak password, by a user clicking on a tempting link (phishing), by corrupting a USB key provided to a user (the Stuxnet virus attacking Iranian nuclear centrifuges in 2010), etc. They continue with the injection of malicious programs (malware), of which there are already nearly a hundred thousand: Trojans, which provide permanent points of entry, creation of hidden channels for the theft of data, information, or viruses that will reproduce throughout the network. They enable theft, destruction or modification of data, program corruption, complete encryption of computers (ransomware), takeover of physical systems, etc. It often takes only a few tiny holes to take control of a machine and a network.

A common point of attack is the detection of a user's weak password. A good password is a long string of random characters, but it is impossible to remember. Users therefore often use few that they can remember, and often repeat them on multiple sites – a joy for attackers. Rules about the presence of special characters are of little use and often differ from site to site. The best solution is to use a password vault: it asks to know only one password, the one that opens the vault, and takes care of generating and storing the others without you even needing to know them, with automatic connection to sites.

Worse, connected objects often come with admin/admin as user/password, which almost nobody changes. Video cameras and other infected objects have been used to launch attacks on Dyn, an Internet name server, preventing it from performing its essential function of translating domain names into IP addresses (DDoS attacks have already affected entire countries). The malware used was Mirai, originally designed by teenagers to attack the game *Minecraft*.

Another major point of attack is due to the many bugs that still exist in operating systems or other software. When properly exploited, the frequent memory corruption bugs (writing to the wrong place) are formidable. They are often corrected in system update reports.

Other sources of attacks come from obsolete ciphers that are left in place because they are used by machines whose programming cannot be changed, or subtle errors in encryption key exchange protocols like SSL (secure socket layer) for https (see "FREAK and Logjam attacks" on Wikipedia). Users can also be tracked by "trackers" downloaded into browsers to follow what they do on the network, corrupt web pages to detect passwords as they are typed, etc. The list of possibilities is growing all the time.

Defense

Organizing the defense is not simple, and it cannot be absolute because the attackers' opportunities are not known in advance. It is necessary to systematically use strong passwords stored in a good safe, and to keep all software up to date. Outdated systems whose security is no longer maintained, such as Windows XP, which still leads to devastating attacks against many hospitals, administrations and systems (WannaCry, Petya, NotPetya viruses, etc.), must be abandoned; this is not easy as XP and many others are often at the heart of connected machines that cannot be easily modified. Finally, operating system and application manufacturers need to continuously strengthen their software, which is difficult because they were often not designed with maximum security in mind. Even worse, many of the software programs still in use are no longer maintained or, worse still, their manufacturers have disappeared. Knowing what to fix and how to fix it is often a key issue in the industry.

In organizations, strong rules must be imposed and automatic systems must be built to detect attacks by analyzing network traffic; machine learning techniques are beginning to play a major role. The source of the attacks must be sought and, if possible, the attacking networks dismantled, which requires strong international coordination. Defense is now being organized in France, with the creation of a specific industry and the actions of the ANSSI (which publishes useful documents that can be read by everyone) and the CNIL.

Scientific research

Research in cryptology plays a key role. It covers three areas: physical communication, cryptography and security protocols.

Physical communication is currently introducing quantum entanglement transmission techniques, where an attacker can only intercept the transmitted bits by destroying them, which becomes useless. However, this does not exclude the possibility of cancelling all propagation of information from one speaker to the other.

Based on combinatorial mathematics and number theory, cryptography defines and analyzes encryption techniques to make them secure and efficient. It is often based more on difficult conjectures than on proven certainties. The current challenge is to replace existing ciphers, which could be broken by a future large quantum computer (NASEM 2019). Post-quantum ciphers are being studied and standardized.

On a complementary level, security protocols define the set of operations to be carried out by the parties in order to communicate, for example, to exchange encryption keys or vote using a secure electronic system. The mathematical approach is also essential, especially for the analysis and verification of operating system kernels, security protocols or electronic voting systems that can be used in trusted environments (Cortier *et al.* 2018). But meeting all the desired guarantees for large-scale e-voting is not yet achievable and may never be.

Training

A major systemic weakness remains the widespread lack of education about information technology and its security, especially among the public, managers, industry players, doctors, lawyers and politicians, who are too easily susceptible to malicious actors. Although specialized cybersecurity training is being organized, it will take time to solve this problem.

References

Cortier, V., Dragan, C.C., Dupressoir, F., Warinschi, B. (2018). Machine-checked proofs for electronic voting: Privacy and verifiability for Belenios. *Proceedings of the 31st Computer Security Foundations Symposium* (CSF), 298–312. Oxford, UK.

NASEM (2019). Quantum computing: Progress and prospects [Online]. Available at: https://www.nap.edu/catalog/25196/quantum-computing-progress-and-prospects.

Contributory Economy

François Elie
ADULLACT, Montpellier, France

The terms "contributory economy" or "contribution economy" appeared around 2010 and were added to the vocabulary of alternatives or changes in production (co-opetition) or products (social and solidarity economy, circular economy, economy of functionalities). The term is new, but the concept is not. The free software sphere was among the first theaters of these transformations (as was also the case for the debates on the commons and the practices of agile methods).

Shaking up the capital-labor model, the *collaborative economy* uses individualized means of production (Uber, Airbnb). While the *social and solidarity economy* socializes the ends, the *contributory economy* dreams of a model that guarantees that volunteers will not be exploited and contributions will be paid at their fair value.

The contributory economy is characterized by original relationships between actors, a particular interest oriented toward a common good and a horizontal mode of production, or at least one that does not respect the usual symbolic powers (hierarchies are only meritocratic but change according to the shifting skills of the actors involved).

At the frontier of third-place practices, Amartya Sen's capabilities and the criticisms of capitalism, the contributory economy seems to describe this utopia where the individuals contribute to the group *according to their means.*

But this utopia does not come from nowhere! The mode of production of free software allows us to understand the articulation between the different aspects. Eric S. Raymond, in *The Cathedral and the Bazaar*, described a collaborative mode of production that, surprisingly, produces as well as the cathedral or monastic mode of the original developers of Emacs or TeX, and better than the *bunker mode* of the companies that produce proprietary software (the "Halloween" reports admit it). Also in 1999, Eric S. Raymond, in *The Magic Cauldron*, observed that it is in the interest of producers to share! The contributory economy, while sometimes described as producing values that cannot be monetized, is in fact a mode of production with very large *positive externalities*. Instead of investing in producing a good protected by copyright or patents, and aimed at a return, actors seem to have more interest in adding value to a common good that they can share with all because they are the authors (the famous *copyleft*), guaranteeing it security (peer review), durability, possible quick success, interoperability, etc. Next to proprietary software, which remains in the world of things thanks to the law, free software, which takes advantage of the character of *non-rival objects* of what is digital, is the matrix, the model or the best example of the contributory economy. By describing the different stages of its development, which successively involves three communities, we can understand the ingredients of this complex concept.

Identity of the producer and consumer

In academic communities (or in local communities) there are strong traditional solidarities that make sharing natural in order to avoid reinventing the wheel. If we take the example of the Internet (which Bernard Lang said "liberated software"), researchers naturally shared TCP/IP, POSIX, C, etc. The lower (network) layers of computing occupy a population of *sysops* and *adminsys* who collectively produce and share the tools they implement.

This way of producing by working together is not new; it is the way of science. For 25 centuries, the civilization of open knowledge, that of libraries and universities, has produced knowledge by sharing it and for sharing it. The four freedoms of free software are the same as those we have for mathematics.

Strictly speaking, *free software* is the world of benevolent hackers who are outside the commercial one. The computer scientists of this world produce their tools *among themselves*; these tools seem to be outside the commercial sphere.

But the computer will spread rapidly and will tend to separate the producer and the consumer. *In-house* developments will give way to *software packages.* Software publishers will offer their "contribution" to users who will reinvent the wheel: they will mutualize through the offer. If the age of microcomputing was still the age of code sharing, the world is going to change quickly: we are starting to sell software in boxes.

Co-operation between actors

The competitive world of industry and marketing will soon be contaminated by the effectiveness of the contributory model. People are discovering that it is more efficient to develop together, while maintaining competition. The Apache Foundation and OW2 are examples of such consortia that bring together competitors to save on research and development costs, to avoid reinventing the wheel... by working together!

At the same time, the competition divides the world of free software that dreamed of unity: one thinks of the multiple Linux distributions and the graphic universes (GNOME, KDE, Xfce, etc.).

The *open source* movement paradoxically instills an economy of contribution between competitors! On the middleware layer, they develop software bricks, libraries (*bookstores*) that are very open, like duelists who make sure that they will fight with the same weapons.

These bricks are used to build software solutions that are sold to users, as if they were still things.

So things get complicated and two communities develop software: computer scientists (academics, amateurs or freelancers), for fun, as we do mathematics, and others, computer scientists, within companies, sometimes on the same strains of code. We value the work of others. Non-market sharing becomes a commodity. Bees produce honey, and beekeepers sell it. In the contributory economy, a *conflict of interests and values* arises (which can be observed in third places). Is crowdsourcing theft?

The users, on their side, gathered in *user clubs*, are not content to advertise in spite of themselves or voluntarily on the software they use: they try as much as possible to influence the evolution (the roadmap) of the software, often in vain. They wipe the slate clean, wait for bug fixes and continue to pay for future releases. Their willingness to contribute is useless if it does not find an echo.

Demand-driven pooling

A new movement is taking shape: users who pay, primarily in the public sphere, will try to regain control of the software they pay for, upstream. Through subscription-based and/or iterative financing, they buy development and share by putting the core software they create under free licenses. This time, it concerns mainly business software.

The circle is complete: the contribution has successively affected the producer-users, the grouped producers who mutualize through supply and, finally, the grouped users who start to mutualize through demand.

Three communities

The history of free software, successively liberating the lower layers, middleware and business software, allows us to distinguish three communities, three circles of contribution that are linked like Borromean rings.

The logic of the contributory economy is the articulation between these three contributions. A free contribution, but one that sometimes needs to be *paid for* (bug bounty platforms), a contribution between professionals who produce in order to take advantage of co-opetition in the commercial world and, finally, the financial contribution of those who have an interest in the development of the commons. We can recognize the ingredients of third places: makers who are partly volunteers, companies that seek to take advantage of cross-fertilization, and those who have an interest in the development of the commons. We recognize the ingredients of third places: makers who are partly volunteers, companies that try to take advantage of cross-fertilization, institutions and communities that finance and try to orient in the direction of social utility. We can recognize the idea of *forges as marketplaces*, bringing together all those who have different but converging interests around digital production.

Borromean rings

Related aspects would need to be examined:

– upstream investment in individuals through a real contributory income and/or early teaching of code (or even low code in the hope of involving more people);

– forms of cooperative models that avoid the temptations of investment in the product;

– modes of governance (mutualist models that guarantee a risk, etc.) that encourage the emergence of the commons while avoiding the temptation to simply *value* them.

The fact that any digital transformation quickly becomes raw material raises the problem of a contribution that is not really conscious: for example, the traces of our "free" interactions on social networks are monetized ("you are the product").

Economy of "goodwill" contributions

It is perhaps finally this voluntary, lucid and non-alienated character of contribution that is the essence of the contributory economy and that leads each person to wonder where they are, what they can contribute, by asking, when they work for themselves, what difference this makes for others. It can help for producing, correcting, translating, financing, documenting, convincing and organizing, while being attentive to the justice of its retribution and its interest, no doubt, but also and just as much to the positive externalities. It can also contribute to understanding.

[…] it could be one of the greatest missed opportunities of our times – if free software liberated nothing but code (Hannu Puttonen's 2002 documentary on free software, entitled *The Code,* closes with this statement).

References

Benayer, L.-D. (2014). Open models : les business models de l'économie ouverte [Online]. Available at: https://bzg.fr/img/Open%20Models,%20les%20business%20models%20de%20l%27e%CC%81conomie%20ouverte.pdf.

Brandenburger, A.M. and Nalebuff, B.J. (1997). *Co-opetition 1: A Revolutionary Mindset that Combines Competition and Cooperation 2: The Game Theory Strategy that's Changing the Game of Business.* Currency Doubleday, New York.

Elie, F. (2009). *Économie du logiciel libre.* Eyrolles, Paris.

Gosh, R. (2005). Guideline for public administrations on partnering with free software developers. [Online]. Available at: http://joinup.ec.europa.eu/elibrary/case/guidelines-public-administrations-partnering-free-software-developers-2005.

Raymond, E.S. (1999). Le chaudron magique [Online]. Available at: http://www.linux-france.org/article/these/magic-cauldron/magic-cauldron-fr-1.html.

Contributory Governance

Michel Briand

IMT Atlantique, Brest, France

Contributory governance: the changing posture of open cooperation

A society transformed by digital technology

Digital technology, which has become omnipresent in all of our activities, confronts us for the first time in our history with abundance: a text, a photo, a piece of music can be copied at very low cost, without depriving the person who allows it to be copied. One person can produce and put online a video that will have thousands of views. Social networks allow hundreds of people to join forces to create exchange networks.

To manage this abundance, the actors of free software or Wikipedia have invented new ways of doing things based on the contribution of a large number of people and sharing. Very far from our managerial culture of project management, which hierarchizes and distorts, here we are in a "bazaar" that self-regulates: Linux, Firefox, many components at the heart of the Internet, then Wikipedia or the open maps of OpenStreetMap illustrate this new reality of a co-production that aggregates the contributions of thousands of contributors.

Gradually, this understanding of the interest of a new form of organization that relies on the contribution of a large number of actors is spreading. In our daily lives, we practice it by trading on online classified ads, by exchanging music with friends, by spreading useful information to others on a messaging network, by rating our satisfaction with a purchase or service, or participating in a group on a social network.

These approaches are giving rise to new models, which Stiegler (2015) calls a "contributory economy", no longer based on large projects decided from above and compartmentalized into silos, but based on the involvement of actors working in networks. Our society is no longer just that of a public sector alongside a market sector guided by free competition, but is also articulated around commons, as Peugeot (2011) explains, with myriad initiatives from the social and solidarity economy, territories in transition and solidarity associations to respond to the profound transformations of society and the crises we are facing.

Thinking in a contributory way: a change in culture

However, many of our representations and our functioning date from the pre-digital era, where the hierarchical and compartmentalized social organization was based on performance and competition. Selective calls for projects that only retain 10% of responses and "desires to do", without mutualization, are an example. The practice, since 2003, in Brest of a "call for ideas", where *all* projects are retained, shows another way, one where the primary wealth is the involvement of people, which, even more than the budget (which has remained more or less constant over 17 years), is not easily extendable. The absence of competition in the projects (since everyone is selected!) promotes cooperation between the actors with the publication of their project.

Even today, the websites of local authorities or universities reflect the projects of the structure, but not much of what the actors in the territory or the structure are doing. The collection of pedagogical innovations at *Télécom Bretagne* (which in 2015 became *IMT Atlantique*) has made it possible to publish dozens of pedagogical practices that are of interest to other teachers elsewhere. This is how the open website innova-tion-pedagogique.fr came into being, which shares teaching initiatives in French-speaking higher education by (re)publishing two or three articles a day, which are read by no less than 2,500 visitors a day.

In a crisis situation, such as the Covid-19 pandemic, initiatives of solidarity, innovation and sharing are multiplying. For example, there are dozens of webinars and educational resources reported by *Riposte créative pédagogique*. It was also the network of manufacturing visors from digital manufacturing workshops (fab lab) mapped on the collaborative space *Riposte créative Bretagne*, initiated during the first lockdown alongside 450 initiative sheets in a text open to all, like Wikipedia.

Alongside "attention to" and "making visible", "action with" is the third aspect of this contributory dimension which takes into account the abundance of initiatives and the long-term nature of the involvement. Thus, at the beginning of the Internet for the general public (1997), the network of "PAPIs" (public Internet access points) in Brest allowed each neighborhood structure to join the network at its own pace. It took 10 years for the last neighborhood facility to join the network, testifying to the long period of time it took for the cultural change in the use of digital tools, which was not self-evident for social action actors. This "action with" approach at the pace of the actors is frugal (each one participates with their own means and ways of doing things) and robust (20 years later, there are still about 100 "PAPIs" in Brest).

Thinking in a contributory way: a factor of social transformation

This abundance of initiatives, these myriad implications that participate in the transformation of territories exist all around us, without us really seeing them. However, *Transiscope*, a map of alternatives that aggregates some 30 sources, now has more than 30,000 referenced places. For example, there are 297 shared gardens and composters run by the *Vert le Jardin* association in Brest, where thousands of people are involved, most often in social housing areas.

On climate, food resilience, social innovations, new ways of doing things "in common" are emerging and groups of people are organizing themselves to produce shared resources, managed according to their own rules of governance. In line with the knowledge commons (Vecam 2011), the commons movement is laying the groundwork for social organization that can foster transitions in action.

But this attention to initiatives is not yet part of our culture within communities and institutions. In order to support this emerging contributory dynamic, we must learn to move from a culture of prescription to a culture of territorial animation and cooperation. Beyond learning to achieve "action with", "attention to", and "making visible", it is the understanding and the dissemination of a culture of sharing and cooperation that will lead to progress. Open cooperation (Sanojca and Briand 2018) expresses this idea of cooperation beyond a closed group, that is, the actors of the pedagogical commons, transition, third places, free software, knowledge exchange networks, reuse, participatory finance and municipalism.

In a world in crisis, not only in terms of health but also in terms of social, economic and ecological issues, the old ways of thinking are no longer appropriate. The old, hierarchical and compartmentalized modes of governance cannot lead the transformations necessary for the intersection of climate, ecological, health and social challenges. Emerging contributory dynamics, by proposing new modalities for public action and by relying on the commons, can contribute to a liveable and desirable world.

References

Briand, M. (2015). Gouvernance contributive, réseaux coopératifs locaux et communs. *Éthique publique*, 17(2) [Online]. Available at: 10.4000/ethiquepublique.2297.

Cardon, D. (2006). La trajectoire des innovations ascendantes : inventivité, coproduction et collectifs sur Internet. In *Actes du colloque Innovations, usages, réseaux*. ATILF-CNRS, Montpellier.

Peugeot, V. (2013). Les collectifs numériques, source d'imaginaire politique. Millénaire 3 [Online]. Available at: http://www.millenaire3.com/texte-d-auteur/les-collectifs-numeriques-source-d-imaginaire-politique.

Sanojca, E. and Briand, M. (2018). La coopération ouverte, un concept en émergence. Innovation pédagogique [Online]. Available at: https://www.innovation-pedagogique.fr/article3428.html.

Stiegler, B. (2015). *L'emploi est mort, vive le travail ! Entretien avec Ariel Kyrou.* Fayard/Mille et une nuits, Paris.

Vecam (2011). *Libres savoirs : les biens communs de la connaissance. Produire collectivement, partager et diffuser les connaissances au XXIᵉ siècle*. C&F Éditions, Caen.

Course Guidance

Francis Danvers
  CIREL-PROFEOR, Université de Lille, France

Digital technology and course guidance: reflections on Parcoursup

Parcoursup is a computerized application designed to collect and manage the assignment wishes of future students in higher education in France. This national access system for returning to school was opened in January 2018. The Parcoursup platform replaces the admission post-bac (APB) system, which, from 2009 to 2017, was used to assign students, whether new baccalaureate holders or reorienting, to most higher education courses. Parcoursplus, for its part, is presented as a national access system for the resumption of studies.

APB functions as an assignment algorithm, which matches the wishes of applicants with the number of places available on each course. The *Cour des comptes* (Court of Auditors) considered that "the use of an algorithm to support the processing of a large number of applications seemed judicious" (APB, October 2017). However, this system has been challenged as it has been confronted with a significant increase in the number of applicants (increase in births from 1997) and an insufficient number of places offered correlated with the decrease in public spending per student. APB has been perceived as a "huge mess" (Vidal 2017). Research (Clément *et al.*) has since shown the segregational and unequal effects of the reform. Among the criticisms that remain: the lack of consideration of the candidate's motivation and, above all, the obscure nature of the selection algorithm set up by the institutions, which arouses the mistrust of certain high school or university students who fail to pass. The prevailing doctrine is "active course guidance" at university, but not all young people have equal access to useful information. The course guidance platform causes real stress among high school students at the time of their baccalaureate exams and acts as a supplementary exam (*Le Monde*, June 17, 2019).

From the second grade onwards, students are required to project themselves into the future, at the risk of taking the wrong path. This "competition society", which glorifies individual performance, feeds a market of anxiety: "By reinforcing competition between establishments to attract the best students, Parcoursup transforms students into consumers obliged to compare training offers. To help them make this difficult choice, the media will soon draw up lists of the best courses of study, as they do already for business schools, engineering schools or masters' degrees... There is no doubt that former students will suffer when they send in their CVs to find a job" (*Le Monde diplomatique*, April 2018).

In the major recommendations of Pierre Mathiot's report, dating from June 2018, there is the definition of a "*grand oral*", with, in particular, a time that can be devoted to "the student's course guidance project, on the condition that this test does not introduce a form of social censorship according to the origins of students". The director of Sciences Po Lille acknowledges that "the transition to algorithmic logic under exceptional conditions has had effects", some of which have been brought to the attention of the Constitutional Council. The sheer volume of applications requires an agile procedure. How can scores be made comparable according to the criteria of the school of origin? How can we judge the seriousness of a motivation based on a declaration? All the elements of the complexity of a young person's course guidance cannot be brought to the attention of an expert who decides in a compressed timeframe in order to allocate a limited number of available places.

The Parcoursup device questions inequalities in higher education, when we know that in France, the official cost of a year in a preparatory class for the *grandes écoles*[4] is three times higher than the cost per undergraduate student at university. "We are still very far from building an educational norm of justice" (Piketty 2019).

We are witnessing a paradigm shift in post-bac orientation. We are moving toward a "platform state", in the sense of Mazet (2019), whose price to pay is the "dematerialization of the administrative relationship" and its counterpart, the educational relationship in the guidance process.

Mr. Huteau, former director of the Inetop-Cnam Paris, mentioned in 2019 the competition of the public course guidance service, which is undergoing a "regression", particularly with "the development of the private offer in terms of guidance assistance [...] with the implementation of Parcoursup, the private guidance centers have seen their activity increase by about 30%. It should be noted that the cost of a consultation in a private center varies from 200 to 500 euros, depending on the extent of the service". The right to guidance is no longer respected in the context of a public service that is free for all.

The third edition of Parcoursup (April–May 2020) was confronted with the constraint of lockdown and the risk of unequal processing of wishes due to differences in living conditions and housing, which are added to the pre-existing social inequalities usually generated by the device. Indeed, sociologist Annabelle Allouch says: "The operation of Parcoursup, which favors continuous assessment and the series of early tests, will undoubtedly amplify these differences between students and between high schools".

The *Comité éthique et scientifique* (Ethics and Science Committee) (report to Parliament 2020) recalls the four key notions: transparency, efficiency, equity and safety. "The respect of these four requirements should condition the ethical and scientific character of the system". Parcoursup will have to integrate new functions and adapt to new audiences in connection with the reform of the *baccalauréat*[5] (2021).

---

4 Elite French academic institutions.
5 French national academic qualification that is obtained at the end of secondary school.

References

Allouch, A. (2020). *La société du concours : l'empire des classements scolaires*. Le Seuil, Paris.

Danvers, F. (2017). *S'orienter dans la vie : un pari éducatif ? Pour des sciences pédagogiques de l'orientation. Dictionnaire des sciences humaines, tome 3 : De la 601° à la 700° considération*. Presses universitaires du Septentrion, Villeneuve-d'Ascq.

Weixler, F. (2020). *L'orientation scolaire : paradoxes, mythes et défis*. Berger-Levrault, Paris.

Critical Thinking (Education for)

Marie Cauli
Université d'Artois, Arras, France

Thinking about it briefly

Following the uses and observations concerning disinformation and the diffusion of fake news, but also the propensity for new obscurantism and the endangerment of democracy, a consensus has been established on the need to develop critical education at school. However, this recommendation, presented as necessary and urgent, should not allow us to forget that it responds to one of the missions of the republican school[6] and that it is a traditional goal. Moreover, French culture, which has been carried on historically since the Greeks and the skeptics, via Montaigne, Cyrano de Bergerac, Voltaire, etc., has long invited us to oppose the verb "to believe" with the verbs "to know" and "to doubt". Also, it is in the name of individual freedom but also in the name of truth that Montaigne, in his time, exhorted to examine personally and carefully the accepted beliefs: "Let him pass everything through the muslin and let nothing be lodged in his head by mere authority and on credit". Since then, when one might have thought that the truth would prevail, one finds that students and even adults manage to be suspicious of the most established things, while adopting a boundless credulity in the face of the most dubious things. We can therefore believe that the issue is not to develop critical thinking in the face of the digital world, but to understand the difficulty, despite the rise in the general level of knowledge, to distil autonomous thinking and the rudiments of scientific thinking, in short, to think successfully at all in the digital society.

---

6 In the French meaning of this term, it is a free, secular school, teaching the core values of the French republic.

A context of "decline of rationality"

The misunderstanding of digital technology that has been denounced is only the tip of the iceberg in a general context. How many people radically refuse contradiction and claim a way of thinking that is closed in on itself, deaf to all dialogue and self-criticism, while trying to rally others to this bias? To this is added "an epidemic of sensitivity", where simple contradiction becomes "offensive". And all this leads to an imagination that favors collective scenarios or dubious theories, which coexist with information that includes tangible reality. This imagination, forged through "naive amateur thinking", as Serge Moscovici puts it, is no longer sufficiently filtered by interpretive frameworks, be they those of the State, university networks or scientific institutions, which are themselves subject to suspicion. Also, even if these beliefs have always existed, as Edgar Morin tells us in *La rumeur d'Orléans*, they are taking on unprecedented proportions in terms of content and scope, amplified by the digital environment and the networks, territories that are particularly conducive to the development and propagation of the most outlandish ideas. Similarly, news feeds that continuously broadcast and hammer out the same content, giving priority to the latest buzz, to small phrases, generate consequences in terms of disinformation. To this, we can add the lack of reference points in a complex world and the difficulty of conceptualizing changes. Individuals, connected to interactive situations, are both overwhelmed by this "fuzzy" collective thinking and the mass of information to be processed, which they can create as authors, transforming their subjective perception into universal thinking. According to Gérald Bronner, this movement, characterized by a context of "rationality decline" and by the intrinsic limits of our cognitive biases, no longer spares the sphere of the elite. However, it can be modulated by the knowledge and dispositions of each individual, the educational, social or relational context at the source of a scientific approach, a common sense posture, or even intellectual resistance.

Critical thinking: an intellectual survival kit

Thus, faced with the collective consequences of individual behavior, one can only agree with the desire to strengthen critical thinking. However, it is necessary to have a better understanding of what this term covers. However, the definition of critical thinking, which has been fuelled by distinct philosophical, educational and psycho-cognitive traditions, is accompanied by a multitude of related terms that do not mean exactly the same thing. Thus, we speak of critical thinking, critical knowledge, critical rationality, well-formed minds, etc. Moreover, this notion refers to educational interventions, programs, educational objectives, cognitive mechanisms and attitudes. It is therefore difficult to define these constituent elements, whether they are cultural, cognitive or behavioral. Critical thinking is therefore not so simple to develop and teach. This is probably because it cannot be reduced to a discipline, a subject, a skill, or an extra soul, but because it is a culture. Culture in the "anthropological" sense, which the example of Santa Claus studied by Claude Lévi-Strauss reminds us of. He shows us how the child becomes an actor through a process of self-naming and socialization that initiates them into reality. By spotting the signs of "scheming" adults, by grasping their inconsistencies and by looking for clues to the existence of this marvelous character, the child sees doubt creeping in, until tangible evidence is found that will make them leave behind their belief but also their childhood. Culture in the historical sense, that is, marked by a period, when it was mobilized to confront the propaganda techniques of totalitarian states, then when it was developed in the context of consumer society to cope with the development of mass media. Finally, culture of the present time, where digital technology, beyond its technical performance, has become a decisive element in understanding the world around us, given its omnipresence in all human activities.

A "total social" culture

This culture is learned, transmitted, disseminated and appropriated. This objective should not be taken lightly in the face of ignorance of the technological issues at stake and the need to preserve one's autonomy of thought and action. It requires a long path fuelled by knowledge, strengthened by "knowing how to think", equipped with know-how and interpersonal skills. It is an integral part of teacher training and must be carried out not only in schools, in initial and in-service training, but also in the extracurricular world. Nevertheless, it seems an insurmountable goal in view of the broad spectrum covered, and there is no magic formula. The message must therefore be simplified. Disseminating such education requires the support of two pillars: on the one hand, the teaching of computer science, which is part of scientific and technical culture, and, on the other hand, the dissemination of the conceptual repertoire of the humanities and social sciences, to which we can refer in order to think scientifically about the situations of the present that are induced by digital technology. These two foundations can be transformed didactically and reinjected in the form of training programs.

Computer science as a scientific and technical thought, as a method and a concrete construction

This first pillar has a profound relationship with mathematics. It is based on major unifying concepts, such as the algorithm, language, information and machine. At the same time, it evokes a systematic method that transforms a need into a series of elementary operations. Once the goal is set, the computer scientist looks for the means to achieve it, gathers the necessary information and materials, and then performs a series of elementary actions to reach the goal. This process is both a science and a technique, and it builds knowledge as well as objects, whether they are concrete like machines or abstract like programs. It mobilizes intellectual operations, such as demonstration, calculation, reasoning, approach to complexity, hierarchization, reduction of uncertainty, problem-solving, etc. These intellectual operations rely on the transparency of scientific methods, the reduction of the share of subjectivity and participate in building a scientific truth in the face of the tenacity of revealed truths. In this, it participates in the formation of critical education, but it has its limits. Because, thanks to computer science, modeling has become the preferred mode of expression for most projects and collective decisions, locking us into managerial processes and procedures, leading us to think of reality as a problem to be solved and to favor technical solutions to social questions, even relieving us of thinking for ourselves.

Anthropological, social and cultural culture

This is why other aspects of culture, more determined by the use of digital technology in social activities, must not be neglected. To do this, the concepts of the humanities and social sciences must be reactivated more than ever, in order to make room for a social and political interpretation of the implications of the digital age, including in sectors that previously seemed to be excluded. This repertoire is essential to better grasp and put into perspective the civilizational and geopolitical shifts, but also to become aware of the political effects. The fact of having knowledge in this field remains a potential lever for citizenship training. It increases the chances of better perceiving this unprecedented phenomenon and, in doing so, of revealing the best conditions for addressing the ethical, legal, political and social questions that await answers.

In short, one can only be encouraged to become a technophile, to be interested in the workings of computers, to understand the technical terms and symbols, the passion of the technical dimension explaining many things. Being at ease with computers and computer tools should allow for positioning oneself in front of technical problems, or even, at best, to freeing oneself from the technological complexity. But this approach must be accompanied by a social intelligence that promotes the exercise of our citizenship and our rights. These two aspects are necessary to understand what is at stake, the ambivalences of the actors and interests.

With these insights, the school's agonizing indecision on the subject should give way to a real strategy that would assume the fundamentals to be transmitted, conceived not as cognitive acts or manipulative techniques but as social intelligence. Between the challenge of a revolution in mentalities and the abyss of social practices, the operational implementation of this shift is necessary and urgent. The school is still in its infancy.

References

Bronner, G. (2019). *Déchéance de rationalité : les tribulations d'un homme de progrès dans un monde devenu fou*. Grasset, Paris.

Gauvrit, N. and Delouvée, S. (2019). *Des têtes bien faites : défense de l'esprit critique*. PUF, Paris.

Lévi-Strauss, C. (1952). Le père Noël supplicié. *Les Temps modernes*, 77.

Morin, E. (1982). *La rumeur d'Orléans*. Le Seuil, Paris.

Moscovici, S. (2013). *Le scandale de la pensée sociale*. EHESS, Paris.

Crowdsourcing

Laurence Favier
 GERiiCO, Université de Lille, France

Crowdsourcing: when the digital crowd supplants the machine

What is there in common between the participatory construction of an archive on the First World War by the *Bibliothèque nationale de France* (French National Library)[7], the classification of images of galaxies according to their shape for a scientific[8] program, the classification of consumer tweets to find out what is being said about a product, or the transcription of various types of handwritten documents? These approaches are part of what is known as crowdsourcing: an open call to the crowd, that is, to an undefined set of voluntary individuals, not selected beforehand because of their particular qualities or knowledge, to accomplish tasks or provide information (or documents) on an online platform. These tasks are, most often, either content contributions, or operations of classification, labeling (adding keywords describing images, for example), transcription of handwritten documents or content moderation (sorting according to ethical and legal criteria).

The main reason for the use of the digital crowd is the non-automatable nature of the proposed tasks or the poor performance of the machine on such tasks. Moreover, the work performed by the crowd is essential for "machine learning", which is at the heart of artificial intelligence (AI): a lot of data previously processed by human intelligence is essential to develop a model that will allow a machine, because of an

---

7 Available at: europeana1914-1918.eu/en.

8 Available at: https://www.zooniverse.org/projects/zookeeper/galaxy-zoo?language=en.

adapted algorithm, to identify its own criteria that will allow it (almost) certainly to recognize new data (e.g. to identify a dog, a cat or a giraffe in a new photo according to the photos already processed) and trigger operations. Thus, CrowdFlower is a human-in-the-loop AI platform that aims to simplify the creation of training data to continuously improve machine learning algorithms.

Moreover, crowdsourcing can feed a content-sharing platform: this is the case, for example, with archives (Europeana, *op. cit.*, or the American Citizen Archivist project), information on pollinating insects (The Great Sunflower Project) or information on car traffic provided by drivers and map editors to contribute to the real-time guidance of car drivers (Waze)

The contribution of the digital crowd can be more or less creative and more or less open to the "crowd". Creativity is what can sometimes happen when people contribute to scientific projects, such as the famous "Foldit", an experimental video game on protein folding developed by the University of Washington's computer science and biochemistry departments. Amateurs had accomplished the feat of remodeling in 3D the structure of an enzyme (called M-PMV) present in a virus similar to AIDS in monkeys. The same game is now being used to fight Covid-19. Using logical skills and without scientific expertise in the field, players find the solution to a problem that computers cannot solve. Another mechanism of creative crowdsourcing is that claimed by open innovation platforms (such as Inno Centive or Praesens), which also offer problems to be solved (of a technological nature) but without a game interface, to a group of people likely to be able to find a solution. However, in the latter case, that of open problem-solving, the crowd is limited to that of registered experts: openness is therefore limited.

From the "wisdom of crowds" to the "productive crowd"

In one of its facets, crowdsourcing appears as an illustration of the "wisdom of crowds" (or collective wisdom), a theme taken up by James Surowiecki to describe our contemporary society in his 2004 book, *The Wisdom of Crowds: Why the Many Are Smarter Than the Few and How Collective Wisdom Shapes Business, Economies, Societies and Nations.* According to the "wisdom of crowds", the quality of the judgments, decisions and knowledge produced by individuals isolated from a group would be worse than the average of those in the group. In other words, the multitude working on a common project gives rise to a "collective intelligence", an emerging property of the group that is not the sum of the particular intelligences. The multitude here is a set of individuals who, without consultation, can turn the collective intelligence into a creative crowd. The idea is not new and, from Aristotle to Condorcet, there have been attempts to establish the possibility of collective competence in political and aesthetic matters, a competence that is all the stronger because it is based on the number and diversity of individuals. This tradition is opposed to the conception of the crowd dominated by emotion which, impulsive, dangerous and credulous, engulfs the individual and annihilates their rationality (Gustave Le Bon).

Three main variables determine the "wisdom of crowds": their size, their diversity and the motivation of individuals. Digital technology gives a new dimension to these three variables. The size of the crowd (a determining factor in crowdsourcing content) and cognitive diversity are decisive in forming a "collective intelligence" of such quality that it can offer an alternative to expertise. Thus, "diversity takes precedence over competence". In social debates, the crowd could be an advantageous substitute for the community of experts, by opposing the variety of points of view to the univocal knowledge of specialists. It is in the area of motivation that crowdsourcing reveals another facet of the digital crowd than the one claimed by the "wisdom of crowds". The crowd solicited in scientific projects, which is at the heart of what is called "citizen science", advocates scientific education, the interest or intellectual passion of volunteer amateurs whose commitment can sometimes even be rewarded by association with a scientific publication. It is notable that citizen science projects in a wide range of disciplines (genetics, earth sciences, environment and humanities) make use of crowdsourcing. But this type of individual motivation does not sum up the motivations of the digital crowd.

Another facet of crowdsourcing is therefore that of task-based work. A platform offers tasks to free and willing individuals who perform them for a fee. This "click work" (Casilli 2019) is the raison d'être of platforms such as Amazon Mechanical Turk, Foule Factory or CrowdFlower, to name but a few, which make the crowd perform "human intelligence tasks" (HITs). According to Howe (2006), who is credited with coining the term, crowdsourcing is a process of "out-sourcing" by the crowd in the same sense as that which consisted of exporting industrial production to low labor cost countries.

In this respect, crowdsourcing differs, according to some authors, from open source (free software), both of which use digital "collective intelligence". According to Gram (2010), open source, as a collective and collaborative process of software development and an alternative model to proprietary software publishing, is based on the principle of "many contributors for many beneficiaries", whereas the basis of crowdsourcing is "many contributors for few beneficiaries". Indeed, at the origin of crowdsourcing, there is always an order from a few (a group of researchers, a company) and a crowd willing to serve their particular interests. It is the link of subordination defining the paid relationship between the commissioners and the crowd that is then questioned. The crowd may find a selfless or financial benefit. It is therefore the use of the irreplaceable human substitute to feed automation and the tension between the paradigm of digital sociability conveyed by "Web 2.0", on the one hand, and the economic model of online work, on the other hand, that are at the heart of crowdsourcing.

References

Casilli, A. (2019). *En attendant les robots : enquête sur le travail du clic*. Le Seuil, Paris.

Elster, J. and Landemore, H. (2012). *Collective Wisdom: Principles and Mechanisms*. Cambridge University Press, Cambridge.

Favier, L. (2016). Humanities crowdsourcing. *ZIN*, 54(2), 7–21.

Grams, C. (2010). Why the open source way trumps the crowdsourcing way [Online]. Available at: http://opensource.com/business/10/4/why-open-source-way-trumps-crowdsourcing-way.

Howe, J. (2006). The rise of crowdsourcing. *Wired Magazine* [Online]. Available at: http://archive.wired.com/wired/archive/14.06/crowds_pr.html.

D

Data Economy

Bruno Deffains
Université Paris 2 Panthéon-Assas, France

Data economy and regulation of digital giants

The data economy has become a thriving industry. A century ago, the strategic resource was oil, and its control by certain large companies, such as Standard Oil, gave rise to "antitrust" laws. Today, similar concerns are raised by the giants that collect and process big data, the oil of the digital age. Alphabet (Google's parent company), Amazon, Apple, Facebook and Microsoft have steadily increased their power in this new economy. Their profits have not stopped growing. In the decade from 2010 to 2020, Amazon was able to capture 50% of every dollar spent online by American consumers. Google and Facebook, meanwhile, absorbed most of the growth in online advertising revenues.

Such dominance is a real challenge to the functioning of economies. Size alone is not a crime and the success of the digital giants has benefited consumers. Few people can do without Google's search engine, Facebook's network or Amazon's same-day home delivery. These companies have long flown under the radar of competition authorities, as traditional tools are not always adapted to their practice. Indeed, most of the services offered seem to be free of charge, while the consumer often pays indirectly through the data they provide to the companies.

However, there is reason to wonder. The control of data by the Internet giants gives them considerable market power. The classic approach to competition law, conceived in the era of the oil industry, seems outdated in the face of the data economy. Smartphones and the Internet have made data abundant, ubiquitous and far more valuable in terms of its economic value. All online activities create a digital trail and, therefore, raw material for data processing factories, fed by an ever-increasing number of connected objects. At the same time, artificial intelligence techniques, such as machine learning, enable more value to be extracted from the data collected.

This abundance of data changes the nature of competition. The digital giants have benefited greatly from network effects: the more users there are, the more people tend to sign up for Facebook and the more attractive it becomes for other users to sign up. Moreover, with data, there are additional network effects. By collecting more data, a company has more room to improve its products, which can attract new users, generating even more data and so on. For example, the more data Tesla collects from its autonomous cars, the better it can make them drive. That is part of the reason why the company, which sells far fewer vehicles than traditional car manufacturers, has become the world's largest automotive market capitalization at more than \$700 billion by 2021. Large data pools can thus act as protective barriers for companies with a first mover advantage.

The surveillance systems of these giants cover the entire economy: Google can see what people search for, Facebook what they share and Amazon what they buy. They have an overview of activities in their markets and beyond. They are the first to know when a new product or technology appears, allowing them to acquire it before it becomes too big a threat. Facebook's purchase of the messaging app WhatsApp for \$22 billion in 2014 (the company had barely 60 employees at the time and was not generating high revenue streams) provides a good example of such a strategy to stifle any form of competition.

The nature of the data makes the "antitrust" remedies of the past less appropriate. Splitting a company like Google into multiple entities would not necessarily solve the problem in the long run. Over time, one of these entities would likely become dominant again as network effects would not disappear. It therefore seems essential to radically rethink the solutions. A first solution is for market authorities to turn the page on the industrial era of the 20th century. Most often, in a merger case, for example, they traditionally use size to determine when and how to intervene. Today, the volume of data held by the companies should be taken into account when assessing the consequences of the merger. Second, the purchase price of a target company is a signal as to whether or not the company is trying to neutralize an emerging threat. In the same sense, the analysis of market dynamics, using simulations to detect the use of possible price collusion algorithms or to determine the best way to promote competition, appears to be more and more useful.

Another solution is to loosen the grip that online service providers have on data. Governments could encourage the emergence of new services and open up their own databases as essential public infrastructure. They could also make it compulsory for companies to share some of their data, with the consent of users, in a commons approach (an approach that Europe has favored in the area of financial services, by requiring banks to make their customers' data available). It is not easy to revive "antitrust" in the era of the data economy, but it is a major concern. One of the difficulties is that increased data sharing can threaten privacy. If governments do not want an economy dominated by a few giants, they must regulate them.

The end of 2020 was marked by a series of judicial events that could mark a turning point in American antitrust. The Federal Trade Commission (FTC) and 46 American states launched a double action against Facebook for abusing its dominant position and illegal monopoly on social networking services. The case concerns the acquisition of Instagram and WhatsApp, as well as anti-competitive contractual conditions imposed on the use of its APIs (Application Programming Interfaces). The use of user data is being challenged and the FTC is seeking prior approval for future acquisitions. These actions could lead to major lawsuits and force Facebook to dismantle and divest itself of its two subsidiaries.

Alongside another complaint against Google, this is the main offensive by the US authorities since their 1998 lawsuit against Microsoft. The FTC accuses Facebook of "stifling threats" through the acquisition of competitors. The complaint sets out to make a strong case that the acquisitions of Instagram and WhatsApp were part of an overall strategy to maintain a monopoly. The legal battle is at the heart of the FTC and state prosecutors' complaints, which require Facebook to show that its economic success would have been the same without WhatsApp, while justifying the high prices for the acquisition of both companies.

There has been a clear shift in the rationale for these complaints, reflecting a return to the origins of antitrust. At the end of the 19th century, the time of the Sherman Act, there was a fear of the concentration of economic power and its influence on political power and democracy. In recent decades, however, an approach has been adopted that requires proof that the monopoly penalizes the consumer in terms of price. The problem is that this approach is ill-suited to the digital giants, whose business model is based on characteristics that are now well-known (network externalities, data control, advertising and free access). A new form of competition has emerged in which firms compete more "for the market" than they do "in the market". The temptation to take advantage of one's dominant position to consolidate one's power is strong, just as the robber barons did shortly before the birth of "antitrust".

Is this a desirable turning point? The question must be discussed without any ideological consideration. The issue is the effectiveness of the rules governing the competitive process. When, in June 2020, Facebook was fined $5 billion for violating the privacy of its users in the Cambridge Analytica affair, the stock value rose. This was the largest penalty imposed by the FTC on a digital giant, but the decision was not seen as a challenge to the business model. It is therefore difficult to conclude on the economic and social efficiency of a system that does not encourage the company to change its practices once the fine has been paid. Many commentators therefore agree on the need to rethink the boundaries of "antitrust" by analyzing barriers to entry, conflicts of interest and data control. This is the case for Lina Khan, who believes that it is essential to look at the value of the target's data in the event of an acquisition, rather than just considering its turnover.

It should be noted that the opponents of this more political vision of "antitrust" evoke a rapprochement with the European Commission, which raised its tone at the end of 2020 in the face of the digital giants and changed gear with new regulatory frameworks (the Digital Services Act (DSA) and the Digital Market Act (DMA)). On the American side, as the laws do not change, it is the judges who must adjust their assessment of the contested practices. How will they explain, on objective grounds, particularly economic ones, that the previous tests no longer apply, that the previous decisions taken for infringements or mergers are not good benchmarks? For its part, Europe can adapt the rules, as it is doing today with the DSA and DMA projects. It is probably simpler to start on new bases. One example is the category of gatekeeper operator (controller of access to a given market), which is making its appearance in competition law, accompanied by specific obligations. This does not mean that the new guidelines are effective per se in the face of the challenges posed by the practices of the digital economy, but that the market authorities have become aware of the need to revisit the concepts and tools of "antitrust" in order to better supervise the activity of the players in the data economy.

References

Khan, L.M. (2018). Amazon's antitrust paradox. *Yale Law Journal*, 126(3).

Monerie, N. (2018). Les défis de la commercialisation des données après le RGPD : aspects concurrentiels d'un marché en développement. *Revue internationale de droit économique*, 4(32).

Pouvoirs (2018). La datacratie. *Revue Pouvoirs*, 164.

Data, Information, Knowledge

Serge Abiteboul
INRIA, Paris, France

Data have been with us since the Sumerian tablets. Stored, processed and exchanged, they are at the heart of our personal and professional lives. Computer systems help us manage these data, represented in digital form, as sequences of bits. Data management is the basis of three of the most successful computer applications: database management systems, web search engines and conversational assistants. They deal with three distinct facets of data that overlap so closely that they are difficult to separate: data (in a more precise sense), information and knowledge.

An example from (Abiteboul 2013) will illustrate the distinction. Temperature measurements taken each day at a weather station are data. A curve showing the evolution over time of the average temperature in a place is information. The fact that the Earth's temperature is increasing as a result of human activity is knowledge. Let us try to clarify these three concepts (Floridi 2011).

A piece of data is the elementary description of a reality. It is, for example, an observation or a measurement. It will be represented numerically by a sequence of bits. From collected data, information is obtained by organizing it to make it understandable to humans. By understanding the meaning of the information, we arrive to knowledge, that is, "facts" that are considered to be true in the universe of a speaker and "laws" (logical rules) of that universe.

The following table can represent a bit sequence.

Manon	Imperial College	London
Pierre	ENS	Paris-Saclay
Jérémie	Mines de Paris	Paris
Marie	ENS	Paris-Saclay
Myriam	Sorbonne University	Paris

A two-dimensional table representing, for students, their name,
the name of their university and the name of the city where they study

An alien probably would not understand this sequence of bits, but we do; a program or text editor has been able to parse it and present this table in a form we are familiar with.

From data to information

The entries in this table are strings. For now, they are data. We can now specify that the first column contains the first names of PhD students at a summer school in Cargese, Corsica, the second their university and the last the city where their university is located. By obtaining a meaning, these data have become information, such as "Manon, a student at Imperial College London, attended this summer school". This sentence, which we can read and which for us has a meaning, is information that we can extract from this table.

From information to knowledge

This information turns into knowledge when we introduce it into a logical universe. Each line becomes an assertion and we can reason with these assertions or meta-knowledge, such as: this table contains the complete list of all PhD students in computer science from ENS Paris-Saclay who attended this summer school. We can deduce that either Philippe is not a PhD student in computer science at ENS Paris-Saclay or he did not attend this summer school.

We have moved from data to information, and from information to knowledge. Obviously, the boundaries between these concepts are blurred, and the world we seek to model with knowledge is complex and can partly elude us. For example, the fact that Myriam is listed as a student at Sorbonne University could have many explanations: she left the university at the beginning of the month, she filled in the form incorrectly, etc.

We are accumulating more and more data. Digital data helps us in our daily lives, just as it contributes to leading to harassment and facilitating the mass surveillance of citizens. In the world of digital data, the exponential has its place. It is said that the amount of data doubles every 18 months. A few examples will give orders of magnitude (Abiteboul and Peugeot 2017): a dozen of Chopin's "Nocturnes" on a cell phone occupies 75 megabytes; a holiday video a few gigabytes; the raw text of all the books ever written a few hundred terabytes; the quantity of data produced by the Large Hadron Collider at Cern in one minute is of the order of a hundred petabytes; the representation of all the sentences ever spoken a few exabytes, etc. Finally, the zettabyte is the order of magnitude of annual traffic on the Internet today. Let us get back to the three great computing successes mentioned earlier.

Data

Database management systems (Garcia-Molina *et al.* 2009) manage structured data for us in, for example, the form of two-dimensional tables that may be in large quantities. They allow us to share these data, to query them (by asking "queries") while guaranteeing their safety (against failures) and security (against hostile people).

Information

Web search engines (Brin and Page 1998) manage masses of much less structured information for us – typically text, images and videos. Their achievement is enabling billions of people to find the information they are looking for, a needle in a haystack, in a few tenths of a second.

Knowledge

Conversational agents answer our questions, allow us to play music, do our shopping, etc. A first feat consists of being able to understand the user's speech and synthesize a voice to answer. However, they do much more than that. More and more, they are true knowledge base engines. They accumulate knowledge on many subjects: music, history, sports, etc. and reasoning allows them to choose an answer to each question. If they still often get it wrong, it is because the subject is far from simple.

To conclude, let us note that two of the latest big buzzwords in the field, Big Data analysis and machine learning, are both based on the idea of learning something from data. This can be statistical properties, for the former, or standard behaviors from training data, for the latter.

References

Abiteboul, S. (2013). *Sciences des données : de la logique du premier ordre à la Toile. Leçon inaugurale au Collège de France*. Fayard, Paris.

Abiteboul, S. and Peugeot, V. (2017). *Terra Data : qu'allons-nous faire des données numériques ?* Le Pommier, Paris.

Brin, S. and Page, L. (1998). *The Anatomy of a Large-Scale Hypertextual Web Search Engine*. Stanford University, San Francisco.

Floridi, F. (2011). *The Philosophy of Information*. Oxford University Press, Oxford.

Garcia-Molina, H., Ullman, J.D., Widom, J. (2009). *Database Systems: The Complete Book*. Pearson, Hoboken, and Prentice Hall, Hoboken.

Digital Commons

Sébastien Shulz
LISIS, Paris, France

Emergence and definition of the "digital commons" concept

The concept of the "digital commons" emerged in the United States at the turn of the 1990s. Debates about the regulation of the digital environment were in full swing. Web technologies were leading to the increase of information and collaborative practices around digital resources. Industries that were beginning to invest in the Internet wanted to protect their online assets by extending intellectual property rights (IPR) legislation. Liberal governments, in line with the dominant economic theory of law and economics, supported the privatization of digital resources but wanted, at the same time, to regain control over the regulation of cyberspace. A group of free Internet activists and committed academics then formulated a critique of this dual dynamic. They opposed the maximalist vision of digital capitalism and the authoritarian centralism of states that constrained informational practices of sharing and contribution. They highlighted the economic, political and social virtues of the latter, drawing on the successes of free software, Wikipedia and peer-to-peer file sharing (Benkler 2006). To distinguish themselves from the state (public goods) and the market (private goods), they reinvented the old concept of *commons*, which traditionally referred to land whose ownership, management and maintenance practices were shared. The parallel also lies in the fact that these common lands underwent a phenomenon of enclosure through physical and legal barriers from the 13th century onward, which is analyzed by the Marxist tradition as the beginnings of capitalism. Legal scholars interpret the extension of IPRs as a second wave of enclosure, enabling the advent of information capitalism (Boyle 2003).

It is within this critical matrix that the digital commons were conceptualized. They are defined as digital resources for which: (1) free access is protected against attempts at exclusive appropriation; (2) the rules of governance are democratically determined by the producers and/or users; and (3) the aim is to enrich the resource and encourage practices of sharing and contribution over time. In reality, this generic definition conceals differences in interpretation that have not yet been stabilized and that relate to each of these three points.

Open access digital resources – open digital commons

The first characteristic of the digital commons is an ownership system that guarantees open access. For some economists, open access is antithetical to technological progress because it does not encourage entrepreneurs to engage in innovation by producing a resource that everyone can use without paying. In the 1980s, computer scientist Richard Stallman was one of the first to criticize this proprietary perspective. Drawing on his own experience as a developer, he showed the technical and social benefits of giving users free access to software source code. In the following decades, a range of actors – librarians, hackers and academics – followed this free software movement and defended the free circulation of digitized information, ranging from scientific knowledge to public domain works. In particular, some lawyers from the largest American universities called these open access resources *commons*. They turned the conclusion of Hardin's famous "tragedy of the commons" on its head, which states that an open-access resource will always end up being overexploited. They showed that, on the contrary, digital resources are theoretically inexhaustible because their reproduction cost is close to zero and their open access stimulates creativity and innovation, leading instead to a "comedy of the digital commons" (Lessig 2001).

Far from an absence of IPRs, the first characteristic of the digital commons is an alternative system to exclusive IPRs. It is translated into legal licenses, and, more rarely, into laws, which have the common thread that they guarantee the right of access to digital resources. Therefore, a *continuum* of different access rights systems is emerging. The most permissive is the public domain or the "free license" chosen by most governments to make their public data available (open data). Other systems that are more protective of access rights, such as the General Public Licence and the Creative Commons licence, guarantee freedom of access but, to avoid exclusive re-appropriation, either force derived versions to be placed under the same free licence (these are known as "contaminating licences") or restrict certain commercial uses.

Democratically managed digital resources – self-governing digital commons

The second structuring characteristic of the digital commons is their democratic governance by producers and/or users. In the 1990s, governments sought to regain control over the regulation of cyberspace. They were confronted with criticism from part of the Internet world, which had historically been constructed through anti-hierarchical self-organization practices and values. The communities of people who contribute to the governance of the Internet, to free software or to collaborative projects such as Wikipedia all share the common feature that they seek to establish their own rules to organize the production and management of their shared resources. For some researchers, this characteristic of the digital world evokes the seminal work on common-pool resources by Nobel Prize-winning economist Elinor Ostrom, who was interested in studying the way in which citizens self-organize to manage a shared natural resource outside the hierarchical framework of a company or a state (Ostrom 1990). These two perspectives were finally brought together in the 2000s by translators, including Elinor Ostrom herself, who described these collaborative and democratic projects for producing digital resources as self-regulating commons (Hess and Ostrom 2007). It should be noted that while the aim of governance of a natural resource is to avoid its overexploitation by limiting its use, that of a digital resource seeks to ensure its enrichment by increasing and improving contributions.

Here again, the democratic criterion for the governance of a digital commons has several levels. Some projects, such as the open source software Linux, are managed by a "benevolent dictator for life" and most of the contributors are salaried employees of digital capitalism companies, but the democratic character is nevertheless ensured by the possibility, for those who wish to do so, to *fork* the project. Other communities, such as Debian or Wikipedia, have developed extremely refined and theoretically egalitarian rules of governance, which, however, still lead to forms of authority between old community members and newcomers. While in most digital commons, governance is not established by all users but by a minority of contributors, more open forms of governance are emerging, especially around cooperative platforms like Fairbnb.coop. Finally, the formalization of governance can vary, from informal rules to the statutes of a foundation, through the writing of "constitutions" to organize the powers within a project.

Digital resources produced through contributory practices – digital commoning

The third structuring characteristic for the digital commons concerns the mutual contribution practices that are at the heart of the production and maintenance of these shared resources. In economic theories of rational choice and collective action, humans are driven by selfish and calculating choices. At the turn of the millennium, a group of theorists, such as Michel Bauwens, used the example of peer-to-peer sharing practices to highlight the altruistic motivations that exist on the Internet. For these authors, what counts in digital communities is commoning.

This concept can be characterized as the practices of actors working together toward a common goal. Commoning in the digital context includes contributing to a digital resource, participating in its governance and sharing it. The authors who defend this perspective seek to rethink the digital commons as structured by an alternative praxis to the individualism and proprietorship characteristic of neoliberal capitalism (Bauwens and Lievens 2015).

The uncertain future of the digital commons

To conclude, we would like to point out some of the political issues raised by the digital commons. On the one hand, they carry divergent political projects, some more liberal and reformist, others more anarchist and radical. One can legitimately wonder to what extent the notion of the digital commons can designate objects that are sufficiently homogeneous to fall under a single analytical category. The distinction we have drawn between open digital commons, self-governed digital commons and digital commoning would, in our opinion, allow for a better distinction. On the other hand, while the digital commons all designate alternative forms of social organization to capitalism and the state, we see the emergence of hybridization phenomena between these three terms. Digital capitalism seems to have incorporated the open source critique and is making full use of some digital commons. Google, which uses Wikipedia data to enrich its search results, recently donated $3 million to the Wikimedia Foundation, and also allowed it to use its machine learning tools. These connections have led some activists to fear that digital commons will become the "commons of capital" if they do not provide themselves with the means to reproduce autonomously, without depending on the calculated generosity of large corporations. On the other hand, the digital commons movement has an ambiguous relationship with the state. While some commoners want to see the advent of a partner-state and some political actors have taken a stand in their favor, many fear forms of commons-washing. The future therefore remains uncertain for the digital commons.

References

Bauwens, M. and Lievens, J. (2015). *Sauver le monde : vers une économie post-capitaliste avec le peer-to-peer*. Les Liens qui Libèrent, Paris.

Benkler, Y. (2006). *The Wealth of Networks: How Social Production Transforms Markets and Freedom*. Yale University Press, New Haven.

Boyle, J. (2003). The second enclosure movement and the construction of the public domain. *Law and Contemporary Problems*, 66(1), 33–74.

Hess, C. and Ostrom, E. (2007). *Understanding Knowledge as a Commons – From Theory to Practice*. The MIT Press, Cambridge.

Lessig, L. (2001). *The Future of Ideas: The Fate of the Commons in a Connected World*. Random House, New York.

Ostrom, E. (1990). *Governing the Commons: The Evolution of Institutions for Collective Action*. Cambridge University Press, New York.

Digital Humanities

Joana Casenave
GERiiCO, Université de Lille, France

Digital humanities is a relatively recent field of research, even though it has its conceptual and methodological roots, at least partially, in already established disciplines. It was in 2004 that the notion of digital humanities was introduced through the publication of the *Companion to Digital Humanities* (Schreibman *et al.* 2004). The term was unprecedented, since, until then, researchers had relied on *humanities computing*, which qualified the scientific approach of using computer techniques not only for the analysis of linguistic corpora, but, more generally, for research in the humanities and social sciences.

Born in the bosom of linguistics and literary studies, the digital humanities movement has continued to open up to various disciplinary horizons. However, above all, the development of this field of research marks a paradigm shift: digital technology has gone from being a working tool to being a vector and also the very object of research.

In the introduction to the *Companion to Digital Humanities*, the authors highlighted three elements that they felt were structuring this new field of research: interdisciplinarity, which brings together theorists and practitioners in a common movement and project; the constitution of this field as a scientific discipline in its own right; and the diversity of the links and exchanges that this new discipline forges with the humanities and social sciences.

As with any newly created research field, it takes time to stabilize. Its scientific and methodological outlines are therefore still being developed. However, it is already significant that the digital humanities invites researchers to reflect on their methodology and reinterprets the conceptual structures through which they organize their knowledge.

Generally presented as an application of digital methods and tools to research in the humanities and social sciences, the digital humanities is collaborative, transdisciplinary and involves a wide range of actors (Burdick *et al.* 2012). Dominique Vinck indicates that they form "the set of scientific disciplines that capture, analyze and restore, through tools and computer calculation, the cultures and social dynamics, past, present and emerging" (Vinck 2020, p. 13). They concern very diverse objects: constitution and processing of corpora, analysis and exploration of masses of data, visualization of data, digital editions and editorialization and organization of knowledge, to name but a few.

Defining digital humanities: between conceptual reflection and technical experimentation

The epistemological reflection currently underway in the digital humanities, developed, above all, by theorists, is also nourished by practitioners. It is in this spirit that the *Day of DH* was inaugurated in 2009. Under the impetus of *CenterNet*, an international network of research centers dedicated to digital humanities, researchers, engineers and students who define themselves as digital humanists are invited each year to present their work and to express what they believe the term digital humanities covers.

The resulting definitions are very eclectic; they show a wide range of views, which illustrate the disparity of definitions in the field of digital humanities. In the book *Defining Digital Humanities* (Terras *et al.* 2013), a selection of the definitions proposed between the years 2009 and 2012 has been taken up to illustrate this eclecticism. It appears that these definitions relate to the methods used, the theoretical foundations, the actors involved, the sense of community and even the results of research and the products developed. The ways of defining this field are therefore multiple and correspond to the various approaches taken by researchers interested in this disciplinary field.

For Lincoln Mullen, the digital humanities is in fact akin to a spectrum: "Digital humanities is a spectrum" (Mullen 2013, p. 237). All scholars use digital tools in the course of their work to varying degrees. There is therefore no firm boundary between traditional researchers and digital humanists. What is therefore important is that digital humanities cultivates an ethics of inclusion (Mullen 2013, p. 238). The ethical aspect of digital humanities is also prominent in the definitions given by researchers. Indeed, this disciplinary field is characterized by a set of values that it promotes in the research world. In the book *Debates in the Digital Humanities* (Gold 2012), Lisa Spiro made a contribution devoted to this theme: "This is why we fight: Defining the values of digital humanities." She summarizes these values as "Openness, collaboration, collegiality, connectedness, diversity, experimentation" (Spiro 2012, pp. 23–30). One of the most obvious expressions of these values is collective work, which is particularly valued by digital humanities scholars.

Finally, the feeling of belonging to the digital humanities community appears to be essential. Thus, a digital humanist is someone who first defines himself as such. Stephen Ramsay, in his article "Who's In and Who's Out," asked the question of what characterizes a digital humanities researcher. While the community and collaborative aspect of digital humanities is essential, it is not enough to define the field. It is necessary to go beyond a simple inventory of the work, practices and tools used in the field. Digital humanists can be identified, above all, by the fact that they create and develop products for other researchers. The set of individuals who enters the creation process then corresponds to this definition: those who theorize, those who design the product, those who develop it and those who supervise the projects (Ramsay 2013, p. 241).

Progressive institutionalization of digital humanities

The task of defining the digital humanities is therefore complex. In various monographs, the metaphor of a big tent is often used to illustrate this disciplinary field: the digital humanities is similar to a big tent that welcomes researchers and actors from a wide range of academic backgrounds (Gold 2012). Above all, the adoption of the big tent metaphor avoids the fragmentation of the digital humanities which is linked to the hyperspecialization of the sub-disciplines of this fragmented disciplinary field (Terras *et al.* 2013). However, this inclusive definition is not without drawbacks and questions the notion of disciplinary cohesion. The theoretical and practical aspects of the digital humanities lead some researchers to focus the definition of this disciplinary field on this dual purpose (Piotrowski and Xanthos 2020). Above all, the epistemological grasp of this field of research requires an analysis and a critical attitude towards the digital that can be nourished by the conceptual approach of the information and communication sciences in particular (Cormerais *et al.* 2016).

Several studies and publications have been instrumental in the development of this disciplinary field and have marked important theoretical milestones. These include monographs that have focused on theoretical and conceptual reflection: *A Companion to Digital Humanities* (Schreibman *et al.* 2004), *Debates in the Digital Humanities* (Gold 2012), *Digital Humanities* (Burdick *et al.* 2012), *Defining the Digital Humanities: A Reader* (Terras *et al.* 2013), *A New Companion to Digital Humanities* (Schreibman 2016), *Digital Humanities* (Vinck 2020), to name the main ones. Digital humanities also benefits from editorial structures. Several journals specialize in the field, such as the *Digital Humanities Quarterly*, published by the *Alliance of Digital Humanities Organizations* (ADHO), or the more recent journals *Digital Studies/Le Champ numérique,* published by the *Canadian Society for Digital Humanities* and *Humanités numériques*, which is published by the Francophone association Humanistica. Some publishing houses have developed specific collections: such as the "Digital Humanities" collection at the University of Illinois Press, or "*Parcours numériques*" at the Presses de l'université de Montréal.

Finally, the collaborative nature of the digital humanities is further underlined by the fact that they are built around associative structures. Thus, in 2005, the Alliance for Digital Humanities Organizations (ADHO) was created, which organizes the annual *Digital Humanities* conference. The first French-speaking association, named Humanistica, was created in 2014. Today, digital humanities has dedicated research infrastructures and an international network of training courses that mark their progressive institutionalization.

References

Burdick, A., Drucker, J., Lunenfeld, P., Presner, T.S., Schnapp, J. (2012). *Digital Humanities*. The MIT Press, Cambridge.

Cormerais, F., Le Deuff, O., Lakel, A., Pucheu, D. (2016). Les SIC à l'épreuve du digital et des humanités : des origines, des concepts, des méthodes et des outils. *Revue française des sciences de l'information et de la communication*, 8.

Gold, M.K. (2012). *Debates in the Digital Humanities*. University of Minnesota Press, Minneapolis.

Mullen, L. (2013). Digital humanities is a spectrum, in *Defining Digital Humanities: A Reader*, Terras, M., Nyhan, J., Vanhoutte, E. (eds). Routledge, Abingdon.

Piotrowski, M. and Xanthos, A. (2020). Décomposer les humanités numériques. *Humanités numériques*, 1.

Ramsay, S. (2013). Who's in and who's out, in *Defining Digital Humanities: A Reader*, Terras, M., Nyhan, J., Vanhoutte, E. (eds). Routledge, Abingdon.

Schreibman, S., Siemens, R.G., Unsworth, J. (2004). *A Companion to Digital Humanities*. Blackwell Publishing, Malden.

Schreibman, S., Siemens, R. G. and Unsworth, J. (2016). *A New Companion to Digital Humanities*. John Wiley and Sons, New York.

Spiro, L. (2012). 'This Is Why We Fight'. Defining the values of digital humanities, in *Debates in the Digital Humanities*, Gold, M.K. (ed.). University of Minnesota Press, Minneapolis.

Terras, M., Nyhan, J., Vanhoutte, E. (2013). *Defining Digital Humanities: A Reader*. Ashgate Publishing, Farnham.

Vinck, D. (2020). *Humanités numériques : la culture face aux nouvelles technologies*. Le Cavalier bleu, Paris.

Digital Inclusion

Vincent Meyer
Université Côte d'Azur, Nice, France

This expression has become a vulgate in public discourse since 2017. This is in order to qualify forms of illiteracy in the context of dematerialization of a (future) "digital everything". It must "feature" support for the use of digital technologies for those for whom they still constitute an obstacle or who reject them. In the way that France is governed at the beginning of the 21st century, it is also a question of making digital inclusion a lever for social and economic (re)integration and proposing practical solutions (training, tutorials, third places where public service activities – or substitutes – are developed), and labels to allow the greatest number of people access to rights and citizenship. Digital inclusion, in a way, puts a political will and highly differentiated situations of use on an equal footing. In this sense, it remains a presupposition, because such inclusion (as well as its associated expressions: acculturation or maturity) is based on an observation that is difficult to quantify (the number of people who are far from the digital world: 14 million in March 2018, according to a CSA Research study for the *Syndicat de la presse sociale*). Let us first look at "inclusion", a "fashionable" or politically useful term, before detailing its translation, its uses and challenges in a digital society.

The noun "inclusion" is defined, in the Cambridge Dictionary, as both "the act of including something or someone as part of something" and/or "the act of allowing many different types of people to do something and treating them fairly and equally". The notion of inclusion is, above all, "social" and it is all the more appealing because it is almost entirely removed from all the previous notions: integration, insertion, readaptation, etc. A possible common denominator of these notions is the capacity of an institution or an environment to adjust – to provide a response – to the specific situations and needs of people (from infancy to old age) in a situation of disability, fragility and/or vulnerability (almost a slogan, like "My body is not maladjusted, your society is").

Inclusion sounds like an injunction (all our practices must be inclusive), but it remains an expectation. It is disseminated in professional circles through announcements, experiments, calls for projects, roadmaps or via gray literature. In the economic and political rhetoric, inclusion can be found in any human organization: schools and companies must be/become inclusive, and the same applies to the urban environment with smart cities (Meyer 2019), which bring together, among other things, the requirements of ecology, mobility versus accessibility, security, etc. The current situation is therefore – almost to the point of excess – one of inclusive practices (at school, in employment, in public places, for mobility, etc.). Even writing is becoming inclusive in order to "determine" (culturally, even ideologically) our interactions, texts and discourses. To take the example of people with disabilities alone, who experience and feel – on a daily basis and with good reason – various forms of discrimination, inclusion is the subject of numerous political and legal translations in areas as varied as public space, housing, employment, education and training, culture, etc. In other words, the promise of an inclusive (vs. accessible) society ensures the possibility of full participation of all citizens. However, there is still much room for improvement. Inclusion is a subsidized term, that is, a "polysemous term that is, if not an argument of self-righteousness, at least has the capacity to influence and orient a demand in a given political context" (Meyer 2021).

Thus, digital inclusion can be defined as a desire on the part of the public authorities to make digital technologies accessible to as many people as possible, mainly via equipment such as computers, telephones, the Internet (100% of sites should be accessible), the Internet of Things (from watches to television, via home automation; connected cities between cameras and sensors), and, on the other hand, to enable everyone to appropriate (if not master) the possibilities offered by these technologies. This is particularly the case through training and/or mediation mechanisms that enable people to acquire the basics of a "general digital culture" (for job searches, data protection/security via a "safe", online learning, gamification). It is the 2005 law "for equal rights and opportunities, participation and citizenship of disabled people" which, in a way, lays the foundations for digital accessibility, while a decree of July 24, 2019 aims, in the era of all-digital technology, at equitable access to online public communication services. Consequently, it is a question of rethinking the penetration and social uses of digital technologies for these "remote" people, and, in particular, those who are in charge of or in the care of social systems and supported by volunteers or professionals in the social and medico-social field. Digital inclusion would therefore be a factor, if not of progress, at least of societal commitment, beyond crisis situations. The health campaign which was begun in 2020 even "authenticates" this essential access to digital technologies in terms of training and mastery of them in areas as diverse as teleworking, coworking, job hunting, carrying out administrative procedures, including social rights, etc.

A (first) French National Plan for Digital Inclusion[1] (*Plan national pour un numérique inclusif*) was presented in September 2018 with four objectives: to detect the public who are struggling with digital technology; to offer human support in dealing with the process; to train those who wish to do so, thanks to the "digital pass"; and to consolidate the actors of digital mediation. The year 2018 was thus promoted as the year of digital inclusion to "ensure the equality of citizens and territories within the framework of our republican pact, but also to participate in a virtuous economic strategy". At the end of August 2020, the National Consultative Council for the Disabled (*Conseil national consultatif des personnes handicapées*) held its first Summer Universities of Inclusion with contributions from the *Conseil supérieur de l'audiovisuel*, the *Autorité de régulation des communications électroniques et des postes*, as well as the *Conseil national du numérique* to initiate joint actions around accessibility.

The essential device for digital inclusion today is the platform. With and through it, it becomes possible to manage computerized pathways and cooperation in an increasingly dematerialized and remote work environment. For the above-mentioned groups, the challenge is also to interconnect institutions and public decision-makers, and thus to get social and medico-social professionals to work differently, in

---

1 https://societenumerique.gouv.fr/plannational.

interoperability (to overcome institutional breakdowns, ensure continuity of care or management, etc.). This formalization of professional activities is accompanied by training to learn how to use software or IT "solutions" and thus aim, if not for autonomy or maturity, at least to pursue an acculturation to digital technology. Browsing platforms and producing data are the beginnings of inclusion, and these activities will require more and more time (personal and professional) in the digital world.

Digital inclusion also means an increase in the collection and use of data – in real time – that will go beyond the data stored in the various software programs used to manage "user" files. There is no longer any doubt that the computerized management of social assistance files will be combined with that of a digital health record or the shared medical file, leading to a "single" file containing data on assistance as well as those produced by those responsible for helping. In the context of digital transition, the public health code, like that of social action and families, will also be subject to other codes and scripts, those of information technology and management by algorithms opening up to "individual", "personalized" data processing. The ability to process a considerable amount of personal data simultaneously (activity and situation of individuals, physical and/or psychological health, family history, etc.) has always been a challenge for public decision-makers.

In 2020, however, digital "inclusion" was still something that was not consensual. It is growing through experiments bearing the stamp of innovation (another subsidized term) with, at best, commonplaces tinged with technophilia (it is not a generational issue, it is technology that must adapt to humans) and, at worst, an incantation (the inclusive society). It is because of a lack of resources, of solutions (which are not limited to toll-free numbers), of interest (which is a right) or, more simply, of equipment that millions of French people are in difficulty or far from the digital world. In this sense, digital inclusion remains a technical and political artifact. Thus, we must carefully observe the efficacy – both declared and effective – of measures such as the "digital pass" or this new signage: the (free) "inclusive digital" label (*Journal officiel* of March 27, 2019) or the initiative of the France service houses (now spaces)[2].

Due to the lack of usage studies and specific training to date, social intervention professionals – despite being involved in the digital transition – still get lost in the numerous and pervasive connected devices and objects. They are challenged by various injunctions and questions from users, who are sometimes already equipped and connected, or those among them who want to give up or declare themselves to be in a situation of incapacity. In the testimonies collected, or the serial experiments (Meyer and Pitaud 2017; Meyer *et al.* 2020), these professionals are aware of the needs and demands of a digitally inclusive society, but still struggle to make sense of it.

---

2 https://www.gouvernement.fr/2-000-maisons-france-service-d-ici-a-2022.

References

Meyer, V. (2019). Smart city : quel lien social pour la personne vulnérable ? In *Smart City : une autre lecture de la ville*, Cyrulnik, B. (ed.). Ovadia, Nice.

Meyer, V. (2021). Si tu te connectes, loin de me léser, tu m'inclues. In *Les enjeux de l'inclusion en protection de l'enfance*, Dautigny, S., Mahier, J.-P., Stella, S. (eds). Érès, Paris.

Meyer, V. and Pitaud, P. (2017). Transition digitale et médiations numériques dans les institutions sociales et médico-sociales. *Revue thématique du CREAI PACA et Corse*, Special issue, December.

Meyer, V., Bouquet, B., Gelot, D. (2020). L'avenir du numérique dans le champ social et médico-social. *Vie sociale*, 28, 7–19.

Digital Skills Repositories

Jean-Yves Jeannas
AFUL, Université de Lille, France

Digital skills repositories

Competency frameworks have existed for decades: they have been introduced to guide learning and assessment in different disciplines, and sometimes refer to formal knowledge, sometimes to informal or interdisciplinary knowledge.

Without entering into the debate on the definition of a skill, their main interest lies in the possibility of segmenting learning into specific elements, allowing for a more detailed evaluation, but they also allow for exhaustiveness. Indeed, they make it possible to give a complete description of what the learner must master, thus giving the trainer a checklist so as not to forget anything.

In the field of digital technology, repositories have an important place that should not be neglected, but should not be overestimated. Indeed, it is tempting to summarize digital technology as a series of manipulations, of varying degrees of complexity, which allows operations in the various digital environments that we can encounter to be performed.

In France, the national education system used this principle to list the manipulative skills that were thought to be sufficient to deal with the digital world.

Europe has done the same with the famous ICDL (International Computer Driving Licence), translated into the European Computer Skills Passport, a slightly more respectable name than European Computer Driving Licence (ECDL), but it does not make the content any better. France has invented the B2i (*brevet informatique et Internet*), but the result is not what was hoped for, because the repository is only a guide, which should not be used as it is, but must absolutely be part of a more global system which integrates the discipline in its entirety.

For example, can we imagine learning to drive by learning how to change speeds only, then another day by only learning how to brake, then yet another day by only learning how to accelerate? This inconsistent division has unfortunately been the usual way of using digital repositories in France because, while the contents have evolved, with the C2i (*certificat informatique et Internet*) and then Pix (which is a proper noun), the method often remains the same, taking the skills one by one, forgetting that only the context (preferably an authentic one) will allow a real assimilation and, above all, a real understanding of the concepts and their stakes. Another risk is that the evaluation can be prescriptive and disrupt independent and enlightened learning if the "system" allows itself to choose in place of the user, for example – is not it said that "the measuring device disturbs the measurement"?

Here, for example, is a B2i (*brevet informatique et Internet*) skill addressed in high school: "I can navigate effectively in a document". The objective may seem coherent but it does not oblige the training to consider what a document is: is it an open and interoperable format or, on the contrary, is it proprietary and closed, making its user captive to proprietary software? Thus, the societal issues of digital culture may not be addressed at all, while each skill in the repository will be taken into account and validated.

Reducing repositories to the use of computer tools and manipulative strategies does not allow us to train learners to stand back and critically analyze, a necessary condition for them to become truly autonomous in their choices. Unfortunately, we continue to train button-pushers ("clickers"), who know how to use a tool without knowing the real issues at stake. For example, the use of a "social network" managed on the other side of the Atlantic may seem positive for many things, but it compromises digital privacy and digital sovereignty, not to mention the algorithms of the GAFAM platforms and companies with the same economic model, which amplify extremism and can manipulate voters (see, e.g. the Cambridge Analytica scandal, concerning the potential manipulation of 90 million voters).

Worse still, the ECDL, which has been seized upon by private organizations to sell certification courses, has been nothing more than a Trojan horse for the GAFAMs, who have penetrated the European business environment through opaque and intrusive partnerships and funding.

On the other hand, the European repository DigComp (Digital Competencies for the Citizen) is sufficiently flexible (it could also be criticized for not being precise and exhaustive enough) for users to be obliged to integrate it into an integrated and global system, as they cannot use it as is. Thus, it can be used for training in handling, like the French repositories, but it can also be a guide to identify missing skills when building a training system based on authentic situations, modeled on real life. It can then be used to ensure that all fields of the discipline have been covered, in particular, the issues of sustainability, interoperability and the societal stakes of these associated concepts.

However, even in the case of DigComp, the repository, if it is legitimate for a diagnostic or certifying evaluation, must also pave the way to the self-appointment of the user in their choices and uses of digital technology, which the GAFAM world does not allow. This is shown by the fact that the conditions of use of the licenses are not readable and that the users' signatures are mechanical.

In conclusion, digital skills repositories can be good tools if they are in the hands of people who master the completeness and the challenges of digital technology. They can have a margin of flexibility to provide, in addition to digital skills, a real digital culture that is necessary to enlighten citizens on the issues of their choices and uses. In particular, this culture must raise awareness of the issue of the fairness of online services and the impact of digital technology on the climate and health of all.

In this context, the repository can also have a reflexive vocation, so that each person can measure the state of their strengths and limitations in relation to digital tools. This is because, although manipulation seems to be mastered by a large number of citizens, digital culture is lacking for the majority. Thus, the European working group that produced DigComp counted 48% total or partial illiteracy in the European population, all ages and all socio-professional categories combined. This notion is defined as the difficulty, or even the incapacity, that a person encounters in the informed use of tools due to a lack of knowledge of their functioning and their impact. Paradoxically, digital natives, who have been "connected" from a young age, fail to take a distanced and critical look at a constantly changing environment. This is why they are now often called digital naives. As for the other generations, they are sometimes passionate about learning but remain dependent on the tools they use. These "digital immigrants" are in fact "digital dependants" of GAFAM, often without being fully aware of it.

References

Colloque C2i (2011). C2i niveau 1 et culture numérique. JYJ, Montpellier [Online]. Available at: https://slideplayer.fr/slide/11309429/.

DigComp V2.1 [Online]. Available at: https://ec.europa.eu/jrc/en/publication/eur-scientific-and-technical-research-reports/digcomp-21-digital-competence-framework-citizens-eight-proficiency-levels-and-examples-use.

Wikipedia (2022). Certificat informatique et internet [Online]. Available at: https://fr.wikipedia.org/wiki/Certificat_informatique_et_internet.

Digital Sovereignty

François Pellegrini
LaBRI, Université de Bordeaux, Talence, France

The attributes of sovereignty

Sovereignty refers to the capacity to exercise power over a geographical area and the population that occupies it. It implies, for the political system concerned, being able to carry out, with full autonomy, all its missions, and in particular that of ensuring the essential needs of its population. In modern mechanized societies, in addition to guaranteeing food sufficiency, it implies ensuring the supply of energy, raw materials and goods that are necessary for the realization of the various human activities. The inequality of the geographical distribution of various resources has precipitated many major conflicts and significantly directs the strategy of powers.

The digital revolution is opening up new areas of exchange in the form of digital services, collectively referred to as "cyberspace". The Internet, a global public digital inter-network, has gradually become a common good to which access has become a fundamental right. More broadly, the generalization of digital tools in conducting human activities has made them dependent on the proper functioning of computer systems.

Digital sovereignty can therefore be defined as the ability to exercise power over the digital spaces necessary for the activities of organizations and the population of a geographical area.

The obstacles to digital sovereignty

Digital spaces, although allegedly immaterial, exist only because the data that represent them are made available and processed on hardware that is rooted in the physical universe.

Digital sovereignty is therefore based on the ability to implement and maintain the IT systems used by organizations and individuals. It requires mastery of technological chains, including the design and manufacture of processors and hardware, the production of software and the implementation of IT hosting, as well as mastery of cross-disciplinary skills such as cyber-security. To carry out these activities successfully, the necessary scientific, technological, design and financial capacities must be available. However, the digital economy is characterized by the pre-eminence of economies of scale and network effects, which favor the dominant players.

On the hardware side, the cost of a modern processor foundry is in the order of tens of billions of dollars, and the volumes of processors to be manufactured must be large enough to amortize the costs of production. Control of the processors is central, as it is illusory to implement cyber-security measures at the system and software level when the processors themselves have irreparable security flaws, as is the case with almost all of those currently available.

The same is true for hosting infrastructures (the cloud), even if on a smaller scale. The dominant players are investing more and more in infrastructure that allows them to provide these services at the lowest price, as well as in software deployment services that maximize customer comfort in terms of functionality and usability.

In terms of software, user habits and opaque file formats create closed rent ecosystems, from which users and "knowers" are reluctant to leave. The bundling of pre-installed software helps to convince buyers that there is no alternative to the dominant players.

In terms of services, network effects are also at play. Social networks are only attractive when the majority of users' contacts are present, as are the content platforms with the largest catalogs. This greatly reduces the possibility of competition, especially since users' available time, which is an eminently competitive commodity, becomes scarce and limits the number of tools used (three or four social networks on average).

On the legal front, a number of mechanisms also have the effect of limiting the competition that could be presented to the dominant players. This is the case of patents on algorithmic methods (improperly called "software patents") which, although irrelevant or even counterproductive from a macro-economic point of view, and even illegal in Europe, are promoted by the offices that make a lucrative trade in them. This is also the case with "digital locks" (DRMs, Digital Rights Management devices), which limit access to distribution channels and reduce competition in the market for reading tools and the systems that support them. But "who controls the pipes controls the content".

The difficult implementation of digital sovereignty policies

In view of the complexity and intertwining of the various issues (hardware, infrastructure, software, services, etc.), a digital sovereignty approach can only be conducted on the basis of a holistic vision of the strategies to be deployed. Such a vision presupposes, on the part of the political system that implements it, the political coordination of a wide range of resources. This can be observed in the United States, China and Russia, but not (yet?) in the European Union, where the doctrines of economic deregulation and the "free market" have led to an economic vassalization whose effects and causes must be reversed. The non-exhaustive list presented here is intended to illustrate the scope of the subject.

The first of these strategies concerns taxation: many companies in a monopoly position benefit from privileges in calls for tender and tax systems of convenience, which bias the market in their favor. At the European level, the combination of a freely circulating single currency with uncoordinated national tax policies allows for the existence of perfectly-assured tax evasion mechanisms. These tax losses result in a lack of investment capacity, especially in education and innovation.

The second concerns the law. As we have seen, certain legal systems favor rent-seeking behavior to the detriment of innovation and competition. Moreover, if a company is not fully subject to the law of its host country, it may be forced by its parent country to provide it with data in secret or to install back doors in the systems it markets. From this stems the need to make territorial law exclusively applicable to all actors operating in the sovereign territory. An example of this is the General Data Protection Regulation (GDPR) which, in the area of personal data, aims to apply to any actor wishing to interact specifically with residents of the European Economic Area. The apparent extraterritoriality vis-à-vis foreign companies is the counterpart of the sovereignty regained regarding the law that is applicable in full and without derogation in the European Economic Area, regardless of the origin of the service provider.

The third is software. Because it is not conceivable to fully fund the development of alternative software with a limited market, governments must invest in the emergence of local ecosystems of publishers and service providers for application portfolios that are available under free licenses. Investment in these products, in terms of functionality and design, aims at making them more widely used in order to reduce the total cost of ownership. In the same way, the use of such software and open formats must become the rule in administrations, and particularly in education, in order to break the bias that leads to the perpetuation of captive markets from the start. For this model to work, a significant fraction of the savings made must be reinjected as investments in the ecosystems concerned.

The fourth concerns hosting platforms. Having software is not enough if one is not able to implement it in a secure way. It is therefore necessary to have shared platforms, governed solely by local law, with ergonomic deployment services that are subject to specific investments. The public–private partnership model can be a virtuous one here, if strict compartmentalization of activities is ensured.

The fifth concerns components, especially processors. It is in this sector, as in the aeronautics and space sectors before this, that the most concentrated investments must be made. These investments must be accompanied by actions to encourage the creation of sufficient markets, because an industrial tool cannot be maintained if its products do not find buyers. It is also in this field that partnerships must be the most extensive, as with Africa, a political powerhouse and a market in the making.

The sixth concerns the regulation of platforms. The banning of Donald Trump from social networks illustrates the power of these players, which is highly concentrated, over the circulation of ideas and the creation of opinions. The methods of promoting or silently filtering content must be made explicit, in the interests of transparency and in order to reduce the considerable asymmetry that exists between, on the one hand, these platforms and, on the other hand, their users and the regulators. While the creation of "French-style" competitors is illustrative of the good old days, the promotion of the use of decentralized alternative tools can be an achievable political objective.

The sample presented above fully illustrates, for political systems wishing to ensure their digital sovereignty, the need to assume the strategist role, both in vast technological fields and in those of law and regulation of uses. This requires, in particular, the definition of long-term industrial strategies and a global vision at the highest level, as a strategy for leadership in digital spaces and their underlying technologies.

References

Degans, A. (2019). La sécurité économique de la France dans la mondialisation : une stratégie de puissance face aux nouveaux défis du XXIe siècle. PhD Thesis, Université de Reims Champagne Ardenne.

Latombe, P. (2021). Report no. 4299 made on behalf of the Information mission on "Building and promoting national and European digital sovereignty". Report, French National Assembly [Online]. Available at: https://www.assemblee-nationale.fr/dyn/15/rapports/souvnum/l15b4299_rapport-information.

Longuet, G. (2019). Le devoir de souveraineté numérique. Report no. 7, Commission d'enquête sur la souveraineté numérique, French Senate [Online]. Available at: https://www.senat.fr/notice-rapport/2019/r19-007-1-notice.html.

Morin-Desailly, C. (2013). L'Union européenne, colonie du monde numérique ? Report no. 443, Commission des affaires européennes, French Senate [Online]. Available at: https://www.senat.fr/rap/r12-443/r12-443.html.

Digital Transition

Vincent Meyer
Université Côte d'Azur, Nice, France

Explosion, mutation, revolution, transformation: summoning these terms to qualify the (great) deployment or folding of the digital world makes one dizzy, so much so that they intersect in the accompanying discourse and set the bar extremely high. These terms are widely used in scientific and gray literature, in the daily and specialized press, without the authors going back over their meaning or proposing a distinction between them or their temporality. For "transition" and, moreover, "digital", it is first and foremost a question of qualifying an "atmosphere" of progress for all, where we breathe more easily, in the sense given to progress in the work of Pierre Teilhard de Chardin (1881–1955), but not free of turbulence due to always being subject to accelerations in terms of innovation. According to the *Cambridge Dictionary*, the noun "transition" designates the change from one form or type to another, or the process by which this happens. This is the case of the so-called industrial and/or ecological transitions, which are generally included in "energy transition", which also includes the deployment of very high speed broadband throughout the territories (particularly in rural areas, for the benefit of agriculture and telemedia). Second, the use of such a conjunction and the emphasis on "digital" owe nothing to chance, to the reproduction of a hype vocabulary. With this conjunction, it is a question of taking into account an effervescence in the making and of giving another meaning to the conventional formula of the moment, namely, "don't say that you don't know anything about computers, say that you are in digital transition". There are several reasons for this choice (Meyer 2017, pp. 19–21).

Work in the sociology of professions, on the emergence of sociomedia at the end of the 1990s (Meyer 2004) to those emerging around a sociotics, has led to a great deal of caution about the impact of technological innovations in a field or a professional group as well as on their qualifications, a fortiori today for the "new" digital employees. This is still confirmed, unsurprisingly, by work in the sociology of digital technology (Boullier 2016). Talking about digital transition makes it possible to express this caution – which is sometimes, or already, intuitive – without distorting it or thinking that we have grasped all the issues at stake in this transition "from one state to another" (especially given the pervasive nature of these technologies). Thus, talking about digital transition allows us to begin by thinking about the devices (places, infrastructures, equipment, etc.), then about/with the public and their uses. Thus, transition is a process that is never imposed in a brutal manner (even with a test like the 2020 health crisis), it is indeed a time of passage that is worrying and/or exhilarating for some; full of pitfalls (but without repulsion or rejection) for others. The so-called demographic transition is a good illustration of this; our living conditions are improving, leading to an annual growth rate, which is certainly progressive, depending on the country, but continuous for the world population. Concerning the "digital" transition, it is, more certainly, a time of action where the abundance of these technologies (they are now everywhere...) helps us to think and formalize their possible consequences and/or impacts. Consequently, talking about digital transition tempers the dizziness of "explosions/mutations/ revolutions/transformations" linked to digital technologies, whatever the professional field (and more broadly of a "this will kill that"). It supports the effective digitization of all professional and/or personal human activities, by referring to their traces or marks. It does not omit the passive, cautious, even suspicious state of individuals with regard to managerial strategies and procedures that will be imposed by a communication that "organizes" through digital technologies.

Finally, the "digital", coming back to the hand (to the fingers, to the traces and signatures that each individual leaves), is also what one brings of oneself (from one's manipulation of a technical object to the contents of the messages that it allows) to others in one's capacity as an Internet user (or surfer) without a priori or direct profit, and this is already the case with different generations of Internet users. Of course, digital natives no longer have to demonstrate any agility in the use of digital technologies that is not innate, even if, as mentioned in a report by the French National Digital Council (*conseil national du numérique*, CNNum[3]) in 2013: "the non-connected [had already] become a minority". This council rightly insisted on the need to avoid approaching digital accessibility/access solely from the angle of a digital divide, specifying that, in such a transition with these objects, everyone is "at risk", since everyone is in a permanent learning situation. These issues call for an overall ambition to make special efforts to help people who are far from the digital world, and to address the problems of equipment and territorial coverage. This ambition is that the digital devices on which we are going to depend (platforms in the lead and search engines in support) must be increasingly easy to access and use and that the generational aspects, which are only "human biases", will erase themselves.

This transition also involves a political awareness that is being consolidated in reports (making digital an accelerator of diversity, in September 2020) and in proposals (45 proposals for an inclusive digital technology, in October 2020) to offer solutions or issue recommendations, such as the one on digital technology as

---

3 https://cnnumerique.fr/.

an accelerator of "diversity"[4]. This report formulates 15 recommendations that are organized around three accompanying speeches aiming at: guaranteeing information and access to training on digital opportunities for profiles coming from diversity; rethinking and measuring the recruitment policy in order to respond to the lack of diversity in the digital sector; and reinforcing the policy of revitalization of territories. Thus, such a transition allows us to support a postulate that some may find absurd: we are all in a situation of disability, vulnerability and/or fragility when it comes to digital technologies and their development, which we must first access, then appropriate and then use. This exposure and visibility of a diversity of uses already shows that any transition can reinforce social inequalities, as well as produce other norms, and also allow for an awareness.

The relationship with living environments and spaces will also be impacted by this digital transition. Beyond the (sustainable) environmental impact of digital technologies for everyone, our research fields – social and medico-social establishments and services (ESSMS) – are concerned with the issue of de-institutionalization and, with it, increasingly technologically-equipped and connected and/or virtual home help that already goes beyond the stage of "teleconsultation". Certainly, in their design of places and services, the ESSMS will not soon be transformed into a coworking space and/or equip each reception or living space with a smartroom or "Google home assistant", these non-social but vocal assistants that speak to us and are able to capture and store personal and professional data, or assistance robots with "responsible" algorithms. Beyond the ESSMS, as "closed" environments, which are often still disconnected, this transition will also affect the territories, the so-called city policies. Here we are thinking of the so-called intelligent city (or metropolis) projects, equipped with third places where public service activities, or substitutes, such as those linked to the social and solidarity economy, are developed. More broadly, environmental and climatic factors and those linked to the preservation of biodiversity – that is, reducing the environmental footprint of digital technology – will be essential elements of the future digital ecosystem, both urban and rural.

Far from the universality imagined by the pioneers of the Web, and in the face of competition from the "giants" and/or key players of the moment, who are – becoming more and more like proto-states – concentrating all the power and capital of the digital world via high-performance computers, we are indeed living in a world in transition, where we have to permanently learn to mobilize "resources", to be satisfied with "solutions", to work alongside/with increasingly intelligent socio-technical devices (and with, one hopes, their designers or developers), whose uses

---

4 https://www.vie-publique.fr/sites/default/files/rapport/pdf/276196.pdf.

are far from "neutral", to accept being assisted or even supervised by these devices. Being in digital transition means managing a connected present that is different at every moment. This digital transition will progressively be equipped with notional entities that researchers will have to define, such as acculturation, inclusion, transparency, self-nomination, maturity and/or temperance or sobriety. All these notions, or "solutions" that are already available, allow for the (re)qualifying of digital technology and its impacts, including environmental ones. Finally, this transition is also about time, time spent in front of screens, objectified/connected time. Being in transition will also mean a temporality of ownership for the ever-new or eternal learners that we are, and our membership of a community in the face of the proliferation of technological innovations (if I am a geek and I think I am more efficient that way, I am not sure my colleague is…).

The digital transition is therefore not a technological tsunami in the sense of a wave that would submerge everything, coupled with the proverbial tension of "this will kill that". It prefigures a world in the making where everything converges via a computational approach, with cumulative effects linked to the interoperability of basic digital services with connected objects with increasingly personalized functions. It will also be a transition of knowledge allowing a capture (and mainly that of our attention), a monitoring as well as a data visualization (of all our professional as well as personal activities), which will go further and further back in time. We can think that such a memory will also serve as an accelerator of social cohesion, of the advent of low tech, of open data, of digital commons where, in addition to the input of serendipity, the question will be asked of what, in the end, is indispensable data to (re)create social links. This must be at the "center" of the digital transition (Meyer 2017), while respecting the fundamental public liberties of all.

References

Boullier, D. (2016). *Sociologie du numérique*. Armand Colin, Paris.

Meyer, V. (2004). *Interventions sociales, communication et médias : l'émergence du sociomédiatique*. L'Harmattan, Paris.

Meyer, V. (2017). *Transition digitale, handicaps et travail social*. LEH, Bordeaux.

Meyer, V., Bouquet, B., Gelot, D. (2020). L'avenir du numérique dans le champ social et médico-social. *Vie sociale*, 28.

Stiegler, B. (2016). *Dans la disruption. Comment ne pas devenir fou ?* Les Liens qui Libèrent, Paris.

Disability

Nathalie Pinède
MICA, Université Bordeaux Montaigne, Pessac, France

Digital technology is becoming an increasingly important part of our lives, our societies and our cultures, whether by choice (in our leisure activities, in our relationships with others) or by injunction (dematerialization of administrative procedures or job-seeking processes). The Covid-19 pandemic, with its accompanying periods of lockdown and social distancing, is contributing to a sudden acceleration of this digital pervasion. However, this omnipresent digital reality, through the multiple tools and connected techniques that give it form, is not self-evident for everyone. Beyond the promises of opening up, substitution and even re-enchantment, there are many risks of fragility, vulnerability and even exclusion. In this context, the question of disability, in conjunction with that of digital technology, poses major challenges. Two dimensions should be highlighted in particular: the first, which is fundamental, concerns the notion of "disability" in the face of the digital environment and the collective responsibility for the environment in which we live; and the second, which is linked to the diversity of people, highlights the sensory and cognitive involvement at the heart of digital mediation and the fractures that can result from it.

Representations of disability have changed considerably in recent decades, moving from the "crippled person" of the 1930s to the "disabled person" in the early 2000s, that is, from an individual approach based on illness and health to a focus on the environment as the source of the person's disability. This is a major paradigm shift, which no longer makes disability a mainly medical problem but rather a real social responsibility, requiring adapted public action methods, such as those deployed around accessibility. Two important milestones can be highlighted here, firstly the new classification proposed by the WHO in 2001, the "International Classification of Functioning, Disability and Health" (ICF), which, inspired by the model of the Quebec classification (Fougeyrollas *et al.* 1998), goes beyond a vision based on deficiencies alone to take into account, in a multidimensional way, the person's environment, in connection with personal factors or with the context of the activities in which they operate. In France, the law of February 11, 2005 on "equal rights and opportunities, participation and citizenship of people with disabilities" defines disability as "any limitation of activity or restriction of participation in society due to a substantial, lasting or definitive impairment of one or more physical, sensory, mental, cognitive or physical functions". This change in perspective and a broader vision of disability that includes permanent or occasional, emergent or sudden situations, as well as vulnerabilities or weaknesses that may arise with age, are fully in line with this new approach. Senior citizens are particularly affected by significant sensory and cognitive changes. The World Health Organization (WHO) estimates that more than 1 billion people, or about 15% of the population, live with some form of disability, and that this proportion is tending to increase, particularly due to the aging of the population and the increase in chronic diseases. In France, approximately 12 million people are said to be disabled, although the vast majority of these situations are socially invisible: 80% of disabilities are invisible (hearing and psychological disabilities and chronic diseases, for example). In a broader vision of disability situations, we can include the phenomenon of "illectronism", that is, the persistence of difficulties in browsing the Web or handling digital-related services, despite very good rates of equipment ownership (CSA Research 2018).

This shift from an individual issue to a social responsibility for disability fully concerns digital technology, since it is becoming a major component of our current ecosystems. From the moment that digital technology concerns the social, professional and cultural participation of each and every person, based not only on personal choices, but also on compulsory access to certain services, resources and means, the question of equity of access to digital technology is legitimately and necessarily raised at the level of society.

In addition to the social responsibility in the situation of disability in relation to digital technology, it is important to underline another dimension, that of the physical and cognitive commitment through which we put the digital into action. While the latter can be interpreted as a "milieu" (Bachimont 2015), as we have seen previously, it is embodied by our daily manipulations in a multiplicity of devices, tools, services and media, which implicitly suppose a certain level of skills and competences, both physical and cognitive. Digital technology is a matter of "sense", in two meanings (mobilization of our senses and ability to produce meanings from digital content). Sight, hearing and touch are constantly mobilized in the manipulation of digital access interfaces. The problem of disability is therefore also raised with regard to the "scripted" dimension of the technical object, the choices in the organization and architecture of digital services and documents, and also the affordances of interfaces. These elements and their arrangements may come into conflict with sensory (sight, touch and hearing), bodily (gestural) or cognitive specificities, with regard to which, as individuals, we are not all equal.

The link between disability and digital technology therefore poses real challenges, particularly in terms of inclusion, challenges which, moreover, are by no means totally without ambiguity, with both positive effects and risks. Undeniably, we must emphasize the possibilities opened up by digital technology for people with disabilities (enabling technologies), such as facilitating access to information resources or the fluidity of communication processes (Casilli 2010), and also the preservation of social and family ties, particularly for the elderly. Nevertheless, the risk of a digital divide remains constant for people with disabilities on three different levels: the instrumental divide, that is, the difficulty or impossibility (physical or sensory) of accessing services via media and interfaces; the cognitive divide, resulting from difficulties in understanding content; and finally, the social divide, the third level and a direct consequence of these multiple obstacles. These divisions are mirrored by symmetrical levels of vulnerability for people with disabilities: vulnerability of the body, in the face-to-face encounter with the material and sensitive devices of the digital world, a face-to-face encounter that presupposes a certain level of commitment; cognitive vulnerability, in the face of complex content and arrangements, which can prove to be a source of exclusion; and finally, social and educational vulnerability, a rebound effect that all too often accompanies situations of disability. The implicit image of digital technology as "self-evident", together with the vision of an "ideal digital user", therefore represents fantasies which still alienate many achievements and conceptions. This means that we forget somewhat the "technical constraint" and the "vulnerability of the subjects" (Voirol 2011), two extremes between which flaws related to disability situations can be created and reinforced.

References

Bachimont, B. (2015). Le numérique comme milieu : enjeux épistémologiques et phénoménologiques. *Interfaces numériques*, 4(3), 385–402.

Casilli, A. (2010). Technologies capacitantes et "disability divide" : enjeux des usages du numérique dans les situations de handicap. In *Vers la fin du handicap ? Pratiques sportives, nouveaux enjeux, nouveaux territoires*, Gaillard, J. (ed.). Presses universitaires de Nancy, Nancy.

CSA Research (2018). Enquête sur l'"illectronisme" en France [Online]. Available at: http://sps.fr/wp-content/uploads/2018/06/Rapport-CSA-pour-SPS_Illectronisme_Mars-2018_120318.pdf.

Fougeyrollas, P., Cloutier, R., Bergeron, H., Côté, J., Saint-Michel, G. (1998). *Processus de production du handicap, classification québécoise*. International Network on the Disability Creation Process, Quebec.

Voirol, O. (2011). L'intersubjectivation technique : de l'usage à l'adresse. Penser une théorie critique de la culture numérique. In *Communiquer à l'ère du numérique : regards croisés sur la sociologie des usages*, Granjon, F. and Denouel, J. (eds). Presses des Mines, Paris.

Diversity

Samia Ghozlane
Grande École du Numérique, Paris, France

Women and the digital world

Under-representation of women in the digital sector...

Women are under-represented in the digital sector: they represent 30% of employees in this sector, in all the professions combined, and the figures are not improving. For example, the number of female tech graduates (higher education, digital and engineering) fell by 6% in France between 2013 and 2017, and the proportion of female graduates fell by 2% over the same period in the digital sector (Gender Scan 2019). Moreover, when we look at the hierarchical level of women in the digital sector, we note that the more responsibilities there are, the less women are represented: only 18.5% of managers are women, according to the study conducted by the organization AnitaB.org in 2019. As for women entrepreneurs in the tech sector, they face barriers when launching their startups. The 2019 study by SISTA and Boston Consulting Group shows that startups founded by women are 30% less likely to raise funds than those founded by men, and in France, only 2% of funds are raised by female startups.

...with damaging societal and economic consequences

The under-representation of women in the digital sector represents a societal regression that leads to inequalities and has an economic cost. The digital sector is one of the driving forces of the economy today, with companies that are experiencing significant growth and transforming society. It is the sector that will generate the most new jobs in the years to come: employment in the digital sector is growing 2.5 times faster than in other sectors, according to Pôle emploi. Excluding women from this sector is harmful, as it would deprive them of employment opportunities in a fast-growing sector with more stable employment status and prestigious and highly-paid career prospects (Stevens 2016). Similarly, from an economic point of view, according to the European Commission, if women held as many jobs as men in the digital sector, there would be a gain of around 9 billion euros per year for European GDP (Stevens 2016). As for France, parity in digital would generate 10% more GDP by 2025, according to a 2016 McKinsey study.

Women, pioneers in the history of technology

In 1843, Ada Lovelace created the first computer program. In 1941, Hedy Lamarr filed a patent to secure telecommunications, which is still in use in Wi-Fi and Bluetooth links. Grace Hopper created the concept of the "compiler", developing the first model of a computer in 1952, and Mary Keller submitted the first thesis in computer science in 1965. In the 1950s, half of the computer workforce was made up of women, and they remained the majority until the late 1980s. By the 1980s, 40% of computer science degrees in Europe and the United States were awarded to women (Collet 2019).

In the 1990s, two phenomena were decisive in the fall in the number of women in the digital sector. During this period, computing grew in power and became a strategic issue for companies and governments. As a result, it gained in prestige and men rushed into it en masse. Moreover, the appearance of personal computers has made it possible to equip almost exclusively men (fathers and their sons) in the home. In fact, history shows that when a field of knowledge gains importance in the social world, it becomes masculinized (Collet 2019). In the 1980s–1990s, when computer security was discussed, 20% of professionals in this field were women. In 2020, as cybersecurity has now become strategic, women represented only 11% of the workforce in this speciality. In 1990, the "computer employee and operator" profession was mainly made up of women data entry operators. These positions have gradually become "computer operations operators"; much more qualified and gaining in importance, they are now mainly occupied by men. Finally, until 1960, computer "coding jobs" in Britain were almost exclusively female (Morley 2019). In 1965, in the United States, women represented 30% of programming specialists. In 1982, 35% of computer science jobs in France were held by women. Today, the importance of data processing and analysis and the advent of artificial intelligence mean that women now account for only 12% in this field in France (Collet 2019).

How can we restore gender diversity in the digital sector?

For centuries, gender stereotypes have been embedded in the collective unconscious, with social and cultural beliefs about women's unfitness for scientific fields and the attribution of specific skills to women based on pseudo-scientific studies. Restoring gender diversity in the digital sector will require much more than a change of mentality. Numerous voluntary initiatives, supported by various political, economic and associative players, are working to improve gender diversity in the digital sector: communication campaigns with regard to training and professions have highlighted female "role models" with exceptional careers, awareness-raising actions in elementary and high schools, quotas in educational establishments, etc. Indeed, digital professions must be presented, from school onward, in a sufficiently inclusive and attractive way so that young women can see themselves in them. These presentations by female role models are useful but, in reality, it is the trivialization of women in the digital sector that will ultimately demonstrate that gender diversity has been achieved (Collet 2019). The introduction of a quota policy has also proved effective. For example, in Norway, after quotas were introduced at the University of Trondheim, the percentage of women in computer science rose from 7% to 40%. Nowadays, quotas have been removed there (Collet 2019).

In France, many initiatives exist to develop the presence of women in the digital sector. These include the *Femmes@numérique* foundation, the *Journée de la femme digitale* (JDF) and the *Grande École du Numérique* (GEN). The latter works within the framework of a public policy through its GEN label and the granting of money to improve access to tech professions for women. The goal of the GEN is to train at least 30% of women in its accredited courses (Key figures 2020).

References

Collet, I. (2019). *Les oubliées du numérique*. Le Passeur, Paris.

Gender Scan (2019). Formation, emploi, satisfaction au travail et évolution de carrière des femmes dans le secteur des technologies et du numérique [Online]. Available at: http://www.global-contact.net/wordpress/wp-content/uploads/2019/11/Synth%C3%A8se-Gender-Scan-19.pdf.

Grande École du Numérique (2021). Chiffres clés 2020 [Online]. Available at: https://www.grandeecolenumerique.fr/sites/default/files/2021-06/GEN_ChiffresCles_2020%E2%80%93A4-web.pdf.

Morley, C. (2019). Pourquoi les filles ont délaissé l'informatique ? *The Conversation* [Online]. Available at: https://theconversation.com/pourquoi-les-filles-ont-delaisse-linformatique-110940.

Stevens, H. (2016). Mais où sont les informaticiennes ? *Travail, genre et sociétés*, 2(36), 167–173 [Online]. Available at: https://www.cairn.info/revue-travail-genre-et-societes-2016-2-page-167.htm.

E

Eco-digital Responsibility

Jean-Yves Jeannas[1] and Marie Cauli[2]
[1] *AFUL, Université de Lille, France*
[2] *Université d'Artois, Arras, France*

The use of digital technology is now an integral part of our daily lives. If we can see the beneficial aspects of this evolution, we can also see the undesirable effects on our lives and on our environment. By optimizing manufacturing or management processes, digital technology has long been presented as allowing environmental gains, such as the "zero paper" objective linked to digitalization. However, the figures have toned down these claims. They show that the surging wave of digitization and the exponential growth of digital uses, accompanied by an inflationary supply of consumption, have accelerated and amplified energy consumption and environmental impacts. However, they are ignored as such by the general public, who have difficulty understanding the lifecycle of the infrastructures that condition the technical equipment with which everyone is equipped today. Thus, we all too often forget, when we handle them, that dematerialized exchanges only exist by relying on material infrastructures, consisting of terminals, computer centers and networks that are far from the image of an impalpable sector. What is certain is that there is nothing immaterial behind the operation of the digital sector and that energy consumption is very real, with the exploitation of natural resources and the impacts linked to the manufacture of equipment, transport, waste treatment, uses, etc.

Resource and energy prices

Energy consumption and the resources required are difficult to measure, as they are intertwined in the production cycle of equipment and its operation, as well as in its use. Energy is essential not only for the electrical operation of computers, laptops and connected objects, but also for computing, storage and routing operations, as well as for accessing emails, sharing videos and all types of documents, not to mention the equipment used in cars and household appliances. In view of the increase in usage and the renewal of technologies, digital energy consumption is increasing by 9% per year and accounts for 4% of global GHG (greenhouse gas) emissions, with digital usage responsible for 55% of energy consumption compared to 45% for equipment production, according to the 2018 Lean ICT report, "For digital sobriety". As an example, and to present an enlightening order of magnitude, 600 kg of raw materials are needed to produce a computer, a box consumes between 150 and 300 kWh, as much as a large refrigerator, the distance traveled by a digital data is about 15,000 km on average.

Materials, precious and rare metals

The production of such equipment requires raw materials, in particular metals, including certain precious or rare metals, and substances such as cobalt. It involves the extraction of these resources, the production of components and the assembly of equipment. Networks, for example, are major consumers of copper, while devices need rarer metals to improve their performance. In a smartphone, for example, no less than 40 metals are present, ranging from a few milligrams to a few dozen grams (indium for touch screens; silver, copper, tin, platinum, tantalum, tungsten for electronic cards; cobalt for batteries, magnets, etc.). Similarly, with the penetration of digital technology in sectors such as the automotive industry, the demand for metals involved in the manufacture of electronic components is becoming exponential. Thus, the connected world and digital uses could not exist without the exploitation of energy resources. Among these are rare earths. These include 17 different metals that have been described as rare for three reasons: their difficulty of extraction; their low production because they are mixed with other minerals; and their chemical properties. Their unique properties are coveted for developing wind energy technologies, and also fiber optics, touch screens, miniaturization of equipment, etc. Only a few players are involved in exploiting them, and as a result they have a virtual monopoly. According to a USGS report, "International Minerals Statistics and Information", China produces 44% of the world's indium, 55% of its vanadium and 95% of its scandium and yttrium. Other countries are also involved – Brazil (90% of niobium) and Congo (60% of cobalt) – and also Vietnam and Russia. The demand for such minerals, in the face of accelerating technological developments, places non-producing countries in a situation of dependence, and raises questions about geopolitical, economic and social balances, given the conditions under which the minerals are exploited, as well as the scarcity of these resources. It may lead to the risk of depletion of deposits, as well as tensions over supply.

Equipment and uses

On the usage side, the increase in environmental impact can be explained by the multiplication of peripherals, the acquisition of smartphones, the rise of the Internet of Things and the explosion in data traffic. In the energy consumption balance, data consumption is responsible for almost half of the global impact of digital technology. These flows are mainly generated by the consumption of services provided mainly by the GAFAMs (Google, Apple, Facebook, Amazon, Microsoft) and companies with the same business model. These are divided into different uses, the relevance of which can be questioned in terms of human progress, or simply in terms of quantity. Among the most energy-intensive contents, videos are those that require a large volume of data. They account for nearly 60% of global data traffic. Similarly, the cloud, which gives the impression of having infinite storage space, encourages the multiplication of seemingly innocuous actions that use infrastructures of global proportions and require a substantial amount of energy to operate. All of these information transfers trigger a complex routing mechanism, activating algorithms all along the path of our data and inducing a consumption of energy that is not negligible (e.g. a 10-minute video is equivalent to five hours of emails sent in a row).

Moreover, we are increasingly equipped and renew our equipment while it is still in working order, which has consequences in terms of carbon footprint disparity, for example. The carbon footprint of an American with a much higher level of equipment is 16 times higher than that of a person living in a developing country. This effect is justified by the benefits that a more recent model is supposed to provide, by making it easier to use, more fun, more powerful or faster. The most successful example is that of telephony, which is subject to a commercial campaign extolling the prowess of the latest models and exacerbating the ever-growing temptation to buy. The aim is to encourage consumers to buy the latest model, even though they often only use a small part of what is possible (two to two and a half years on average for a smartphone before it is renewed).

Overconsumption

In addition, the proliferation of equipment is compounded by its programmed obsolescence. This process, which is planned and integrated into the design of equipment, often shortens its lifespan and leads to its premature replacement while it is still in working order. There is no shortage of examples. For example, some operating system or software companies have designed their software and updates in such a way that the increase in power is necessary and makes a computer in perfect physical condition unusable from a software point of view. In addition, there is the difficulty of repairing increasingly complex devices, because they cannot be dismantled or are incompatible with new developments. In the same way, it was thought that digitization would lead to savings and gains in environmental protection, but this was without taking into account what is known as the "rebound effect". This term is used when a gain in efficiency of a product or service results in an increase in demand, canceling out the expected benefits. For example, the substitution of products and services by digital equivalents has long been presented as a factor of energy saving or work saving, but dematerialization processes have run up against rematerialization effects. Examples include users printing administrative documents from digitized contracts or invoices, etc.

At the other end of the chain, the growth in equipment has an impact on the growth in waste production, which continues to increase. Landfilled, burned, illegally traded or difficult to recycle, the end-of-life management of all this equipment has become a huge issue.

Thus, in the energy field as in all others, digital technology is accelerating the systematization and intensification of production and resource exploitation techniques. It amplifies environmental impacts in inverse proportion to the level of public awareness of these issues, which have become tangible.

Developing eco-digital responsibility

This is why, in view of the importance of this issue, it is essential to raise awareness of the dangers of pollution and exploitation of natural systems, and to apply the recommendations for the preservation of the planet and the protection of future generations in the digital sector. These recommendations, applied to the environment, are beginning to have an effect, as shown by the emergence of new behaviors aimed at reducing, for example, the consumption of certain food products, or favoring more ecological modes of transport. In the same way that we clean up nature or plant trees, we need to encourage eco-digital responsibility, to disseminate a culture of sobriety, or even frugality in the digital domain, which, contrary to a general trend, encourages over-consumption. To achieve this, regulatory levers must be identified at different levels. First, this means questioning the sustainability of the social and economic model supported by digital companies, while the demand for sobriety and the fight against waste are spreading. This implies setting up a concrete collaboration between regulatory, political and judicial actors, service providers and users, and inventing the most judicious modes of regulation. This is a vast question that researchers can address by identifying the obstacles and the conditions for removing them, and a difficult question for researchers who cannot dissociate the results of their research from their consequences. For users, that is for the majority of people, it is a question of facilitating the transition from an intuitive and instinctive digital world to an informed digital world. This is a difficult task, as it requires a real deconditioning process. This can begin by raising awareness of the collective impact of individual behavior and unintentional clicks, and by giving preference to equipment, operating systems and software that do not practice programmed software obsolescence. This approach can only be efficient by developing training in the fundamentals of digital technology and technical gestures, such as data compression and sorting, to detect the real relevance of uses, to break away from automatisms generated by an addictive design, which implies understanding and mastering technological complexity as much as possible, but also becoming aware of the systemic impacts of digital technology on society.

References

ADEME (2014). Internet, courriels : réduire les impacts. Report, ADEME.

European Commission (2020). Best environmental management practice in the telecommunications and ICT services sector. Report [Online]. Available at: https://susproc.jrc.ec.europa.eu/product-bureau/sites/default/files/inline-files/jrc121781_final_bemp_report_telecom-ict_1.pdf.

France Stratégie (2019). La consommation de métaux du numérique : un secteur loin d'être dématérialisé. Working document, France Stratégie.

Pitron, G. (2018). *La guerre des métaux rares, la face cachée de la transition énergétique et numérique*. Les Liens qui Libèrent, Paris.

The Shift Project (2020a). Déployer la sobriété numérique. Report [Online]. Available at: https://theshiftproject.org/wp-content/uploads/2020/10/Deployer-la-sobriete-numerique_Rapport-complet_ShiftProject.pdf.

The Shift Project (2020b). The Shift Project a-t-il vraiment surestimé l'empreinte carbone de la vidéo ? Report [Online]. Available at: https://theshiftproject.org/wp-content/uploads/2020/06/2020-06_TSP-a-t-il-surestime%C3%A9-lempreinte-carbone-de-la-vid%C3%A9o-en-ligne_FR.pdf.

Educational Digital Technology

Valèse Mapto Kengne
Université de Yaoundé I, Cameroon

Solving teaching-learning problems from digital perspectives

Following the Education for All (EFA) goals from 2000 to 2015, and owing to the ongoing Sustainable Development Goals (SDGs) from 2015 to 2030, planet Earth is going through the digital age accelerated by Covid-19. Covid-19 has been a trigger in terms of uses and is pushing us to look at the potential of educational technology, information and communication technology (ICT) and digital technology. In terms of innovation, it is the coherent and systematic adoption of tools in pedagogical perspectives in line with the missions and realities of large-scale education that is targeted. How did Covid-19 drive the process of transformation and innovation, leading to its gradual implementation in the field of education in Cameroon?

The 2030 Agenda for Sustainable Development contains a goal for education (SDG 4), which aims to ensure equal access to quality education for all and promote lifelong learning opportunities. It also contains a goal for industry, innovation and infrastructure (SDG 9) and has several targets including innovation, research and development. The goal is to increase access to ICTs and ensure that all people in the least developed countries have affordable access to the Internet by 2030.

In other words, the 2030 SDGs recognize the importance and power of ICTs in the progressive changes, dynamic and sometimes accelerated growth of human societies. Still, it is necessary to measure their impact on the ground. According to Temkeng (2012), in Cameroon, the appropriation of ICTs in Cameroonian schools is limited by the lack of effective and efficient educational policies, the means of implementing knowledge transmission devices and the lack of willingness of some actors to train teachers accordingly, given the economic and financial contexts.

Yet the health crisis has prompted political and educational actors to make eloquent speeches about the need for effective use of ICTs in schools for health safety and accountability.

In recent years, educational research teams have attempted to measure the level of integration, introduction, appropriation, penetration, innovation and development of ICT and digital technology in elementary schools, high schools and universities (Tchameni Ngamo and Karsenti 2008; Temkeng 2012; Beche 2017; Roland *et al.* 2017). The information collected shows that several policy actions have been undertaken to support and increase ICT in learning, teaching and assessment processes in sub-Saharan Africa, and in Cameroon in particular. School curricula offer computer courses, and distance education is no longer rare. But the pedagogical activities put in place are not sufficient in a constrained environment for learners to acquire, build and develop information, computer and digital skills. These difficulties have an impact on the quality of training provided in education systems.

Thus, the key skills required for literacy, schooling and lifelong learning must now integrate and include digital competence, with digital pedagogues, but without the conditions being met. Moreover, the task is complex, because it is about developing digital competence, which is beyond skills, abilities and aptitudes. Digital competence is a combination of information, knowledge and skills that, under the cognitivist paradigm, is linked to the connectionist approach in a perspective that links know-how, action knowledge, experiential knowledge and self-knowledge.

Digital competence is also tactile knowledge. In order to teach, actors of the education profession posess and mobilize technological knowledge in a given context, in an educational situation and when faced with tasks requested or to be accomplished, to teach, learn and evaluate.

In addition, collaborating, communicating, storing, building, cooperating, exchanging, evaluating, interacting, obtaining, sharing, participating, presenting and producing information via an Internet connection make digital education a complete system. This system is at the heart of transdisciplinarity in the Piagetian sense of the term. Beyond interactions, there are applications and associations between disciplines, knowledge and the various relationships with knowledge.

The results achieved and expected from digital pedagogues are linked to this combination of real-time action capabilities. Problem solving is articulated, so to speak, to the use of collaborative networks, critical distance from negative effects and innovation in relation to ICT in the evolution of society (UNESCO 2018).

However, the new coronavirus, SARS-CoV-2, is breaking through and not only putting an end to physical and social interactions, but also mandating physical and social distancing within the education and training systems, and also in the public sphere. Digital tools and equipment have made it possible to combat the spread of Covid-19 in Cameroonian society, in certain educational environments and families.

However, in developing countries, in sub-Saharan Africa and in Cameroon, urbanization rates, education rates, health coverage rates and connectivity problems are low. Access to education is still restricted by constraints such as distance, rurality and household poverty. Classrooms and lecture halls are crowded. The availability of infrastructure, and the acquisition of educational resources and materials are growing relatively, but are far from meeting the needs. Consequently, the indicators necessary to intensify the processes need to be constructed, educational policies reformulated and strategies put in place in these contexts where the health crisis is one with the crisis of the educational systems.

The digital age has transformed the transmission and sharing of knowledge. The development of digital technologies through the Web has become the main tool for the dissemination of information by some educational and academic institutions, including universities and private colleges. Websites, blogs, Moodle, emails, social networks, Facebook and WhatsApp are now central to the communication, teaching, learning and even assessment strategies deployed by institutions and learners (Karsenti *et al.* 2014). The use of these dissemination platforms often boils down to responding to numbers, depositing and making accessible courses, exchanges and scientific literature.

Moreover, digital tools have the advantage of having multiplied the channels of dissemination, circulation, navigation and free access to articles, school and university works, but they do not necessarily facilitate the task of teachers, because they must not only integrate ICT, NICTs, on the one hand, but also exercise critical thinking, on the other hand, and lead learners to develop the higher levels of the mental activity in learning, as highlighted in Bloom's work (1956), namely, analysis and synthesis for the purpose of constructing and producing knowledge from the discovery of more than abundant documents.

In short, Covid-19 has had a potential and palpable influence on the development of digital education in Cameroon and has established a digital education generation. The facts, objects, phenomena and technological and digital educational realities are gradually being implemented in the Cameroonian educational system. They are pushing all of the actors in the education sector to adjust, invest and change their outlook on the needs of schooling, education and distance training with the Internet.

The observation of innovative approaches in education and training environments confirms these trends. The borrowing and adoption of educational resources that can be used through Internet networks has, in some cases, allowed for educational continuity in educational institutions. However, the issue of training teachers and learners in the use of digital education on a large scale for the purposes of ongoing justice, equity, equality, inclusion, schooling, education and training remains problematic. Teaching ICT, according to the UNESCO report (2018), consists first of all of integrating it into pedagogical approaches, and helping students and learners to use it, but also in inserting it into a collaborative, cooperative, dynamic and interactive learning process of creativity and problem solving.

References

Beche, E. (2017). Conceptualizing the implementation of distance learning system at the Higher Teacher's Training College of Maroua (Cameroon). *International Journal of Humanities Social Sciences and Education* (IJHSSE), 4(5), 51–62.

Karsenti, T., Dumouchel, G., Komis, V. (2014). Les compétences informationnelles des étudiants à l'heure du Web 2.0 : proposition d'un modèle pour baliser les formations. *Documentation et bibliothèques*, 60(1), 20–30.

Roland, N., Stavroulakis, M., François, N., Emplit, P. (2017). MOOC Afrique : analyse des besoins, étude de faisabilité et recommandations. Research report [Online]. Available at: http://hdl.handle.net/2013/ULB-DIPOT:oai:dipot.ulb.ac.be:2013/254265.

Tchameni Ngamo, S. and Karsenti, T. (2008). Intégration des TIC et typologie des usages : perception des directeurs et enseignants des grandes écoles secondaires du Cameroun. *Revue africaine des médias*, 16(1), 45–72.

Temkeng, A.E. (2012). Types de cadrans et appropriation des TIC dans les écoles camerounaises : du manque de moyens au manque de volonté. In *La formation de formateurs et d'enseignants à l'ère du numérique : stratégies politiques et accompagnement pédagogique, du présentiel à l'enseignement à distance*, Karsenti, T., Garry, R.-P., Benziane, A., Ngoy-Fiama, B.B., Baudot, F. (eds). Réseau international francophone des établissements de formation de formateurs (RIFEFF)/Agence universitaire de la Francophonie (AUF), Montreal [Online]. Available at: https://depot.erudit.org/id/003772dd.

Electronic Voting

Chantal Enguehard
LS2N, Université de Nantes, France

Definition

The terms *electronic voting*, *digital voting* and *e-voting* refer to the use of one or more digital tools during the voting period, from its beginning to the end of the counting, whether it is an election, a referendum or a consultation. For example, it may be the *expression of the vote* by an elector, or the *counting of votes* that establishes *the number of votes* attributed to each voting option.

Depending on the vote, several properties are at stake. Here are the most important ones:

– *Uniqueness*: each elector has a unique right to vote.

– *Sincerity*: the candidate who wins is the one who has received the most votes.

– *Anonymity*: there is no link between a voter and their vote.

– *Confidentiality*: each voter votes alone, out of sight.

– *Transparency*: ability to gather evidence and testimony (with probative value) of violations of the fairness of a vote.

Confidentiality can be guaranteed when voting takes place within the confines of a polling station and under the supervision of its members: in France, the latter must ensure that each voter publicly takes at least two different ballot papers and then, in a polling booth, places the ballot paper of their choice in an envelope. On the other hand, when voting is done by mail, confidentiality cannot be guaranteed.

Anonymity and *confidentiality* protect the *secrecy of the vote*. In order to ensure that the *freedom to vote* is complete and that each voter votes without being *pressured*, the secrecy of the vote must not only be respected, but voters must also be fully aware of it.

Transparency is a prerequisite for the possibility of filing an electoral dispute that could lead to the annulment of the voting operations by an electoral judge.

There are specific features for certain digital devices. Here are two of them:

– A *signature management* takes place to guarantee the respect of uniqueness – some e-voting devices record votes as well as signatures. Others are limited to the recording of votes, while signatures are embodied in those written by voters in a signature book[1]. The recording of the signatures, independently of the electronic voting system, allows studies to be carried out on the accuracy of vote collection (Enguehard and Noûs 2020).

– *Verifiability* – when voting is anonymous, it is not possible to check, even approximately, the accuracy of the election results provided by electronic voting devices operating without a ballot paper. So-called *End-to-end Voting Systems (E2E) have* been proposed, such as paper-based voting computers or some Internet voting systems (where verifiability relies on encryption processes). However, these systems suffer from limitations (Chevallier-Mames *et al.* 2006) and from an insufficient legal framework to guarantee the implementation and effectiveness of verifications in case of electoral disputes (Enguehard 2015).

Diversity of electronic systems

There is a wide variety of electronic voting systems, and also different terms for them (which we indicate in italics). Here are models of the most common devices. We also point out some variations.

---

1 In France, when a member of the public wishes to vote, the members of the polling station consult the *cahier d'émargements* (signature book). In this book is a list of all the voters and a place to collect their signature. If the signature space is empty, the person is allowed to vote. They then cast their vote and sign their name in the signature book. This ensures that each voter has only voted once.

Voting computer (voting machine)

These devices are most often placed in polling stations. They directly record voters' voting intentions in their memory (usually without printing ballots) and then provide some election results at the end of the voting period. In some countries, they can be linked to the Internet to transmit the results. However, the electoral code and the rules applicable in 2020 do not allow this in France.

In general, ballots are recorded independently of the voting computer; however, there are models that also handle ballot counting, as in Venezuela, where the voting machines perform biometric fingerprint scanning.

For political elections, voting computers are used by some voters (France, Belgium) or in the entire country (Brazil, India).

There are voting computers advertised as verifiable, despite the limitations identified (see above), that print ballots in addition to recording voters' choices in their memory (*Voter Verified Paper Trail*, *Voter Verified Paper Audit Trail* and *Voter-Verifiable Paper Record*). They are used in some counties in the United States.

Finally, there are rare cases of atypical use, such as in 2018 in the Democratic Republic of the Congo, where voters used voting computers to consult the list of candidates, make their choices and then print them on the ballot paper provided to them. The counting was then done manually (Lesfauries and Enguehard 2018).

Optical recognition system (scanner)

Each voter has a ballot paper on which they tick boxes (with a pen or using an electronic system) to express their choices. The ballot papers are then read by an optical recognition system which establishes the number of votes obtained by each voting option. This method of counting is partially used in political elections in several countries, such as the United States and Russia. Voters either have their ballots read directly by a ballot box equipped with a scanner, or the polling station staff performs this operation at the end of the voting period, possibly centralizing the ballots from several polling stations.

Many cases of ballot papers bearing a code identifying the personal identifier of each voter have been noted. This code makes it possible, if necessary, to break anonymity by revealing the voter's identity.

In France, many professional elections by correspondence are managed via scanners. This is the case, for example, for the election of members of the CNRS scientific councils.

Internet voting (e-voting, online voting)

Voters express which candidate they wish to vote for using a platform accessible via the Internet. This is a form of remote voting.

For political elections, Internet voting has been tested in Switzerland since 2001 and is authorized in Estonia. Several countries have opened up this new voting channel to their expatriates (occasionally in France, for certain cantons in Switzerland, etc.). However, tests carried out in many countries (the United Kingdom, Japan, the Netherlands, Norway, Canada, etc.) have not led to the development of this voting method.

Voting by e-mail

This is the modality proposed to US armed forces based abroad. In this case, neither confidentiality nor anonymity can be guaranteed. The same applies to *voting by fax*.

From uses to controversies

There is also a wide variety of uses, whether for political, professional or associative voting. It is necessary to examine the technical devices proposed and the functionalities that will be used, their compulsory or optional nature, their possible coexistence with other voting methods and their legal framework.

Thus, in France, the use of a voting computer is the only voting option in polling stations where they have been authorized and are installed. The use of Internet voting was allowed for French citizens abroad in the 2012 legislative elections, but voters could choose to vote in a polling station or by mail using ballot papers.

In Estonia, Internet voting is allowed for all political elections, but an Internet vote can be canceled and restarted several times, including by voting at a polling station. This would be illegal in France, as it implies maintaining a link between each voter and their voting expression, which may threaten the respect of anonymity.

The use of electronic voting devices in political elections has been controversial in several countries, resulting in the limitation of the deployment of voting computers in France, the cessation of their use in the Netherlands and Germany, and their non-use in Ireland.

Proponents of e-voting point to many advantages, such as the ease of voting for all voters (including people with disabilities), speed, protection against fraud and modernity. However, these arguments are rarely based on scientific research.

On the other hand, all electronic voting systems reduce transparency, and therefore the ability to cancel an election in the event of an infringement of sincerity. Moreover, there are strong questions about security, respect for the secrecy of the vote or even ease of use (Sénat 2014). In Germany, the Constitutional Court of Karlsruhe prohibited their use because of their violation of the constitutional principle of the public nature of elections (Seedorf 2015). There have been several reports of the vulnerability of Internet voting systems, from bugs (Teague and Halderman 2015) to demonstrations of fraud (Grégoire 2012).

Conclusion

The single term "electronic voting" covers a multitude of electronic voting systems that try to fit into heterogeneous electoral traditions, governed by very different legal rules. This situation limits the scope of possible comparisons or transfers (of technology, knowledge). Each case of use must be carefully examined.

Moreover, electronic voting remains a difficult and resistant problem, due to the conjunction of anonymity, uniqueness and dematerialization. It requires examination at the confluence of several scientific disciplines (computer science, political science, legal science, sociology, etc.), which further complicates the understanding of all its dimensions.

It is the subject of intense research published in several conferences and scientific journals, including:

– *International Conference on E-Voting and Identity* (E-VOTE-ID);

– *Journal of Election Technology and Systems* (JETS);

– *European Conference on e-Government* (ECEG);

– *The Research Committee RC 10 on Electronic Democracy of the World International Political Science Association* (IPSA).

References

Chevallier-Mames, B., Fouque, P., Pointcheval, D., Stern, J., Traoré, J. (2006). On some incompatible properties of voting schemes. *IAVoSS Workshop on Trustworthy Elections (WOTE)*, 191–199.

Enguehard, C. (2015). Les dispositifs de vote électronique dits vérifiables. In *Le Vote électronique*, Guglielmi, G. and Ihl, O. (eds). Lgdj, Paris.

Enguehard, C. and Noûs, C. (2020). Some things you may want to know about electronic voting in France. *E-Vote-Id 2020*, September 28.

Grégoire, L. (2012). Comment mon ordinateur a voté à ma place (et à mon insu). *Les Inrockuptibles*, 27 May.

Lesfauries, V. and Enguehard, C. (2018). L'introduction de machines à voter en République démocratique du Congo. *Les Convergences du droit et du numérique*, October.

Seedorf, S. (2015). Germany: The public nature of elections and its consequences for e-voting. In *E-Voting Case Law*, Laurer, A.D. and Barrat, J. (eds). Routledge, London.

Sénat (2014). Rapport d'information n° 445 fait au nom de la commission des lois constitutionnelles, de législation, du suffrage universel, du règlement et d'administration générale sur le vote électronique. Anziani, A., Lefèvre, A. (Senators). Sénat, 9 April.

Teague, V. and Halderman, J.A. (2015). Security flaw in New South Wales puts thousands of online votes at risk. *Freedom to Tinker*, March 22.

Empathy[2]

Serge Tisseron
Université de Paris, France

Empathy for objects and machines

It is customary to distinguish two aspects of empathy. The first is strictly emotional: it is feeling, in front of someone who has an emotion, the same one as they had. This impression can be misleading, in that we sometimes have a false idea of what another person is feeling. The second aspect of empathy is cognitive: it allows us to understand what the other person is feeling and thinking, and this may be different from what we feel and think ourselves. This skill relies on observation, memory, knowledge and reasoning, and is commonly associated with the "theory of mind", which emerges around the fourth grade. But these two aspects are insufficient to speak of complete empathy. A "motivational" factor must be added: complete empathy is a mental construct that enables not only understanding the experience of others, but also representing it subjectively (Tisseron 2010). It is this component that determines whether the person, after experiencing emotional and cognitive empathy, will act or not, especially if it is to prevent the suffering of others.

---

2 This text is a reworked and translated version of a more complete version of my text that originally appeared in Tisseron and Tordo (2021).

However, empathy is not only a condition of social life with our fellow human beings. It also plays an essential role in our relationships with objects because of the fundamentally anthropomorphic nature of human beings (Tisseron 1998). Far from being a disadvantage in our relationships with the environment, it is even at the origin of the formidable human capacity to domesticate the world. But with non-living animate agents, the errors of judgment that it entails are likely to create new problems.

Two systems

In 1966, Joseph Weizenbaum designed a computer program written to simulate a Rogerian psychotherapist, called ELIZA. He found that some users, while convinced that it was a machine, thought it gave them the same quality of attention as a human being. Weizenbaum referred to this as "cognitive dissonance" between the user's awareness of the programming limits and their behavior respect to the information given by the program.

In the 1990s, the anthropocentric attitude of the user of a digital tool was studied within the CASA (Computers Are Social Actors) paradigm. It applied to both natural and laboratory environments, even though users agreed that their machines were not humans and should not be treated as such.

In 2011, Daniel Kahneman showed that human beings use two modes of reasoning to manage their relationships with their environment. The first, which he calls "system 1", is fast and intuitive, while the second, which he calls "system 2", is on the contrary slow and reflexive. In the case of our relationships with objects, system 1 leads us to adopt the same behaviors as we do with our fellow creatures for convenience. For example, if my computer breaks down, I can say to it, "No, you're not going to do that to me. Not today!" But if I can reprimand my computer, I do not expect it to respond to me, and I do not worry about it being angry. Indeed, this is where system 2 comes in, taking into account the fact that only living things have specific goals that they pursue according to their own logic. However, the more objects are able to simulate human skills, the more we will integrate objects into our network of relationships, and the more the cognitive biases of system 1 will increase.

It has already been shown that the imagined suffering of a mishandled or damaged robot is very poorly experienced by many observers, and may even be unbearable for some (Rosenthal-von der Pütten et al. 2013). This is hardly surprising, since imaging has revealed that the emotional displays of a human and a robot are perceived relatively similarly by an observer. Similarly, a person talking to a robot spends as much time looking into the eyes of the robot as they spend looking into the eyes of a human interlocutor, even though the robot's "eyes" are reduced to two shapes drawn on the sphere that serve as its "head". An experiment conducted at a pain clinic in Japan showed that when a robot was introduced as an observer in a consultation, 33% of patients were rather reassured by its presence, while only 6% said they would prefer it not to be there. Moreover, when the robot smiled and nodded in sync with the doctor's words, 40% of patients said they were reassured by its presence, and this percentage rose to 50% when the robot smiled and nodded in sync with the patient's attitudes (Takano *et al.* 2008). Finally, in a disaster simulation, some people would even prefer to sacrifice humans who seem useless to them than robots who seem useful (Nijssen *et al.* 2019).

More research is needed to better understand the personal, cultural and environmental factors that influence our empathy for robots, and what can increase or decrease it. But already, three risks must be taken into account: forgetting that these machines are permanently connected and impose the solutions of their programmers on us; thinking of them as equivalents of humans capable of emotions when for a long time to come they will only be machines to be simulated; and finally preferring them to humans or thinking of them as a desirable image of the human. Their development poses legal, ethical, educational and psychological problems that are essential to consider today.

References

Kahneman, D. (2011). *Système 1/Système 2 : les deux vitesses de la pensée*. Flammarion, Paris.

Nijssen, S., Müller, B., Van Baaren, R. (2019). Saving the robot or the human? Robots who feel deserve moral care. *Social Cognition*, 37(1), 41–52.

Rosenthal-von der Pütten, A., Krämer, N., Brand, M. (2013). Investigation on empathy towards humans and robots using psychophysiological measures and fMRI. *Computers in Human Behavior*, 33, 201–212.

Takano, E., Chikaraishi, T., Matsumoto, Y., Nakamura, Y., Ishiguro, H., Sugamato, K. (2008). Psychological effects of an android bystander on human-human communication. In *Humanoids 2008 – 8th IEEE-RAS International Conference on Humanoid Robots*, 1–3 December.

Tisseron, S. (1998). *Comment l'esprit vient aux objets*. PUF, Paris.

Tisseron, S. (2010). *L'empathie au cœur du jeu social*. Albin Michel, Paris.

Tisseron, S. and Tordo, F. (2021). *Comprendre et soigner l'homme connecté : manuel de cyberpsychologie*. Dunod, Paris.

Ethics

Gilles Dowek
INRIA, ENS Paris-Saclay, France

Information technology ethics is placed in the general framework of the ethics of technology: information technology, like any technology, gives humans greater power to act well or to act badly. It is, for example, the same image recognition algorithms that are used in medical imaging and mass surveillance. The purpose of the ethics of technology is therefore to inform our individual and collective choices, so that we can use technology to do good and not evil.

Information technology ethics is, however, different from the ethics of technology on several points.

Values under construction

The ethical approach is often based on a set of values, which makes it possible to qualify an action as good or bad according to whether it respects them or not. For example, biomedical ethics is based on the values of personal autonomy, benevolence and justice. The ethics of information technology brings about new values, such as transparency, respect for privacy, explicability, loyalty, security and oblivion.

This corpus of values is still under construction and is already being constantly called into question by the global, and therefore transcultural, nature of information networks. The question of the values to be respected in an exchange between people located at geographical and cultural antipodes creates tensions, which can only be resolved by seeking compromise. For example, in their hierarchy of values, Europeans place respect for the memory of the victims of crimes against humanity before freedom of expression, while the United States does the opposite. This creates a tension between the hierarchy of values when a website publishes a text on US soil that glorifies crimes against humanity and this text is accessible in Europe. For some, it should be removed, for others, it would be an infringement of freedom of expression. The tension can only be resolved if each side takes a step toward the other.

Information technology ethics in the face of anthropomorphism

Because they complement some of our intellectual faculties – reasoning, speech, memory, the ability to learn, etc. – computer objects induce a risk of anthropomorphism that looms or locomotives that do not – but which, paradoxically, teddy bears also induce. Faced with this risk of blurring the line between human and machine, information technology ethics vigorously reaffirms the unbridgeable nature of this boundary.

A tension appears here between ethics, which values the human, and science, which tends to reify it. This movement of reification began with the materialism of the 18th century, developed with the rise of biology and was accentuated by information technology, in particular by the Church-Turing thesis and the Turing test, which ingenuously questions the way to evaluate whether an entity, human or not, thinks. This tension is resolved by a distinction between the scientific and ethical fields, the latter introducing a boundary between the human and the machine, and postulating, for example, that it is more serious to injure a human than to break a machine. And that it would remain so even if they had the same faculties.

Ethics of everyday life

Information technology ethics questions every moment of our existence, whereas biomedical ethics, for example, questions moments that are certainly crucial, but rare: birth, death, procreation, etc. This specificity makes information technology ethics, ethics of everyday life.

Ethics and institutions

Political institutions are essentially a place where information is exchanged. It is therefore natural that they should be disrupted by the development of computers – as they were by the invention of writing or printing.

Information technology ethics must therefore consider changes in political models, for better or for worse, the abolition of distances and therefore of borders, new forms of inequality, access – once again, for better or for worse – for all to the public voice, etc.

Open questions in information technology

Ethics readily postulates that we know the difference between good and bad actions, and that we know how to act well, even when we act badly.

Information technology ethics puts this idea into perspective. For example, in the interest of privacy, it may recommend that some data be anonymized. However, data anonymization is a research problem that has not yet been completely solved: algorithms exist, but no algorithm is perfect and researchers are still improving them. So while we need to anonymize some data to do the right thing, we do not know how to do it.

A recent concern

Finally, information technology ethics is a relatively recent concern. While, for example, ethical questions in medicine date back to Hippocrates, ethical questions in information technology only really appeared with the development of the Web at the end of the 20th century. In France, for example, the CERNA[3], whose mission was limited to questions of research ethics in information technology, the CNPEN[4], whose mission is broader, or the more specific ethics committee for educational data[5], only date back to the 2010s.

References

Abiteboul, S. and Dowek, G. (2020). *The Age of Algorithms*. Cambridge University Press, Cambridge.

Devillers, L. (2017). *Des robots et des hommes*. Plon, Paris.

Ganascia, J.-G. (2019). *Le mythe de la singularité : faut-il craindre l'intelligence artificielle ?* Points, Paris.

Grinbaum, A. (2019). *Les robots et le mal*. Desclée De Brouwer, Paris.

---

3 See: http://cerna-ethics-allistene.org/.

4 See: https://www.ccne-ethique.fr/fr/actualites/creation-du-comite-pilote-dethique-du-numerique; https://ai-regulation.com/the-french-national-committee-for-digital-ethics.

5 See: https://www.education.gouv.fr/le-comite-d-ethique-pour-les-donnees-d-education-12146.

F

File Formats

Jean-Yves Jeannas
AFUL, Université de Lille, France

File format: definition

A computer file is either the digital representation of an object (text, image, sound, video, etc.) or a collection, a census of documents, objects, people, places, etc. It refers to the concept of a "list" which existed long before computer science. The latter has changed its scope by making it possible to "translate" the real world into digital data and store a considerable volume of information that can be reused in various fields of application. This process, which is still unknown to a large part of the public sphere, led to the creation of the French "law on information and freedom" in 1978, in order to protect citizens from the liberticidal use of data. Therefore, a few elements are necessary to understand, on the one hand, the technological principle that governs the computer file and, on the other hand, the more complex strategic and societal issues surrounding the notion of computer file format.

Technically speaking, computer files refer to familiar elements that we handle every day and contain our texts, photos, music, etc. We usually think of them as organized documents associated with our office layout. This same principle governs the computer file: it is a matter of arranging documents in files that are themselves deposited in file systems, placed in directories that are themselves organized. They have a name that can indicate the nature of the content and an extension that identifies the software that handles them, and each file has its own specific "instructions for use".

This computer storage technique has a long history. It is explained by the discovery of binary mathematics by George Boole in the 1850s. He invented the "Booleans", numbers and operators that make it possible to calculate using only two numbers: zero and one.

In 1864, John Henry Holmes invented the switch, and in 1867, Charles Sanders Peirce noted the similarities between binary mathematics and the switch, which either activates or does not activate the flow of electric current.

It was not until the late 1930s that physicist Claude Shannon associated the "open" state of a switch with the value 0 (zero in binary) and the "closed" state with the value 1 (one in binary). By the joint meeting of these discoveries, computer storage was born.

What followed was a succession of discoveries, from applied physics to electronics. Switches were miniaturized, from the lamp to the integrated circuit, via the transistor. The latter is an electronically controlled switch, without any mechanical part. It stores the same simple information: open or closed, current flows or current does not flow, zero or one, etc. This coding allows logical operations to be performed using these two states, zero and one, which are bits, representing the smallest units of information physically corresponding to the electronic components. These are also known as binary digits. In the same way, thanks to various physical properties, the juxtaposition of eight wires that will allow the transmission of data between electronic circuits will create the "byte", composed of eight bits.

These units of measurement, written in binary, are still used today and remain at the heart of computing. They feed a specific logic that is different from human logic and language and is not "natural" to us. To understand its mechanism, we can, for example, experiment with converting a byte into a decimal number. To do this, the decoding algorithm consists of summing each of the bits multiplied by the base to the power of the rank. This sounds complicated but it is simply multiplication and addition. Namely: in binary, the base is 2 (two) and the rank is counted from right to left, starting at 0 (zero), which makes it possible to calculate the value of the byte which, in our case, gives for example that of a pair of shoes.

Byte	0	0	1	0	1	1	0	1
Rank	7	6	5	4	3	2	1	0
Base	2	2	2	2	2	2	2	2
Base^rank	128	64	32	16	8	4	2	1
Byte x Base^rank	0	0	32	0	8	4	0	1

Calculating the value of a byte in decimal

The sum gives us: $32 + 8 + 4 + 1 = 45$. The shoes are worth 45 euros. Thus, we can represent "human" numerical values with 0's and 1's, and it is quite simple, if a bit time consuming, to find the decimal value. The same is true for many other objects in everyday life, such as letters, images, sounds and videos. In the case of files, the only difference lies in the computing power and speed of the computer compared to humans. For this reason, computer scientists who design software have devised decoding algorithms for the objects we manipulate on computers. Each algorithm defines a file "format" that allows us to know what it represents. We then simply use the algorithm to find the value of the object we are looking for, such as the price of our shoes, and we are done. In this process, the computer scientists "named" the files to distinguish them. A name and an extension were given to them by convention. Thus, the *name* is in some way equivalent to a first name, the extension is in some way equivalent to a last name. It allows you to know what is in the file (its format). For example, "Report1.odt" describes a text document in a standardized open and interoperable format called ODF (Open Document Format).

Since then, it has been possible to encode very complex data in a very simple way, to store them and to vary numerical values using physical devices that are organized and automatic extensions of the human memory. So everyone travels with billions of bits in their pocket all the time if they have a smartphone. Smartphones contain thousands of files that represent data, the programs (applications) used by each person and all of the files necessary for the system to function, and also files placed on the machine by online services (such as cookies).

Open versus proprietary formats

However, this technical process raises questions about the strategic and societal issues surrounding computer file formats. Indeed, in order to decipher the content of the byte, it is necessary to know the decoding rule. But, even if it is mastered, not everyone necessarily has access to the decoding algorithm of the file. Indeed, this can be "protected", that is registered in a "proprietary format" which is also called "closed", known only by the companies that published the format concerned. Therefore, the user does not have the choice of software and remains dependent on the company. They cannot normally open and modify the file without paying a license fee, materialized by the price of the license of the proprietary software, the only one able to understand the totality of the format concerned. For example, the Microsoft formats .DOC, .XSL and .PPT, used by some people on a daily basis, are the archetypes of these closed proprietary formats.

This mode of operation, which is little known to the general public, has been identified as a risk of dependency. Faced with this danger, actors involved in the implementation of an open digital world, fair and respectful of freedoms, have worked with the ISO (International Organization for Standardization) to define an open and interoperable format, and therefore readable and usable free by all. This format exists and is called ODF, but it does not have enough commercial power to spread. It represents the native format of free software, whose source code has been paid once and for all (by an individual's voluntary work, but also by public and sometimes even private funds), and it is available to all as an *intangible common good*.

It is therefore enough to choose free software to be able to use open formats freely without having to pay the proprietary license fee. On the other hand, any document saved in a proprietary format will require, if we want to be sure of its good use, the payment of the proprietary license.

However, there are barriers to the use of free software. Among those commonly mentioned, there is the difficulty to use it for non-experts, which is much more a problem of change of habit than a real need for new skills. It is therefore important not to reverse the burden of proof: if a file in a proprietary format is not well read by free software, it is because the company that publishes this proprietary format does not provide the necessary information about its format, and not because of a lack of quality in free software. Indeed, formats and software are often of a similar intrinsic quality, but the commercial approach is quite different. Therefore, specific studies are to be pursued in this field, where ethical, philosophical and moral angles should be addressed, in the light of a new economy where the costs of distribution and reproduction of software are almost zero.

References

Wikipedia (n.d.). Office Open XML [Online]. Available at: https://en.wikipedia.org/wiki/Office_Open_XML.

Wikipedia (n.d.). Open format [Online]. Available at: https://en.wikipedia.org/wiki/Open_format.

Wikipedia (n.d.). Proprietary format [Online]. Available at: https://en.wikipedia.org/wiki/Proprietary_format.

Formal Language

Gilles Dowek
INRIA, ENS Paris-Saclay, France

Computers are machines built to execute algorithms, and this explains why the concept of machine and algorithm are among the fundamental concepts of computer science. However, next to these two concepts, there are others, notably that of formal language.

During the genesis of computer science, this concept of formal language took some time to emerge. On the one hand, because the engineers and mathematicians who built the first computers and programs brought with them the concepts of their primary domain: machine, algorithm, etc., but not that of formal language, and on the other hand, because our culture prefers things and always looks at too much interest in words with suspicion. Thus, when the first "high-level" programming languages appeared and, with them, the first compilers who made it possible to translate a program from one language into another, a certain number of computer scientists, such as John von Neumann, looked at these objects with skepticism.

However, the need to write algorithms and design languages to do so allowed the notion of a programming language to emerge. Then other formal languages appeared – specification languages, query languages and Web page description languages, etc. – which consolidated the place of the concept of formal language among the fundamental concepts of computer science.

What is a formal language?

So what is a formal language? And what distinguishes an expression of a formal language created from scratch, such as *ttc = ht + ht * 16.85 / 100.0*, from an expression of a natural language, such as this excerpt from the law of April 10, 1954 introducing the value-added tax in France: "This tax is levied at an ordinary rate of 16.85 p. 100"? First of all, a relative simplicity of the lexicon and grammar: where a dictionary contains several tens of thousands of words, a programming language uses a few dozen key words. Second, a formal language is specialized, whereas a natural language is universal. A programming language, for example, can express algorithms, but not contracts, prescriptions or novels. A natural language, on the other hand, can express everything: contracts, prescriptions, novels and also, even if imperfectly, algorithms. Another difference is that an expression in a natural language is oral before it is written – there are even some natural languages that are exclusively oral. On the other hand, an expression in a formal language, such as a program, is written and can only be read aloud with difficulty.

But above all, formal languages break free from the two most fundamental properties of natural languages: their one-dimensionality and their double articulation. There are two-dimensional formal languages, in which the "words" do not come in a certain order. There are also languages in which each word is expressed by a single symbol, without being broken down into sounds or letters.

Formal languages, from Antiquity to the present day

The central place taken by formal languages in computer science – from the theory of automata to the semantics of programming languages – has allowed us to become aware, in retrospect, of the place that formal languages have silently occupied in our culture, since the invention of writing: from the language of prescription glasses – OD: - 1.25 (- 0.50) 180° OS: - 1.00 (- 0.25) 180°– to that of music scores, from the language of equations – $x^3 + 3x^2 = 20$ – to that of the developed formulae of molecules, we have created dozens of formal languages and this notion has its own history, relatively independent of that of natural languages. This history has its revolutions: the first formal language allowing us to write numbers 5000 years ago, the *The Sand Reckoner* in the 3rd century BC, decimal positional numeration in the 5th century, the language of scores in the 13th century, the language of arithmetic in the 15th century, the language of algebra in the 16th century, chemical nomenclature in the 19th century, predicate logic and programming languages in the 20th century, etc. And this history of formal languages is linked to two others.

Formal languages and writing

First of all, to that of writing. It is an empirical fact that texts written in a formal language – scores, equations, programs, etc. – are often exclusively written and can only be read aloud with difficulty. It is also an empirical fact that the first texts written on clay tablets expressed numbers and were therefore written in a formal language and not in a natural language.

These two observations lead us to the hypothesis that writing was not invented to write a natural language, to transcribe speech, but to write a formal language: more precisely, the protolanguage of numbers expressed in the form of clay envelopes.

Formal languages and science

The history of formal languages is also linked to that of science and technology, even if formal languages are also used elsewhere: to write music and chess games. Each science invents its own formal languages and borrows formal languages from others. Thus, the difference between physics and chemistry can be explained simply if we look at the languages they use: physics, in the 17th century, essentially borrowed the language of differential equations from mathematics, whereas chemistry, in the 18th and 19th centuries, created its own languages to name molecules and describe their reactions.

The sciences and the humanities can, perhaps, be distinguished by the fact that the former innovates by inventing formal languages, while the latter innovates by using natural language with a new virtuosity. Thus, linguistics and economics would be on the side of the sciences, and history and anthropology on the side of the humanities.

Computer science and logic

But, while there are formal languages in music or chemistry, computer science has undoubtedly given them a more central place. It is true that the acquisition of the language of scores – solfeggio – is part of learning music, as is the acquisition of the language of raw and developed formulae in chemistry, but mastery of these languages remains a marginal skill. Chemistry studies molecules more than the formal languages that allow them to be written. Whereas the study of algorithms and the languages that allow them to be written constitute two branches, both central, of computer science.

This reflexivity is essential in computer science, and perhaps the only field before computer science to have given such importance to its languages is logic. This common interest in the notion of language – even if these languages are very different – is one of the reasons that explain the links between logic – the science of *logos* – and computer science.

A formal language factory

However, computer science has undoubtedly brought about an unprecedented transformation: it has democratized the activity of creating formal languages. Today, it is no longer a matter of using a few formal languages created by François Viète, Antoine Lavoisier or Grace Hopper, but of creating our own formal languages according to our needs.

So when the first parallel computers were built, new formal languages had to be created to program them. Then other formal languages had to be created to build Web pages. And today, the creation of programming languages for quantum computers is in full swing, even though these computers barely exist. More generally, each software program more or less tacitly introduces a formal language, since the interactions between the user and the software program involve an exchange of symbols, the succession of exchanged symbols constituting the expressions of a formal language.

Computer science has, in the end, been the site of the creation of an entire universe, exclusively populated by formal language beings, and which seems to be almost as palpable as the material world, today.

Reference

Dowek, G. (2019). *Ce dont on ne peut parler, il faut l'écrire : langues et langages*. Le Pommier, Paris.

Free and Open Source Software[1]

Jean-Yves Jeannas
AFUL, Université de Lille, France

It is the free licenses that define the software as such.

More simply, software is free if it can be used, modified and redistributed without restriction by the person to whom it was distributed. Such software is thus susceptible to analysis, reuse, criticism and correction by all and for all. This characteristic gives free software reliability and responsiveness to the correction of defects, especially security defects.

Mozilla Firefox, Mozilla Thunderbird, LibreOffice and VLC are examples of famous free software. If you have ever used any of these programs, you have already used free software, perhaps without even knowing it.

Specificities of the designations

There are no profound legal and organizational differences between free and open source software, because in practice the licenses defined as free by the Free Software Foundation (FSF) and the Open Source Initiative (OSI) are identical, except for a few anecdotal cases (see the entry "Free licenses").

The open source movement, however, came well after the free software movement, which was initiated largely for the implementation of humanistic values around the development and use of software, whereas open source does not carry these same values. Free software dates back to 1986, with the founding of the FSF, while the open source movement started in 1998, more than 10 years later, following the shift from mainframe to microcomputing in 1995.

If, in practice, free and open source software refers to the same type of software, they do so from different points of view.

---

1 This text, from the AFUL, is licensed under the Creative Commons CC BY-SA.

Free software emphasizes freedom for users and customers, and open source emphasizes technological and business efficiency for customers, service providers and publishers.

The term free software is preferable because it is clearer and avoids semantic slippage such as the statement: "It's open, since you can read the code", used by some actors to make people believe that some of their source codes are free, when they are not free in the sense of the definition of free software.

The term FLOSS, for Free/Libre Open-Source Software, came later, as an attempt to propose a consensus term.

Questions and answers about free software

Is free software free?

Free software is not necessarily free. The ambiguity comes from the original expression, free software, since in English, free means either available or at no cost. In practice, a lot of free software can be obtained for free, but you have to be careful to download it from official sites, to avoid various threats linked to rogue installation by the distribution site. It is necessary to use the right Internet address (URL).

Paid versions, often very cheap, are sometimes marketed by companies, with documentation and a support contract for installation or maintenance. For example, the companies Canonical (UK), RedHat (USA) and Suse (Germany) distribute different versions of the GNU-Linux system in this way. In this way, they are establishing a new business model in which money is generated by the service, which can be geographically local, rather than an unbalanced system in which a single company (sometimes even hegemonic, in the case of GAFAM and companies practicing the same business model) makes money by "renting" the possibility of exploiting the software, even though the cost of reproduction and distribution is almost zero, thus making a profit practically without creating any jobs.

What is the difference between proprietary and free software?

The vast majority of software sold commercially is proprietary software, which is distributed as an "executable" version, whereas free software is provided with its "source code". Source? Executable? A little detour through a musical analogy helps to clarify these terms. We can consider the source code of a software as its score and the executable code as its recorded version. A score can be played on a piano, a flute or by the Berlin Philharmonic Orchestra. However, a recording pressed onto a disk does not allow the music to be changed, the instruments to be changed or the performance to be modulated.

The transition from one to the other occurs by translating the source code, or program, code that is written and then read and modified by a human into executable code (which the computer understands and executes). Free software is distributed in both forms (source code and also executable), whereas Microsoft™ or Adobe™, for example, only sell the "executable" code and hide the rest. In fact, when you "buy" a piece of software, you do not own it, you "rent" it with other restrictions visible in the EULA (End User License Agreement), often signed without being read, because it is written in a way that is specifically unreadable by the lawyers of the companies that practice proprietary licensing. See, for example, Dima Yarovinsky's artwork, originally created to evoke the terms of use of unfair online services, but whose principle is the same for the terms of use of proprietary software, which is increasingly present online in SaaS mode. Called "I Agree", the work allows us, according to him, "to underline how small and defenseless we are against these giant companies".

Who writes free software?

All people (computer scientists, graphic designers, musicians, translators, proofreaders, testers, etc.) wishing to share, benefit from collective creations and diffuse their works for the greatest number (software, texts, images, videos, music, etc.) can write free software.

At the head of each free software project, to coordinate it, there is a more or less formal structure that is composed of private individuals and/or companies.

For example, the FSF (Free Software Foundation), led by Richard Stallman, produces and/or organizes the development of free software. Thus, the FSF's GNU project (whose logo is, of course, a wildebeest) was instrumental in the creation of Linux (whose logo is a penguin).

References

AFUL (Association francophone des utilisateurs de logiciels libres) (n.d.). AFUL [Online]. Available at: aful.org.

Élie, F. (2009). *Économie du logiciel libre*. Eyrolles, Paris.

Free Licenses[2]

Jean-Yves Jeannas
AFUL, Université de Lille, France

Definition of a license

Copyright allows the law to protect all works of the mind. This protection is effective for the author (with copyright) and prohibits any form of use by third parties.

In order to be usable, the work must be made available voluntarily and explicitly by the author, whether or not it is digital. This applies in particular to the Internet. This act of making the work available is done through the addition of a license. The license is a contract proposed to the users of the work, whether they use it free of charge or not, indicating the rights that are granted to them and the possible obligations that are imposed on them in return.

For software to be free, it must have the license that defines it as such.

Definition of a free license

A license is said to be free when it guarantees the user a certain number of fundamental liberties.

For the FSF (Free Software Foundation), there are four fundamental freedoms:

– The freedom to run the software: this is the guarantee that there are no conditions restricting the freedom to run the software. You can use the software as often as you like, on as many computers as you like and for any purpose (within the limits imposed by law).

– The freedom to study the functioning of the software and to adapt it to our needs: this implies being able to access the source code of the software, that is the way it is made.

– The freedom to redistribute copies of the software: you can make as many copies of the software as you want and give them to anyone you want.

– The freedom to improve the software and publish these improvements: this is the guarantee of the possibility to improve the software and the express permission to publish these improvements.

---

2 This text, from the AFUL, is licensed under the Creative Commons CC BY-SA.

There are also the Open Source Initiative's "ten criteria", which are inspired by Debian's "social contract". These criteria are close to the FSF's fundamental freedoms.

Because of this common base, the differences from one free license to another have no direct impact on the use of the software they cover. On the other hand, they condition the modalities of the reuse of the code of these examples of software and the diffusion of what constitutes a derived work.

The different free licenses

Licenses with reciprocal obligation

The obligation of reciprocity is a basic principle. This obligation of reciprocity is achieved through the use of the copyleft principle. The copyleft principle is twofold: first, to guarantee users the fundamental freedoms; second, to guarantee that derivative works of that software will also provide those freedoms. Any software that uses code licensed under a copyleft license must, where appropriate, be released under an equivalent license.

The GPL (General Public License) of the GNU project (GNU's Not Unix) is the most representative license of free software. It is notably the license of the Linux kernel. It is written in English and translations are provided for information purposes. Most lawyers agree on the validity of the GPL on French territory, but there is no official translation, knowing that it was written with concepts of English law, which can be an obstacle in some cases. This is how the CeCILL license was born, developed jointly by the CEA, CNRS and INRIA, in order to transpose the GPL into French law.

It takes up the letter and the spirit of the Convention, and explicitly mentions its compatibility with it (section 3.4 of article 5). The drafting of version 2 was carried out after consultation with the FSF, AFUL and APRIL.

Permissive licenses

These licenses are similar to the public domain in that they only impose very weak constraints on the release of derivative works, such as the original copyright notice. It is therefore possible to make proprietary software from code released under such a license.

This family of licenses is sometimes referred to as the "BSD-like license" or the "MIT-like license", after two particularly common representatives. Among the projects using this type of license are the BSD Unix systems (OpenBSD, FreeBSD and NetBSD).

According to the same principle that guided the CeCILL license, the CeCILL-B license is an equivalent in French law of these permissive licenses.

Software component licenses

There is a third category of licenses between the two previous ones: copyleft licenses.

They allow you to link a third-party program, regardless of its license, to the programs they cover. This is sometimes called "weak" copyleft, as opposed to the "strong" copyleft of GPL-like licenses. This type of license is most often used for function libraries, but can be applied to other types of programs as well. The LGPL is the most common. The GNU project, the author of this license, has over time changed the meaning of the acronym from Library General Public License to Lesser General Public License. As for the GPL, there is no official translation, only an unofficial translation is available, which led to its adaptation into French law through the CeCILL-C license.

Multiple licenses

It is possible to distribute a work under several licenses, we speak then of multiple licenses.

A software can thus be distributed under several licenses. This is, for example, the case for Perl software, which is distributed under the GNU GPL and Artistic licenses.

In the case of multiple licenses, the author can choose to distribute their work under several licenses, even if these licenses are incompatible with each other. The Qt software, for example, was distributed by its author under the GNU GPL for free projects and the paid and non-free licenses for non-free projects.

The user of the works can only compose and associate them together if the licenses of the works are compatible with each other.

For works other than software

Because software development involves other related elements, specific licenses have been created that combine the principle of free software licenses with the specifics of the elements to which they apply. However, some non-software resources can still be distributed with free software licenses.

Documentation

– Free Document Dissemination Licence – FDDL version 1;

– GNU Free Documentation License Version 1.1 (March 2000) by the Free Software Foundation.

Artistic creation

Artists have been inspired by the free software movement to develop licenses that allow their works to be distributed under comparable conditions:

– Free art license;

– EFF Open Audio License;

– GNU Art, for the application of the GNU/GPL to artistic works.

Creative Commons licenses

The number of licenses for non-software works has grown rapidly and uncoordinatedly, causing confusion for both the user and the creator when choosing a license. The Creative Commons (CC for short) initiative was born out of the desire to provide an organized family of licenses that offer flexibility and readability.

Some CC licenses are not free, in the sense of freedom as defined for free software. This is because choosing a CC license with a NC (no commercial use) or ND (no modifications or derivative works) clause restricts the possible frameworks for further use and distribution.

In order to remain in the domain of free licenses, we can for example choose, within the framework of the creation of resources related to education, the Creative Commons CC BY-SA license (attribution and sharing under the same conditions).

Based on the CC BY-NC-SA license (authorship protection, non-commercial use and preservation of the license on derivative works), the BBC has created the Creative Archive Licence, with an additional clause that does not allow promotional or defamatory uses (which is important for the BBC's image).

References

Clément-Fontaine, M. (1999). La licence publique générale GNU. M.Phil Thesis, Université Montpellier I/Laboratoire CNRS-ERCIM, Montpellier.

Geraud, D. (1999). Le "Copyleft" ou l'état des interrogations quant à l'impact des NTIC en tant qu'élément déstabilisateur des règles de propriété intellectuelle, 12 November.

Séguin, L. (2013). AFUL : libre ou open-source ? [Online]. Available at: https://aful.org/blog/ 2013_05_16_libre-open-source.

Soufron, J.-B. (2002). La licence publique générale : un système original de protection juridique pour les créations issues des systèmes de développement coopératifs. Postgraduate Thesis, Université Strasbourg III Robert-Schuman, Strasbourg.

Working Group on Libre Software – Information Society Directorate General of the European Commission (2000). Case study of a non open source licence: SCSL. Étude de la Sun Community Source License (SCSL).

Free Software (in French National Education)

Jean-Pierre Archambault
EPI, Villejuif, France

The French educational institutional context with regard to free software was defined in October 1998 in a framework agreement signed between the Ministry of National Education and the AFUL (*Association francophone des utilisateurs de logiciels libres*, French-speaking association of free software users), an agreement that was regularly renewed thereafter. In essence, it indicated that there are quality alternative solutions for schools in the form of free software at very low cost, with a view to technological pluralism.

Open source/proprietary software

Free software is the opposite of proprietary software. When you buy proprietary software, you buy the right to use it under very restrictive conditions. For that, only the executable code, object code, is provided.

On the other hand, with free software, you have the following four freedoms. You can:

– use it for any purpose;

– study its operation and adapt it to its own needs (access to the source code is a necessary condition);

– redistribute copies without limitation;

– modify it, improve it and release derived versions to the public, so that all may benefit (access to the source code is still a requirement).

These freedoms are granted only on the condition that others benefit from them, so that the chain of "virtue" is not interrupted, as is the case with public domain software when it is privately appropriated. The GNU-GPL (General Public License), the most widespread license, translates this original approach that reconciles the rights of authors and the dissemination of knowledge to all, into law.

Issues in society

Very quickly, three types of issues emerged:

1) Computer science itself: costs, quality, security, independence, diversity, regulation of the general public computer industry whose structure favors the constitution of quasi-monopolies, open standards. In the French education system, as in companies and government agencies, the growth of open source software began with the infrastructure: in the academic departments and the central administration of the Ministry, Linux was installed on almost all of the thousands of servers that host the major information systems of the French education system. Tens of thousands of "ready-to-use" Linux servers have been deployed in schools. Open source on the desktop has progressed more slowly, despite the success of OpenOffice and Firefox.

2) On resources, the question quickly arose of the degree of transferability, of the free approach to the realization of informational goods in general, and educational goods in particular. Licenses such as Creative Commons are often adopted. With educational resources, we are "at the heart" of the teaching profession.

3) Social issues, as shown by the debates that accompanied the transposition of the European directive on copyright and related rights in the information society (DADVSI) in 2005 or the "Hadopi" law in 2009. Security and personal data protection issues, as revealed in 2013 by the documents disclosed by Edward Snowden concerning the participation of GAFAM and their major platforms in US intelligence surveillance programs.

Free, computer education and general culture for all students

Open source has concretely shown its relevance in producing quality knowledge assets, global information commons. It is also a "conceptual tool" for understanding and thinking about the problems of the immaterial. John Sulston, winner of the Nobel Prize for Medicine, speaking in December 2002 in the columns of *Le Monde Diplomatique* about the risks of privatizing the human genome, said that "the basic data must be accessible to all, so that everyone can interpret, modify and transmit them, following the example of the *open source* model for software".

This aspect of things is very important in the educational context. The stakes of free software are fully in line with the fundamental missions of the school, that is to train men and women, workers and citizens, to give all students the general culture of their time. In a natural way, the actors of free software, first and foremost APRIL, AFUL and Framasoft, found themselves with the EPI in the steps in favor of the creation of a computer science discipline as such in high school (in the final year of high school in 2012, 10th or 11th grade in 2019; we can refer to the section: "Computer science (educational status)" in this dictionary), and more generally, in academic education. In fact, the convergence is real between computer science and education, and it is not surprising that there is a lot of interest in this subject. Indeed, there is a real convergence between the principles and values of free software and the objective of a general scientific and technical computer culture for all students. Of course, a "for" loop or a recursive procedure are neither intrinsically free nor proprietary, but the teaching of computer science in high schools will, in the necessary diversity inherent to learning, make room for free software, and will propose projects that include free software issues, methods and answers in terms of copyright. Whether it is in general training, giving the fundamentals of computer science, or in professionalizing training, free software is an essential component of the curriculum.

Free educational resources

In education, not only is the software free, but also (and perhaps especially) the educational resources. Indeed, we are on the same level as the educational publishing industry and the market of 10 million students. That free software meets resistance is not surprising, especially because of the strong economic stakes.

Teachers have always made documents in preparation for their lessons. This activity is at the heart of their profession. The landscape of academic publishing has been profoundly transformed by the advent of computers and networks, and of open source software, which has rapidly become transferable to the production of other intangible resources, both from the point of view of working methods and of the responses provided in terms of copyright.

It is necessary to ensure a fluid circulation of documents, to allow their reuse, modification and appropriation by all. Pedagogy is therefore synonymous with free licenses, such as Creative Commons. There are thousands of authors, and the Sésamath association has been a reference since 2000.

Free software is a form of "educational exception". The "educational exception", that is the exemption from copyright for works used in the context of teaching and research activities, and libraries, potentially concerns productions that have not been produced for educational purposes. It is a demand of certain sectors of the educational institution and of teaching opinion, put forward with greater acuity in the context of digital technology. The teaching activity is disinterested, and the whole society benefits from it. The challenge is to legalize a "fair use" of cultural resources by teachers for the benefit of students, within the framework of their profession.

A little history

Since 1999, open source software has been a "project" of the CNDP's (*Centre National de Documentation Pédagogique*) "technology surveillance" mission. At the beginning of 2002, on the initiative of this mission, the following was created, the Scérén open source software skills center, bringing together 23 of the 26 CRDPs (*Centre Regional de Documentation Pédagogique*). In fact, this center will be the only institutional educational structure to carry out action in this field at the national level. It will inform the educational community in order to help them make their choices. This will be done in a variety of ways: by organizing and/or participating in days, seminars, symposia and fairs; writing texts and articles; and creating Websites. It will unite initiatives, skills and energies, cooperating with many actors, institutions or partners of the National Education, as well as local authorities, companies and associations, etc. It will coordinate actions of advice, assistance, expertise, production, publishing and support.

Scérén's free software skills center has contributed to free software becoming a valued and fully fledged component of educational computing. The "identity of view", between, on the one hand, the principles of free software and, on the other hand, the missions of the educational system and the teaching culture of dissemination and appropriation of knowledge by all constituted a solid support point. The cause was heard for a certain part, even if it was not always a "long quiet river", with the development of free software meeting resistance, notably within the national education system itself. If the opposition was necessarily muffled, it was no less stubborn. The path is chaotic but the reality remains.

References

Archambault, J.-P. (2012). Enjeux éducatifs des logiciels libres et des standards ouverts [Online]. Available at: http://www.framablog.org/index.php/post/2012/06/19/vincent-peillon-enjeux-libres-standards-ouverts.

Archambault, J.-P. (2014). L'édition scolaire au temps de l'informatique [Online]. Available at: https://www.societe-informatique-de-france.fr/wp-content/uploads/2014/10/1024-4-archambault.pdf; https://www.epi.asso.fr/revue/articles/a1504d.htm.

Archambault, J.-P. and Baudé, J. (2012). Les logiciels libres et l'enseignement de l'informatique au lycée [Online]. Available at: https://edutice.archives-ouvertes.fr/edutice-00826643/file/a1210c.htm; https://www.epi.asso.fr/revue/articles/a1210c.htm.

Association Sésamath (n.d.). Sésamath [Online]. Available at: https://www.sesamath.net/.

Wikipedia (n.d.). GNU General Public License [Online]. Available at: https://en.wikipedia.org/wiki/GNU_General_Public_License.

H

Habitele

Dominique Boullier
CEE, Sciences Po, Paris, France

Inhabiting the digital world

Inhabiting the digital world may seem like a daring formula, but it opens up countless avenues for innovation and allows us to think critically about the spaces created by the digital platforms that shape our daily lives. We often speak of data architectures or digital urbanism to design our eco-system. But we neglect to think about the conditions that allow us to inhabit (live) and not only to lodge (stay). We can take shelter under a porch, and we can stay in a hotel; this does not mean that we inhabit; this elementary human experience (Radkowski 2002) is essential to the construction of a liveable urban world, and the digital world is no exception. The systemic platforms (the GAFAMs) have designed housing for everyone, provided that everything is open to all winds, so that they can collect all the traces they market, on the condition that they impose a rhythm of activity such that it creates an atmosphere of stress that must be shared, if we want to benefit from the reputation effects that are at the heart of this environment. It is high time to redesign these platforms to make them liveable, habitable, and to make it possible to exercise our capacity to inhabit.

The term habitele is formed from the term *habere*, which indicates "having", which should be defining of social entities rather than "being", according to Tarde (2001). But the possessor can also be affected by their possession, to the point of being possessed by it themselves ("I have a car" and "I have it in my skin"). The theory of the person, produced by Gagnepain (1982), has mapped out a series of concepts that make it possible to understand how the human subject can extend their hold on their environment: the habit (beyond the garment/clothes), the habitat (beyond the dwelling) and the habitacle (beyond the vehicle) are "formats of presence in the world" for the subject, who equips their body to make new skins that transform them in what is called an appropriation. These capacities for technical extension of the self are directly linked to the body, to its envelope, to its relationship with the environment. It is in this conceptual lineage that the *habitele* was born, beyond the technical network that ensures connection but not common life. "To connect is not to institute" (Boullier 1988).

The extension of the subject goes beyond these material bubbles, those thought of by Sloterdijk (2005), which constitute so many visible skins: globes, bubbles and foams are part of his spherology and the *habitele* is clearly part of the domain of foams by the plurality of worlds thus associated (family, friends, work, consumption, passions, politics, religion, etc.). It has a material dimension directly linked to social networks, which we want to keep active as close as possible to our bodies, with an increasingly portable and multisensory terminal, which will no doubt end up grafted onto our augmented bodies. But the *habitele* also allows us to envelop pure administrative or legal entities (our civil status, our social security number), for example, of which we are members and which we take into our nets like Leibniz's monads. By extending "habit", habitat and habitacle, the *habitele* expresses the attachment to the body and the dimension of habituation and habit that these attachments entail. By referring to networks, it engages toward the generalized circulation beyond bodies and in a relationalist posture beyond an "egology", to which the reference to an individual and their skills could lead us.

The digitization of the world and the constitution of networks of "portables" (a more accurate term than "mobiles" for the *habitele*) have converged more and more relational functions into the cell phone (payment, access, ID). As McLuhan predicted, the change in scale (two-thirds of human beings have gained access to cell phones since their launch in 1995) provides the building blocks for a change in collective climate and a new kind of shared world. As "connected beings", we inhabit a new "ecosystem of personal data" (Boullier 2014). We maintain connections to vastly different social worlds (from family to the tax department, from soccer fans to political debates, from Tinder to online betting), that we can switch between instantly, whereas moving from one world to another used to require traveling across town or the country (Boullier 2011). But permanent connection leads to a shared state of alertness, of reciprocal vigilance, which manifests itself in constant notifications, shared stress (from SMS to Twitter, which has become the atomic clock of the collective mental rhythm, the tweet per second), a high-frequency life, which is typical of the networks of digital financial capitalism and of a "media warming" that threatens our public space (Boullier 2020). The pace could be different (e.g. Wikipedia) and not guided solely by the imperatives of captology, which designs algorithms and interfaces to make us stay as long as possible and react to anything with a high "novelty score" (Vosoughi *et al.* 2018), totally opposite to the spirit of envelopes and habits. Our environment will need to be better controlled by us to truly inhabit the digital technology, transform it and be transformed, as we do our habitat, but no longer be under its sway.

Identities in the cloud, on the one hand, and permanent physical access, on the other hand, are the elements that make up a universal terminal (which the PC will never be): our digital identities become portable, they dress us up like a new envelope. Service designers should invent a design for this envelope, without piling up applications and functions, but rather giving the users back their hands so that they can produce their own interior and regulate their degree of immunity to their environment, depending on the domain and situation. To do this, we need to escape the advertising drift that has been theirs since the end of the 2000s and which has led them to an unprecedented stock market valuation.

Reclaiming our *habitele* is undoubtedly a political agenda that is taking shape now that the deregulation effects of these platforms are being felt by policy makers. But it will require the cooperation of designers and developers, because it is in the code and interfaces that the *habitele* capacity must manifest itself. And, of course, all of this must be done under the active control of citizens, whose consent will no longer be withheld through an attentional design that favors immediate reaction (everything in one click!). The Chinese and Californian models both confiscate consent, in their own way, but always for the benefit of the people, of course. Now, Europe can make a political and moral demand for people to recover their sovereignty, even if we know that we are woven of all our relationships.

References

Boullier, D. (1988). *Connecter n'est pas instituer. Nouvelles technologies de communication et autres dispositifs pousse-au-jouir*. LARES, Rennes.

Boullier, D. (2011). Habitèle virtuelle. *Revue Urbanisme*, 376, 42–44.

Boullier, D. (2014). Habitele: Mobile technologies reshaping urban life. *URBE*, 6(1), 13–16.

Boullier, D. (2020). *Comment sortir de l'emprise des réseaux sociaux*. Le Passeur Éditeur, Paris.

Gagnepain, J. (1982). *Du vouloir dire : traité d'épistémologie des sciences humaines. Tome 1 : Du signe, de l'outil*. Pergamon Press, Paris.

Gagnepain, J. (1991). *Du vouloir dire : traité d'épistémologie des sciences humaines. Tome 2 : De la personne, de la norme*. Livre et communication, Paris.

de Radkowski, G.-H. (2002). *Anthropologie de l'habiter : vers le nomadisme*. PUF, Paris.

Sloterdijk, P. (2005). *Écumes : sphères III*. Maren Sell Éditeur/Pauvert, Paris.

Tarde, G. (2001). *Les lois de l'imitation*. La Découverte, Paris.

Vosoughi, S., Roy, D., Aral, S. (2018). The spread of true and false news online. *Science*, 359(6380).

Hacking

Éric Zufferey
   Consultant, Fribourg, Switzerland

Introduction

> Our choice is not between "regulation" and "no regulation." The code regulates. It implements values, or not. It enables freedoms, or disables them. It protects privacy, or promotes monitoring. People choose how the code does these things. People write the code. [...] The only choice is whether we will collectively have a role in their choice – and thus in determining how these values regulate – or whether we will collectively allow the coders to select our values for us. (Lessig 2000)

In the collective imagination, hacking may refer to computer piracy or to forms of resistance against large companies or states. What is it really? Etymologically, the term derives from the verb *to hack*, which means "to cut to pieces". In this sense, a child who takes apart an alarm clock or a radio to try to understand how it works is practicing a form of hacking. It therefore refers to the idea of self-education, that is, learning by ourselves outside the school institution.

Hacking thus covers a relationship with technology that is built outside or on the fringe of official institutions. More broadly, it refers to a form of counterculture: hackers approach technology in its social and political dimensions, in opposition to an instrumental perspective that is considered in terms of efficiency or profitability.

The countercultural origins of hacking

Going back to the origins of hacking allows us to understand how it was constructed as a form of counterculture. In fact, hacking was born in the United States from the confluence between a nascent and still not very formalized computer science discipline, on the one hand, and the counterculture of the 1960s, on the other hand.

As early as the 1950s, the first generation of hackers were working in universities and experimenting with calculators, expensive machines normally reserved for scientific activities. These self-taught hackers did not hesitate to bend the administrative rules to develop new uses far removed from scientific calculation: under their hands, the computers could play music or run one of the first video games (*Spacewar!*). They eventually gained recognition from professors interested in their non-conventional approach, such as Marvin Lee Minsky, the founder of the MIT Artificial Intelligence Group.

This first generation was followed by a second generation, which was marked by the counterculture of the 1960s – the equivalent of the May 68 protests in France – and consequently developed a more politicized vision. It campaigned in particular for the democratization of computers and for making them a tool of counter power, as reflected in the famous slogan: *Computer Power to the People!* It was also at this time that the first experiments took place around the idea of "virtual community", itself strongly influenced by the community ideal of the hippies.

A self-taught practice in its early days, hacking has evolved into a counterculture that postulates the command of computer tools as a right that must be guaranteed. Thus, hackers developed the concept of the personal computer as a counter model to the computers owned and operated by universities and corporations. This ideal also had an impact on the design of the Internet – hackers participated alongside researchers and engineers. Indeed, the network of networks adopts a decentralized architecture that places control at the ends of the network, at the users' end, in contrast to the dominant model of its time, namely the centralized architecture of telephone networks (Gillespie 2006).

From ideals to struggles

Hacking proposes an atypical vision of technology that has come up against the lack of regulation by economic and political actors. Not surprisingly, several struggles around the freedom to exchange and communicate have marked the history of hacking.

In 1980, the American government placed computer code under copyright. Researchers and hackers saw their habit of freely exchanging computer code limited, and even attacked in the judicial field in the case of collaborations, sometimes long-standing, between academic and economic actors. The hackers responded by hijacking intellectual property: they created so-called free licenses enabling the reuse of computer code and obliging any software that reuses protected code to use the same license (viral or copyleft clause). This hijacking allowed the constitution of a computer science commons (Mangolte 2013), fueled by numerous free software projects (Linux, Apache, Mozilla Firefox, LibreOffice, etc.).

Hackers have also mobilized in favor of anonymity and the protection of privacy on the Internet. Indeed, they were among the pioneers of computer networks and organized themselves very early on in collectives – so-called virtual communities. In the 1990s, however, this online presence was targeted by the authorities in some countries, such as the United States and France, for police action. While these actions were primarily aimed at online pirates, they also affected militant hackers, and even actors with no direct links to hacking, as illustrated by the court case "Steve Jackson Games, Inc. v. Secret Service". In response to these events, associations for the defense of individual freedoms on the Internet were founded, such as the *Electronic Frontier Foundation* and the Quadrature du Net. More pragmatically, hackers have also developed and disseminated new encryption tools, some for the general public and others for specific audiences, such as whistleblowers in the case of WikiLeaks.

The diffusion of a self-taught computer culture

By promoting personal computing, free software and control of communication tools, hackers laid the foundations for a self-taught computer culture. Since the 1980s, hacking has spread as an amateur practice to a wider public and is particularly attractive to the younger generation. Amateurs can learn from professionals and activists, within groups that advocate social diversity, whether online or in a physical location – we can refer here to the hackerspaces that have been developing significantly over the last 15 years. This social decompartmentalization gives substance to "concrete utopias" (Lallement 2015): hackers seek to concretize a countercultural discourse through new ways of working collectively.

In France and Switzerland, however, this amateur hacking practice does not represent an alternative to the school institution (Zufferey 2018). While amateur hackers seek above all to realize themselves in their work, by importing their atypical ideas and practices, this requires the mastery of legitimate cultural codes, which is mainly acquired at school. For those who succeed academically, hacking is transferred to their professional activity and generally loses its militant dimension. For the others, hacking remains an amateur practice that has symbolic compensatory value and most often feeds a project of professional reconversion. Altogether, militant hacking is only seized upon by a minority of individuals, endowed with the appropriate resources acquired during a classic civic commitment or in connection with a position in the university field.

Hacking: a contemporary issue

Today, computers and the Internet have been "taken over" by economic and political actors: recentralization of the Internet via the services offered by Google, Facebook and others, predominance of advertising targeting and manipulation of opinions (filter bubbles, astroturfing), militarization of cyberspace, etc. As a critical approach to technology and as a vector of alternatives, hacking is therefore more relevant than ever. Hackers remind us that technology is the bearer of citizen issues and that it concerns us all. Their history shows that it is possible, and even desirable, to appropriate technology "from below".

References

Gillespie, T. (2006). Engineering a principle: "End-to-end" in the design of the Internet. *Social Studies of Science*, 36(3), 427–457.

Lallement, M. (2015). *L'Âge du faire : hacking, travail, anarchie*. Le Seuil, Paris.

Lessig, L. (2000). Code is law – On liberty in cyberspace. *Harvard Magazine* [Online]. Available at: https://harvardmagazine.com/2000/01/code-is-law-html.

Mangolte, P.-A. (2013). Une innovation institutionnelle, la constitution des communs du logiciel libre. *Revue de la régulation*, 14 [Online]. Available at: http://regulation.revues.org/10517 [Accessed 2 January 2017].

Zufferey, E. (2018). Changer le travail ou changer la société ? Les *hackers* entre conformation à l'ordre social et volonté d'innover. PhD Thesis, Université de Fribourg/Université de Lille 1, Fribourg/Lille [Online]. Available at: https://doc.rero.ch/record/326663/files/ZuffereyE.pdf.

Health Data

Marie Cauli
Université d'Artois, Arras, France

Health data: a sensitive issue

Faced with digital technologies that govern many aspects of our daily lives and extract personal data from our behavior, every person should have understandable, accurate and fair information on the processing, path, use and future of their data, whether or not their consent is required. But this information requirement is far from being met, particularly in the health field. It is therefore necessary to review the benefits, but also the risks, associated with the collection and use of data, both for individuals and society.

Definition

The new General Data Protection Regulation, which has been applied in all countries in the European Union since 2018, gives a definition of health data understood "as personal data relating to the physical or mental health of an individual, including the provision of health care services, which reveals information about their health status". It concerns information relating to an individual who receives care from the administrative databases collected (registration, invoicing of care, recording of arrivals, discharges, etc.), clinical information obtained during the examination of a part of the body, or paraclinical information, digitized and collected in the care institutions (results of clinical, biological and imaging examinations, genetic tests, laboratory analyses, treatments and drugs supplied), and finally, specific information concerning a disease, medical histories, disabilities, or information from surveys, cohort studies or clinical trials, etc.

An exceptional heritage

Beyond these collection processes, the recent development of digital services, such as online appointment booking or drug sales sites, provides external information. Similarly, the Internet of Things and mobile health, measuring health or well-being parameters at home, have become very important sources of data that can be provided by individuals themselves, whether they are aware of it or not. The interest in these data collected outside the usual medical care procedures comes partly from the fact that they relate to "real life".

As a result, a very large collection of data, from biological measurements, clinical, environmental and behavioral data, have been stored, regardless of their domain (health or other), origin (social networks, mobile applications, computer writings) and medium (image, written, sound).

They make it possible to trace the health care pathway of the entire French population. They constitute a remarkable heritage determined by a series of technological developments: a change of scale, due to the considerable increase in the number of data available and the ability to analyze their volume because of the computing power of computers and algorithms; their durability: using the data does not destroy them, so they can be reused; their rapid dissemination, which allows them to be shared beyond national borders; their ability to generate new information (secondary data) and new hypotheses through their processing.

What about the deduced information?

This last characteristic raises major questions. Indeed, health data can no longer be limited to personal data collected in the context of medical treatment. It makes it possible to obtain sensitive health data secondarily from primary data that are not directly related to health and which, without being qualified as health data by themselves, become so either by their combination with other data that makes it possible to draw a conclusion about the state of health or the risk to a person's health (weight, calorie intake, obesity), or "by destination" (because they are used in a care pathway). Thus, any primary data resulting from a human activity – even if apparently unrelated to health – and apparently harmless, taken in isolation, can contribute – by being crossed with other unrelated data – to the creation of new information relating to a person's health. This newly deduced information, when correlated with a multitude of data can, by cross-checking, give very precise information on the most sensitive individual traits (sexual orientation, lifestyle, etc.). This profiling can be used for commercial purposes, which are presented as being related to the well-being or the good of the patient, and can be used without the knowledge of the holder by insurance companies, which are likely to categorize customers according to risk, by banks, which have an impact on access to financial credit, by companies for employment decisions, etc.

The DMP

This process also raises questions about the DMP (*dossier medical partagé*, shared medical record) which, at the heart of the digital shift in France, is one of the most sensitive elements. This intends to bring together all of the information concerning a patient, to be shared between all health professionals. It aims to promote coordination and continuity of care and is based on a logic of prevention and anticipation of health needs. It contains a lot of information, some of which is protected by medical secrecy. In a vision of networked medicine, the nature of the information and the methods of transmitting it must be questioned, as medical confidentiality remains a pillar of our health system. The digitized electronic health record is part of a worldwide quest to obtain as much information as possible on patients and healthy people in order to offer new services or products. Also, its implementation is not without questions: conditions of access to the file, exact content of the file, conditions of hosting and security, the place of the patient and the patient's rights in this system, so many questions that call for vigilance on the part of the citizen and for them to consider the fate of their data and the conditions of their use. To this end, the CNIL[1] has focused on the regulation and supervision of its uses, so that the patient can be clearly informed of the use of the digital device in their care and the use of an algorithm to base an administrative or medial decision that

1 The *Commission Nationale de l'Informatique et des Libertées* (French National Commission for Computing and Freedoms) is an independent administrative authority responsible for ensuring the protection of personal data contained in public and private files; it plays a warning role but also controls and sanctions.

concerns them, so that medical confidentiality and a principle of human guarantee are preserved.

Data ownership: keeping the upper hand

In addition to the issue of user protection, the DMP is also sensitive because of its economic and geopolitical dimension. A few clues around the rich terminology corroborate the importance of storage locations: warehouses, deposits, bases, banks, in reference to the raw materials to which the data are compared, and notably the *Health Data Hub* platform. Created in November 2019, this device is intended to bring together all the information, facilitate its sharing and promote research. In this gigantic space, we should find health insurance data, hospital data, medical causes of death, data on disability, etc. This is an imposing reserve since the Senate report mentions the reconstitution of the health data of 67 million people over nearly 12 years. This platform went live in April 2020 and its hosting was entrusted to Microsoft, which obtained health host certification in November 2018. Subject to US law, incompatible with the protection of privacy, Microsoft also raises the major question of the transfer of medical power and data to organizations dominated and controlled by global digital giants, strong in their power. The consequences in terms of loss of autonomy of national systems and national and European security are to be feared. In this context, the conditions of access have been partly clarified, making it possible to carry out any activity that is in the public interest, provided that it respects the recommendations for use and the confidentiality of individuals and their private lives. But other questions remain unresolved and deserve to be legislated: the complex issue of data ownership, their scope of use and purpose criteria, the problem of anonymization, the limits of appropriation of our personal data which presuppose the security of computer systems and the choice of the host, as well as a set of guarantees. Beyond the development of technologies, the processes adapted to make them available and the question of infringement of fundamental rights, it is not only the concepts of health but more broadly of society, which are in competition with each other and would take us further away from a democratic society. In the absence of an appropriate response, this would in fact amount to admitting the expansion of a society of surveillance and control of individuals by multiple public or private operators acting in an opaque manner, for the most diverse purposes, whether commercial, political or security related. This challenge necessarily implies setting up a system and processes of regulation and equipping ourselves with appropriate legal instruments, but also increased vigilance by citizens.

References

Avis 130 (2019). Données massives et santé : une nouvelle approche des enjeux éthiques. Comité national d'éthique pour les sciences de la vie et de la santé.

Bourcier, D. and De Filippi, P. (2018). Vers un droit collectif sur les données de santé. *Revue de droit sanitaire et social* (*RDSS*), *Sirey, Dalloz*, 3, 444–456.

CNIL (2017). Comment permettre à l'homme de garder la main ? Rapport sur les enjeux éthiques des algorithmes et de l'intelligence artificielle. Report, CNIL.

Cytermann, L. (2015). Promesses et risques de l'open et du big data : les réponses du droit. *Informations sociales*, 5(191), 80–90.

Léo, M. (2016). Patient connecté et données de santé : les vrais risques. *Information, données & documents*, 53, 65–66.

Human-system

Julien Cegarra[1] and Jordan Navarro[2]
[1] *SCoTE, Université de Toulouse, Albi, France*
[2] *EMC, Université Lumière Lyon 2, France*

Critical contributions of psychology on the use of digital tools

Ubiquitous digital tools

Digital tools are undoubtedly involved in our daily activities and, first and foremost, in our professional activities. In France, the *Conditions de travail* survey shows a continuous increase – since the 1990s – in the use of digital tools at work (Mauroux 2018). In 1998, about half of employees used digital tools at work; this was over 70% in 2013. This type of survey must necessarily be accompanied by more detailed studies to assess the consequences of this development on professional activities. However, digital tools are often considered in terms of their benefits. Without questioning these numerous benefits, the choice is made in this entry to take a more critical look at these tools.

In sociology, Boboc (2017) has thus pointed out the impacts of digital tools, insisting on the simultaneity with profound changes in work in which they participate, such as the intensification of work rhythms, more frequent work reorganizations, restructurings and even a breakup of work collectives. And this author notes:

> Consequently, the impacts of digital technology are to be read as a catalyst and amplifier of organizational mutations already largely underway. (Boboc 2017)

In addition to this work, psychology has highlighted a set of constraints conveyed by digital tools, which are not strictly dependent on work-related issues and which influence our brain functioning.

Lessons from the work in psychology

Psychology has long been interested in the complex relationships between the user, tools and work environment. This so-called ergonomic psychology has historically focused on risky work situations in ultra-safe systems (medical, nuclear, aeronautical). The aim was to better understand the human contribution to industrial accidents and disasters.

While there are countless statistics on "human error", which is the cause of 70% of civil aviation accidents, 76% of medical helicopter accidents and 76% of military accidents, studies in psychology emphasize the importance of putting them into perspective. First of all, analyses of the course of disasters often reveal the poor conception of the interaction between the tools and the user's mental functioning (as well as their expertise). This poor design then contributes to precipitating the disaster. Furthermore, it should be emphasized that if all experts in these situations were replaced by automated tools, there would be no human errors, but many more accidents, given the current poor reasoning capabilities of these tools.

More specifically, these studies show a tension between the search for efficiency, particularly economic efficiency, leading to the use of digital tools in demanding tasks, and the lack of anticipation of the organizational consequences and mental processes at work in the operators. It is generally implicitly expected that users will adapt to the limits of the tools. However, not all adaptations are positive.

Negative adaptations to digital tools

Parasuraman (1997) has looked at the ways in which experts, regardless of the domain, adapt to tools. In particular, he identifies negative adaptations to tools, which can be divided into four categories: overconfidence, complacency, loss of adaptability and loss of expertise.

– *Overconfidence in digital tools*: While the concept of confidence is easily understood between individuals, psychologists have also been interested in the confidence placed in digital tools. For example, a taxi driver may prefer to drive following a GPS rather than using their own knowledge of the road infrastructure; a student may use the calculator on their phone to perform a simple multiplication. These situations illustrate the dynamic aspect of a trust that is built with our representation of the tools and then experience with them. When the experience is positive, the users evaluate the performance of these tools positively. Finally, if this performance seems significantly better than our own performance, we generally prefer to use the tool at our disposal rather than do the activity ourselves. A miscalibration of this confidence can lead to overconfidence where we no longer have the ability to remain critical of the tools and their limitations. This

overconfidence manifests itself in decisions that are visibly inappropriate to an outside observer. The local newspapers often make fun of tourists who follow the GPS to the point of driving their vehicles into a river. However, it is often not pointed out that the tool contributes to this overconfidence by not indicating its own operating limits.

– *Complacency in digital tools*: Old works in psychology show that humans are rather thrifty by nature, especially in the mobilization of their mental activities (e.g. through a confirmation bias that leads to us ignoring information that contradicts our beliefs). This mental economy is reinforced by tools that perform all or part of a task. Let us imagine a student who wants to go deeper into a subject as part of their studies. If they consults the databases, the texts are available online for only part of them. They can then go to the library to find the others. It is likely that they believe that they already have enough resources without having to deal with additional documents. This type of economy is not restricted to people with little expertise in a subject, as it is, for example, recognized that scientific articles whose sources are not available online (and which require more effort to obtain) are on average cited less than available articles. This example illustrates a phenomenon called complacency, which can be strongly accentuated by the limitations of digital tools.

– *Loss of adaptability*: Decision-making, far from being a linear process, is composed of processes integrated in a "decision loop" (from information gathering to action execution). Digital tools are integrated into this loop and can alter the processes that constitute it. When humans are no longer integrated into the loop, they are also no longer aware of the characteristics of the current situation. This is called loss of adaptability. Let us take the example of driving autonomous vehicles. While these vehicles can drive from point A to point B on their own, they can also get into difficulty, for example, if the road markings are blurred. Typically, the vehicle emits an audible signal to inform the passenger to resume driving. However, without being in the loop, the driver can only recover the situation in a very degraded way because they are no longer involved in the driving, no longer aware of nearby vehicles, directions to follow, etc. The recovery of tools that cannot function perfectly is therefore a major source of difficulties and errors.

– *Loss of expertise*: If tools take over tasks done manually, the consequence is also a progressive loss of expertise in the field. If we take the example of our driver, the continuous use of an autonomous vehicle without ever taking the wheel will inevitably lead to the impossibility of ensuring a safe takeover of the vehicle, with the skills linked to driving being reduced. In everyday activities, anyone can see that handwriting is more difficult when it is not practiced. In the school domain, Velay and Longcamp (2012) asked some children, between four and five years old, to perform different writing and reading exercises. The results indicate that if the children did the exercises with a pen, they performed better than if they did them with a computer keyboard. Indeed, the brain areas activated during handwriting are the same as those activated during reading (left premotor cortex) and this facilitates cognitive processing. However, it is not the same brain areas that are mobilized when using a keyboard. Psychological knowledge should therefore encourage us not to consider tools as simple replacements for previously manual actions, but as profound modifications.

The challenges of digital tools

In the logic of economic efficiency, there is often a desire to automate, simplify tasks as much as possible and leave what cannot be done to humans. However, the mental functioning of an individual, especially an expert in a field, cannot be neutral in the face of these changes. Many studies underline the numerous positive aspects of digital tools. It is also important to bear in mind all of the potential consequences, particularly those on mental processes and the maintenance of professional expertise.

This critical look at digital tools should alert us to future technological developments. Thus, in the recent developments of digital tools, deep learning techniques appear ideal to facilitate the work of many professionals. In the field of medicine, a study by Majkowska and colleagues (2020) shows that a tool based on these learning techniques provides a clinical analysis of chest X-rays as good as that of experienced professionals. These tools could therefore help or even replace these professionals. However, the work presented here indicates the necessary loss of expertise that would result. The tool will thus be able to diagnose frequent cases more quickly, while not knowing how to do so in rare or novel cases (for them), while at the same time the professional will have lost the ability to develop a great deal of expertise and, consequently, the know-how to treat the latter.

In fact, the human must be considered as an important part of the human–machine tandem, whose limits must be considered in the same way as the limits of digital tools (Navarro 2019).

References

Boboc, A. (2017). Numérique et travail : quelles influences ? *Sociologies pratiques*, 34(1), 3–12 [Online]. Available at: https://doi.org/10.3917/sopr.034.0003.

Majkowska, A., Mittal, S., Steiner, D., Reicher, J., McKinney, S., Duggan, G., Eswaran, K., Chen, P.-S.C., Liu, Y., Kalidindi, S.R. *et al.* (2020). Chest radiograph interpretation with deep learning models: Assessment with radiologist-adjudicated reference standards and population-adjusted evaluation. *Radiology*, 294(2), 421–431 [Online]. Available at: https://doi.org/10.1148/radiol.2019191293.

Mauroux, A. (2018). Quels liens entre les usages professionnels des outils numériques et les conditions de travail ? *DARES*, 29.

Navarro, J. (2019). Are highly automated vehicles as useful as dishwashers? *Cogent Psychology*, 6, 1575655 [Online]. Available at: https://doi.org/10.1080/23311908.2019. 1575655.

Parasuraman, R. (1997). Humans and automation: Use, misuse, disuse, abuse. *Human Factors*, 39, 230–253 [Online]. Available at: https://doi.org/10.1518/0018720977785438 86.

Velay, J.-L. and Longcamp, M. (2012). Clavier ou stylo : comment apprendre à écrire ? *Donner l'envie d'apprendre*.

I

Indexing

Ismaïl Timimi
GERiiCO, Université de Lille, France

While the table of biblical concordances or the table of commonplaces inspired by Aristotle are often considered as ancestor systems of our modern indexes (Huchet 2010), it was not until the middle of the 20th century that the terms "indexer" and "indexing" appeared within dictionaries, from 1948 onwards (Amar 2000).

Indexing is a very old practice, observed for a very long time in socio-professional environments. And yet, as an object of study, indexing has only recently aroused scientific interest in two very young disciplines – library science and documentation (Timimi and Kovacs 2006). With the documentary proliferation of the Web and the metamorphosis of digital media, its modeling has attracted major interest in search engine algorithms and multimedia indexing.

Indexing, known as a central documentary practice, is a process of analysis and representation of information according to several variables (the nature of the document, the corpus and the field of application, the practices of the users, etc.). It consists of representing, by means of the elements of a free or controlled language, the themes and notions characteristic of the document's content (resource or collection). The aim is to enable the memorization of the content by a distinctive mark (alphanumeric, nominal or other symbol) to easily find the document during a later search (Chaumier 2000). This gives indexing a semantic and also a semiological dimension, in the sense that it aims to extract the meaning of the document and signal it. Indexing can thus be defined by its purpose, as a tool in the service of a function, documentary research (instrumental approach) (Amar 2000).

The difficulty is therefore knowing what relevant approach to adopt and what criteria to retain (procedural approach) to characterize and represent the information present in a document or collection of documents in the best possible way so that, in subjects of investigation, this document is easily visible and accessible. A concise response would be to annotate each document with a set of *metadata* recorded in a bibliographic record (title, author, abstract, keywords, etc.). While this classical approach has the advantage of being formalizable and easy to implement, it remains limited, due to several factors related to the complexity and ambiguities of languages, the multiplicity of document types in the digital age, the volume of the corpus to be processed and the shifting relevance between the author, the indexer and the user.

Depending on the uses and contexts, the applications and the resources available, there are several types of indexing. *Systematic indexing* consists of attaching an index (a numerical or other symbol) to a document in order to situate and classify it in a knowledge organization system. This index comes from a previous classification, encyclopaedic (*Universal Decimal Classification, Dewey classification*) or not. *Analytical indexing* (also known as *subject indexing*) consists of analyzing the document and indicating its subject through one or more keywords from a documentary language, a repertoire of standardized words presenting the subject headings. In the same way, possible combinations of the concepts identified in the document can be represented explicitly, during the indexing phase (*pre-coordinated indexing*), or later, at the time of the interrogation and using logical operators such as union, adjacency, proximity, etc. (*post-coordinated indexing*). In order to reduce the noise of a document search, there is also *role-based indexing*, which consists of associating a role indicator to the various descriptor terms of the document content. The roles (such as action, object, purpose, etc.) are determined according to the domain considered. *Weighted indexing* consists of assigning a weight to the indexing term according to a determined scale, in order to specify its informational importance within the document, or even the collection. Weighted indexing makes it possible to sort and order documents in response to a search. Finally, due to the community effects of Web 2.0, we can also mention *social indexing*, or folksonomy (Le Deuff 2012), a decentralized collaborative indexing practice, based on the natural and spontaneous language of contributors who are not necessarily specialists in the documentation field.

According to Lamizet and Silem (1997), Lancaster identifies seven decisive factors for evaluating the quality of an indexing process: the depth of the index, the choice or not of controlled vocabulary, the size and specificity of the domain vocabulary, the characteristics of the subject and its terminology, the posture of the indexer, the mediation tools and the volume of the corpus.

While indexing has been devoted, and in a privileged way, to textual information where concepts, methods and approaches have been developed (Bachimont 2007), the indexing of images has, for its part, experienced research advances according to two different but complementary processes. The image can be assimilated to its own metadata and we are then in the case of textual indexing, or the image can be reduced to its graphical content (shape, colors, textures), in which case the indexing and search are carried out on the basis of a matching and calculation of similarity between these components (search of the image by the image).

In computer engineering, the first models of automatic document indexing consisted of creating an index of the extracted words or terms, enriched with new data (positions and dispersions, absolute and relative frequencies, weighting, etc.). Depending on the approach, this index could be composed of all the words contained in the text except the stop words, or of the words and terms considered relevant in the text after a lemmatization, or of their equivalents in a document language (*indexing by assignment*). As for current models, they are more complex and remain at the crossroads of computer science and information science, logical-mathematical modeling and linguistic interpretation, and also rely on other disciplines. They are constantly being improved to better understand and organize the digital universe (texts, videos, Web) according to *similarity* criteria (Bachimont 2007). This is the very principle of *search engines*, which make automatic indexing a central component in their referencing algorithms, to the point where we often see a terminological confusion between the two notions of *indexing* and *referencing*.

This confusion is not entirely unjustified, insofar as the indexing of content responding to the levers and prerequisites of search engines in terms of architecture, writing and citation (*sitemap, tags, semantics, tags, internal linking, netlinking*, etc.) is not without major impact on the referencing of a document and, therefore, on its positioning and visibility (as well as on its traffic to remain in the jargon of digital marketing).

Historically, search engines have been inspired by practices and models from the information and documentation sciences (*weighted indexing, Salton cosine, co-citation and the laws of bibliometrics*, etc.), but this rapprochement between intellectual approaches and artificial processes cannot hide major distinctions, in terms of indexing, that are specific to the Web universe, including *paid referencing* through the purchase of indexing keywords and sponsored links, *abusive referencing* (*spamdexing*) to trick search engines, *deindexing* to penalize fraudulent indexing or referencing techniques (*link nurseries, cloaking*, etc.). This distinction leaves a lot of room for questions around indexing and ethical indexing on the web.

References

Amar, M. (2000). *Les fondements théoriques de l'indexation : une approche linguistique.* ADBS Éditions, Paris.

Bachimont, B. (2007). Indexation et archivage de contenus multimédias [Online]. Available at: https://www.techniques-ingenieur.fr/base-documentaire/technologies-de-l-information-th9/gestion-de-contenus-numeriques-42311210/indexation-et-archivage-de-contenus-multimedias-h7500/ [Accessed 8 December 2020].

Chaumier, J. (2000). *Les techniques documentaires.* PUF, Paris.

Huchet, B. (2010). Concevoir l'index d'un livre : histoire, actualité, perspectives. *Bulletin des Bibliothèques de France (BBF)*, 5, 118–119.

Le Deuff, O. (2012). *Du tag au like : la pratique des folksonomies pour améliorer ses méthodes d'organisation de l'information.* FYP Éditions, Limoges.

Silem, A. and Lamizet, A. (1997). *Dictionnaire encyclopédique des sciences de l'information et de la communication.* Ellipses, Paris.

Timimi, I. and Kovacs, S. (2006). Indice, index, indexation. In *Actes du colloque international organisé par les laboratoires CERSATES et GERICO.* ADBS, Paris.

Information Ethics

Widad Mustafa El Hadi
GERiiCO, Université de Lille, France

Introduction

In information science, ethics is considered within the framework of information philosophy, a field that studies the conceptual nature and basic principles of information, including its ethical consequences (Floridi 2013). With his concept of the "infosphere or telematics ethics", Floridi related human beings, information, information technology, society and the interests of individuals from an ethical perspective. In contrast to the classical model, "telematics ethics" in Floridi's sense is primarily concerned with the environment in which information is generated and propagated (also called the "infosphere"). Being a more generic issue than digital ethics, information ethics is concerned with the ways in which all systems, digital or not, organize information in response to users. It can be a library, a search engine or any other computer system for processing, analysis or storage of information. Floridi developed an original ethical framework to describe the new challenges posed by information and communication technologies (ICT), which have profoundly changed many aspects of life – education, work, health, industrial production – and business, social relations, conflicts, leisure, intellectual property, freedom of expression and responsibility. It is a new field of research at the crossroads of epistemology, metaphysics, logic, philosophy of science, semantics and ethics. In this regard, the main areas of concern in information ethics, as highlighted by Bawden and Robinson (2012), are as follows: the contradiction between censorship and intellectual freedom; privacy, confidentiality and data protection; ownership of information and the potential commercial use of public information; universal access, information poverty and the digital divide; respect for intellectual property combined with fair use; issues of balance and bias in the dissemination of information; collection development and metadata creation. In relation to information science, these ethical issues have been identified and typically grouped under the term "information ethics" (Floridi 2013). Ethical issues have historically been addressed by libraries and other cultural institutions, businesses, non-profit institutions, universities, government agencies at all levels, information science research, and the media.

The role of professional and institutional bodies in the development of the ethics of information and knowledge

International institutions and bodies have consistently promoted universal access to all recorded knowledge (Beghtol 2002). From the Belgian visionary Paul Otlet (1868–1944) to UNESCO's World Summit on the Information Society (WSIS), efforts have been made to guarantee and promote this right. Professional institutions and associations have played a crucial role in the emergence and development of ethics. They have organized scientific events, conferences and seminars, and published special issues of ethics journals. A significant number of periodicals, conference proceedings and other scientific productions have served as a basis for this initiative. Some of the scientific events are as follows: Conferences on the Ethics of Electronic Information in the 21st Century, organized at the University of Memphis in 1997; the first UNESCO Conference of InfoEthics in 1997, entitled First International Congress on Ethical, Legal and Societal Aspects of Digital Information; 2003, Karlsruhe, Germany, in 2004, under the auspices of ICIE and with the support of the VolkswagenStiftung.

In parallel with these events, a number of specialized journals emerged in 1992, including the first journal, *The Journal of Information Ethics*. It covered ethical issues concerning the production and dissemination of information and knowledge. Also noteworthy is the 2004 publication *The International Review of Information Ethics* (IRIE). There are currently a number of journals: *Ethics and Information Technology*; *International Journal of Technology and Human Interaction*; *Journal of Information, Communication and Ethics in Society*; and *International Journal of Internet Research Ethics*.

Ethics as a component of knowledge organization: from libraries to the biases of automatic classifications

The ethics of knowledge organization is seen as a process intimately related to language and cultures. From a historical perspective, ethics in KO is rooted in the early criticism of classification systems (SOC). As early as 1973, IFLA (The International Federation of Library Associations) launched the Universal Bibliographic Control Program, asking national libraries to share their holdings and index with them, with special attention to cultural specificities, but this initiative did not succeed. The OCLC (Online Computer Library Center, founded in 1967 as the Ohio College Library Center, is a worldwide non-profit organization serving libraries) and recently Google have taken over control. Most of the criticism has been based on the fact that SOCs only offer a particular representation of language and that their structure modifies our interpretation of language in an unnecessary or false way. It is for this reason that we need to consider the influence of cultures and languages in the design of SOCs, as suggested by Tennis (2012).

Ethics as a component of knowledge organization has been the subject of numerous meetings and publications. Two movements have reinforced these critical approaches: postmodernism and gender studies. The excesses of presumption and confidence observed in the positivist and modernist period triggered a long-term reaction in intellectual environments, which evolved into the post-modern period. The globalization of exchanges has shown how the same subjects can be seen in different ways by different people, just as they are seen by different social classes, different genders, etc. Postmodernism has emphasized the relativity of the manifestations of knowledge and learning, especially since knowledge is no longer seen as an exclusive product of Western culture. The classic examples of the so-called "universal" bibliographic classification systems, developed since the end of the 19th century by American librarians, which have been the subject of criticism, are the Dewey Decimal Classification (DDC) or Amy Cutter's Library of Congress classification.

Although they have been important advances in the field of knowledge organization because of their technical features (the DDC has become an international standard) and are still very useful in libraries around the world, a critical analysis of these systems has revealed many biases, due to the particular Western perspective of their authors, especially for the DDC.

The ethical dimension in knowledge organization: four positions

In addition, the work of four authors specializing in KO has focused on the importance of respecting cultural and linguistic diversity in the design of SOCs. This respect is one of the ethical principles in knowledge organization and is considered one of the foundations of human rights (Universal Declaration of Human Rights). Antonio García Gutiérrez argues for the inclusion of different cultural points of view and their relations, establishing a cross-cultural ethics of mediation. Michèle Hudon advocates multilingualism in knowledge organization. For her, research involving multilingual thesauri in the field of knowledge organization leads to the need to take into account the ethical dimension, which must be respected in the construction of tools, but also in the exercise of professional practices. She advocates multilingual approaches and the consideration of linguistic minorities. In this sense, Clare Beghtol proposes theoretical concepts to support an ethically acceptable knowledge organization system, based on approaches that take into account cultural and linguistic diversity. Beghtol advocates access to information and knowledge as a fundamental human right. In her work, she has analyzed the problem of achieving culturally acceptable SOCs based on an ethical processing of different cultures, as defined by the UN Universal Declaration of Human Rights, which involves ensuring global as well as local access to information and knowledge, in any language, available anywhere, at any time, for any purpose, for the benefit of any individual, from any culture, ethnic group or field. It proposes the concept of "cultural hospitality", a principle that should underlie the construction of SOCs in order to ensure that information ethics are respected.

Hope Olson published her book *The Power to Name* in 2002. According to the author, the power to name directly affects the construction of information and the organization of knowledge. Classifications, subject headings and thesauri reflect the dominant culture of a society and therefore play a key role in the library context. These tools have been constructed and maintained by the mainstream, and librarians have little room to include different points of view. Olson is among the pioneers of critical analysis of SOCs, notably through her book *The Power to Name*, which shows that prejudices against the classes dominated by the dominant classes are often hidden in rubrics and other classifications of knowledge. Discriminated minorities often include women, homosexuals or migrants of various ethnic origins (Asian-American, Black-American, Latinos, etc.). Recently, there has been a debate, involving many American librarians, about the illegal aliens heading, which has been deemed offensive to foreigners cited in immigration materials and has been removed from the Library of Congress Subject Headings (LCSH).

In her seminal article "'Priorities of Arrangement' or a 'Hierarchy of Oppressions'?", Fox (2016) made a major contribution on "intersectionality" generated by classification systems. She describes the transformative, interlocking and conflicting oppressions that occur when humans belong to more than one identity category – with black women (but this is not limited to women) – and has since expanded to different variables beyond gender, race, religion, to sexual orientation, national origin, disability, etc. Oppression can have consequences ranging from inadvertent discrimination to harassment, violence or death, resulting solely from membership in a human group. A broad critical librarianship movement is currently developing in North America, encouraging librarians and users to "decolonize" knowledge organization and LIS (Library and Information Science) in general (La Barre 2017). More attention has been given, for example, to indigenous cultures in North America or Oceania, which may develop their own SOCs, such as the Brian Deer classification. The Brian Deer classification system is a library classification system used to organize materials in libraries with specialized Aboriginal collections. The system was created in the 1970s by the Canadian Mohawk librarian from Kahnawake.

Other examples from the literature show that knowledge perspectives, for example, change not only in space but also in time. A concept developed in the culture of a certain time may change slowly over time, as may the meaning of a corresponding term in an SOC. This evolution has been described as the ontogeny of the subject (Tennis 2001). The example given is eugenics, which can be referred to as "the set of methods and practices aimed at selecting individuals in a population based on their genetic makeup and eliminating individuals who do not fit into a predefined selection framework". The concept and field of eugenics can give us another example of prejudice. Eugenics is a term that first appeared in the DDC in 1911. At that time, it was considered to be within biology, but from the 1950s onwards, following the massacres of Jews, homosexuals and Gypsies by the Nazi governments and their allies during the Second World War, it became impossible for a classifier to place a book dealing primarily with eugenics within biology. After 1945, eugenics became an element of ideology and no longer of biology. The other options are social sciences, applied sciences, philosophy and ethics. And while eugenics currently has a diverse set of related fields, ranging from family planning to anthropometry, we can see its disappearance under certain headings of the Dewey classification. This is especially true since eugenics is still used in population genetics work, even though there is an open debate about what constitutes eugenic work and thought (Paul 1995). Yet even with this debate, population genetics is squarely a biological science, so the erasure/disappearance of the term seems more to avoid a word that could have negative consequences, when in fact it is the term used in the literature. Eugenics has been classified in the DDC, alternately at 575.1 (with genetics prior to the 16th edition) and 363.92 (under "Social and Population Issues") (Tennis 2001).

Thus, many authors argue that all SOCs, whether they focus on one area or encompass several, are biased not only with respect to race, gender, religion, sexual orientations, but also because of the lack of specific, accurate versions.

The biases/prejudices of automatic classifications, an example: the supremacy of a language and its culture

We have drawn a parallel between the control, that is, the "power of naming", exercised by the dominant use and acted through the so-called universal bibliographic classification systems, and the power of the algorithms used by artificial intelligence today. Bowker and Star (1999) consider that all classifications are "powerful technologies" which, once integrated into working infrastructures, can "become relatively invisible without losing any of their power". Our position on control by the power of "naming/designation", *The Power to Name,* that is, classification and categorization and control by AI algorithms, is consistent with Bowker's position.

An example of "naming power" given to classical classifiers is comparable to the control by algorithms of functions such as filtering, automatic translation using English as a pivot language, automatic classification of texts and images, etc. Here, we have translated some elements of the original French text into English: in December 2014, when Google Translate was asked for the equivalent of "This is a pretty girl" in Italian, the program gave *"Questa ragazza e abbastanza"*, which literally means "This girl is enough". What was the cause of this error? The two meanings of the word *pretty* in English, a word that can mean "beautiful" or "rather, to a large extent". The correct choice should have been "This girl is pretty", but the "contextual" program chose the second meaning. It got worse: the query had not been made in English but in French. This meant that the translation algorithms, instead of being closely linked to the initial and final languages, had been transmitted through an intermediary, English, whose status as a universal contact language should not make it an obligatory passage point in a multilingual context. Other queries have produced similar results (Kaplan and Kianfar 2015). For example, an attempt to translate "It is raining cats and dogs" into French, or "It is raining very heavily" in idiomatic English, yielded *"Il pleut des chiens et des chats"*, instead of *"Il pleut des cordes"* in ordinary French.

Conclusion

In order to deal with their ethical problems effectively, library professionals and information-providing institutions need to have a good working knowledge of information ethics. Professional codes of ethics can help provide such knowledge, but they are not sufficient. Unfortunately, there is no universally accepted set of ethical principles that would help. To go further, training in information ethics should be part of the training of information professionals and designers of knowledge organization systems. Such training should enable information professionals and those involved in research to understand ethical principles and how they apply to practical cases. These trainings should also make the link between information ethics and the library professional's mission, research infrastructures, explicit. Finally, a critical approach to the organization of information is needed.

References

Adler, M. and Tennis, J.T. (2013). Toward a taxonomy of harm in knowledge organization systems. *Knowledge Organization*, 40, 4.

Bawden, D. and Robinson, L. (2012). *Introduction to Information Science*. Neal-Schuman, Chicago, IL.

Casenave, J. and Mustafa El Hadi, W. (2019). Developments in ethics of knowledge organization: From critical approaches to classifications to controlled digital communication practices. In *The Human Position in an Artificial World: Creativity, Ethics and AI in Knowledge Organization*, Haynes, D. and Vernau, J. (eds). Ergon Verlag, Baden-Baden.

Floridi, L. (2013). *The Ethics of Information*. Oxford University Press, Oxford.

Fox, M.J. (2016). "Priorities of arrangement" or a "hierarchy of oppressions?": Perspectives on intersectionality in knowledge organization. *Knowledge Organization*, 43(5), 373–383.

Kaplan, F. and Kianfar, D. (2015). Google et l'impérialisme linguistique ; il pleut des chats et des chiens. *Le Monde diplomatique*, 28.

Mustafa El Hadi, W. (2017). Diversité culturelle et linguistique et dimension éthique dans l'organisation des connaissances. In *Sur les sciences de l'information et de la communication : contributions hybrides autours des travaux de Viviane Couzinet*, Fabre, I., Gardiès, C., Fraysse, P., Couzinet, V. (eds). Cépaduès Éditions, Toulouse.

Mustafa El Hadi, W. (2019). Cultural frames of ethics, a challenge for information and knowledge organization. *Zagadnienia Informacji Naukowej*, 57(2), 23–39.

Tennis, J. (2012). Le poids du langage et de l'action dans l'organisation des connaissances : position épistémologique, action méthodologique et perspective théorique. In *Organisation des connaissances : épistémologie, approches théoriques et méthodologiques*, Hudon, M. and Mustafa El Hadi, W. (eds). Université de Lille, 15–40.

Innovation

Serge Miranda[1] *and Manel Guechtouli*[2]
[1] *Université Côte d'Azur, Nice, France*
[2] *IPAG Business School, Nice, France*

Open and Spiralist innovation

We have entered an era in which *mobiquity* (a combination of the mobile phone, which has become a pocket computer, and the ubiquity of the Internet, which has become broadband and wireless) will have spatiotemporal consequences in all sectors of society. All economic sectors are affected and business models are changing. Access to this personal data is a real strategic challenge for companies, a means to create value in a well-established business model. The way companies create value and innovate is clearly impacted.

Innovation is a polysemous term, which designates both a process and its result. Innovation, in the digital economy, is bottom-up and multi-disciplinary, whereas traditional university research is top-down and tubular. Innovation can be seen as an invention that meets a usage. In order to innovate, we must therefore start from the user. This approach to innovation is the basis of the concept of open innovation (OI), which consists of using the internal (*in-formation*) and external (*out-formation*) knowledge of a company to innovate. In an OI system, the initiator shares internal resources with external users and the organization exploits, by absorption, the resources created by these external users. This is a distributed innovation process based on a flow of knowledge that crosses organizational boundaries.

OI reduces the costs inherent in protecting property but, at the same time, innovation becomes a public good with universal accessibility, and this naturally leads to limited remuneration for the organization (Baldwin and Von Hippel 2011).

The implementation of an OI process makes it possible to collectively involve the actors of an entire ecosystem around a common approach. The initiating organization encourages exchange, communication, feedback and practices between all of the stakeholders and supports (often materially) the project. In return, it becomes the (co) owner of everything that results from the process. OI is thus different from open source, because ownership is clearly distributed among the stakeholders who participate in the project.

The process of open and collaborative innovation is nonlinear. This type of innovation can only be constructed in recursive loops or in a "whirling" (Krupicka and Coussi 2017) and spiralist way. It is *spiralist*, as the spiral is made up of successive overtaking; this makes reference to the approach of Frankétienne (the great Haitian writer who created the concept of spiralism in literature). Spiralist innovation allows the network to build and consolidate itself in order to achieve progress in the project (Miranda 2014). Moving from one spiral to the next means overcoming difficulties that appear as the project evolves. Spiralist innovation introduces kinetics on Pisano's innovation quadrants (Pisano 2016); in its economic component (*disruptive business model*), it can be applicable to social sciences (marketing, law, etc.) and the *digital economy* (such as driverless cars).

This spiralist representation enriching Pisano's quadrants makes it possible to identify four types of innovations: disruptive innovations (in this case, technical and economic), routine innovations (leveraging existing techniques or competencies) and architectural innovations that represent a new era in computing (such as the cloud).

We can build the spiral in both directions; let us start with *a roadmap innovation* (e.g. a lighter phone, with a larger screen and a more powerful battery), and then comes a *disruptive technological innovation* (e.g. an OS like Android in the phone, with the NFC standard that makes it a smartphone) and a bundle of personalized services that did not exist before (universities have a major role to play in POC demonstrators and proof of concepts). This innovation leads to new *architectural innovations* (such as downloadable mobile application platforms or the cloud), which lead to new *disruptive business models* for service and application providers (e.g. free business).

The table below illustrates this spiralist dynamic in Pisano's quadrants (2016).

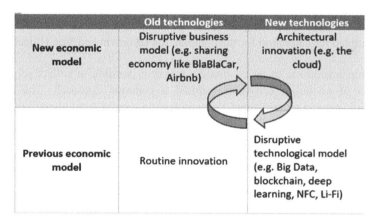

	Old technologies	New technologies
New economic model	Disruptive business model (e.g. sharing economy like BlaBlaCar, Airbnb)	Architectural innovation (e.g. the cloud)
Previous economic model	Routine innovation	Disruptive technological model (e.g. Big Data, blockchain, deep learning, NFC, Li-Fi)

Spiralist innovation and Pisano's quadrants

Communaction and digital "eternity"

Karl Marx was right in his famous adage applied to the informational economy: "From each according to his ability, to each according to his needs". The last sentence he uttered could serve as a guideline for the *communactors* (Miranda 2014) we have become: "If you can't change the world, try changing *your* world".

Communaction encourages us to rethink Marx in the context of a digital economy, a new historicity where one holds ubiquitous access to information and thus autonomy.

Homo mobiquitus will finally become virtually immortal, as Jim Gray had begun to show with his latest research project at Microsoft (digital eternity), via the perpetuation of our digital twin: "I will be able to discuss philosophy, happiness, wisdom with my great-great-great-grandson that I will never meet, from my writings, my Facebook wall, my tweets, my blog bearing the digital traces of my university of life!"

In 30 years, we have moved from an economy centered on products and quantity to one centered on services and quality, and then to one centered on users with smart spaces centered on data and authenticity. Artificial intelligence is not only an economic challenge for companies, but also a competitive advantage; it also induces a profound change in individuals and in society through all the possibilities it offers via robotization and data science. *Homo mobiquitus* is today the bearer of disruptions to their deep nature, their ways of learning, of taking care of themselves, of going through their territory with an increase of their personalized predictive possibilities, for the best as for the worst.

We need to rethink our world in a less linear and more spiralistic way. The dangers induced by all of these changes in paradigms, behaviors and professions cannot be ignored. These are profound issues around data control and protection, intrusive technologies, dependency and smartphone addiction, but also ethics. These issues are and will probably continue to be the major research challenges of the coming years.

References

Baldwin, C. and Von Hippel, E.A. (2011). Modeling a paradigm shift: From producer innovation to user and open collaborative innovation. *Organization Science*, 22(6), 1399–1417.

Krupicka, A. and Coussi, O. (2017). Compréhension d'un cas d'innovation institutionnelle au travers de la théorie de la traduction éclairée par les proximités de ressources. *Gestion et management public*, 5(3), 5 [Online]. Available at: https://doi.org/10.3917/gmp.053.0005.

Miranda, S. (2014). L'Homo mobiquitus : un communacteur pour les nouveaux territoires. In *Devenirs urbains et plissements numériques*, Carmes, M. and Noyer, J.M. (eds). Presses des Mines, Paris.

Pisano, G. (2016). You need an innovation strategy. *Harvard Business Review*, 16–25.

Interoperability

Fabrice Papy
Université de Lorraine, Nancy, France

The deployment of wide area and local area telecommunication networks in the 1990s and 2000s rapidly gave rise to the possibility of data and document exchanges using dedicated applications. Client-server architectures quickly became widespread, proliferating countless access protocols that were often not compatible. The emergence of the Web, conceived by Tim Berners-Lee, opened up new prospects for sharing information by devising a communications protocol called HTTP (Hypertext Transfer Protocol) and a document format called HyperText Markup Language, which made it possible to avoid the overkill of protocol mechanisms in often proprietary computer environments, as well as document formats and their encoding processes.

Because of its simplicity of use and implementation, the interoperable hypertext/hypermedia technology architecture, imagined and designed by Tim Berners-Lee to meet information dissemination needs, has emerged as the best candidate for the instrumentation of our "information and knowledge societies". This change of scale has introduced, however, the need to complement the initial documentary architecture with interactive and functionally collaborative data processing services that the social and participative Web, Web 2.0, has dramatically revealed. The shift from static to dynamic web design, where information, extracted from databases, is laid out on the fly by means of middleware solutions, has been meteoric and this separation of content/layout (background/form) of web pages has become the standard for the design of websites and any online information publishing device (blog, CMS, social networks, etc.).

Under the impetus of the World Wide Web Consortium (W3C), which brings together governmental organizations, standards bodies and powerful international computer companies, the Web is being organized and continues to develop. The principles of openness and interoperability, which are at the heart of the standard-setting proposals adopted by the W3C, have helped to disengage software equipment from the proprietary logic that had previously prevailed in the field of computing and telecommunications. Within the W3C, this functional collaboration between companies competing in a global market, which are also subject to the economic and legal regulation of the countries in which they operate, results in intense activity in terms of technological proposals.

The major developments in the digital document, in particular, are the expression of the intense technological activity of the Internet Society (ISOC) and the W3C, which are organizing the developments of the World Wide Web. With the goal of developing interoperable technologies to drive the Web to its full potential, the W3C produces data description language specifications, guidelines, software and tools. Extensible Markup Language (XML), Resource Description Framework (RDF), Synchronized Multimedia Integration Language (SMIL), Scalable Vector Graphics (SVG), Ontology Web Language (OWL), Cascading Style Sheet (CSS), Document Object Model (DOM), Portable Network Graphics (PNG), Extensible Stylesheet Language (XSL), etc., are just some of the standards (recommendations) initiated by the W3C since 1994.

These standards, which are disseminated by the industrial members of the consortium, are amplified by the technological actions of the International Organization for Standardization (ISO) or of private companies involved in the Web. The MP4/MP3 (MPEG-1/2 Audio Layer 3) audio compression standard, the MPEG-4 (Movie Picture Experts Group) audiovisual object coding standard, the Portable Document Format (PDF) page description language, the ShockWave Flash (SWF) computer file format for multimedia animations (images, sound, 3D, video, Rich Media), etc., are extensions of the W3C consortium's recommendations, which have contributed to the technicalization of the document.

From there, it is easy to correlate the specific recommendations of the W3C with the usual functionalities of the browser, whose data processing possibilities are not limited by these original mechanisms alone ((X)HTML, JPEG, GIF, JavaScript, CSS, DOM, etc.). The extension is infinite, because browsers are designed on the basis of a modular architecture, allowing them to host complementary applications, placed under the control of the browser's core. The logic of interoperability, which is very present in browsers because of the constant dialogue that these complementary modules maintain with the browser's application core, is also predominant in the architectures of heritage, scientific, educational, institutional and market digital libraries.

The profusion of open source software, regularly fed by an international community of developers, enriched by the experiences of users who actively participate in the technical and ergonomic improvement of the devices, testifies to the flexibility and efficiency of these programming models, which are becoming increasingly widespread in the Web development community. While these lean programming models have been instrumental in producing robust and usable software, they have also contributed to the breakdown of proprietary application logic by ensuring the independence of data from processing operations. The coherent multitude of document formats, inspired by XML technologies, has been a catalyst for the application interoperability advocated by the W3C consortium and a significant step toward the perennial separation of data and applications.

Digital documents built from interoperable open formats constitute a vast collection of primary resources that find their organizational coherence in the secondary descriptive data that accompany them. These are no longer a few keywords with approximate syntax, slipped into the header of primary documents to occasionally mislead the indexing engines, but real grammars of structuring and organization, documentary objects in their own right, whose access will be favored in order to effectively control the choices of distribution (for application servers) and consultation (for users) of primary resources.

References

CEN (1999). Interoperability of health care multimedia report systems. CR Report no. 14300, CEN.

Dextre, C. and Stella, G. (2012). ISO 25964: A standard in support of KOS interoperability. In *Facets of Knowledge Organization*, Gilchrist, A. and Vernau, J. (eds). Emerald, London.

Krier, L. (2012). Serials, FRBR and library linked data: A way forward. *Journal of Library Metadata*, 12, 177–187.

Levine, J.R. and Baroudi, C. (1998). Internet : les fondamentaux. *Intern. Thomsom Publicat*, France.

Papy, F. (2016). *Digital Libraries: Interoperability and Uses*. ISTE Press, London, and Elsevier, Oxford.

Intimacy/extimacy[1]

Serge Tisseron
Université de Paris, France

Intimacy and extimacy in the digital age

The 20th century valued the right of each person to privacy. This word, which is defined in reference to the public and the private, designates what one does not show to anyone, or only to a few very close people, while the private space is mixed with the family. But intimacy also includes another dimension, which we call "intimate", or sometimes "interiority": it is what is still too confused about oneself to be named, even by oneself. A permanent movement unites intimate and intimacy. The more the intimate passes from the domain of the informable to the domain of the formulated,

---

1 This text is a reworked version of my original French text that appeared in Tisseron, S. and Tordo, F. (2021). *Comprendre et soigner l'homme connecté : manuel de cyberpsychologie*. Dunod, Paris.

the more it enriches intimacy as a possible space for inner narratives and discoveries about oneself. And then, the more intimacy is shared with others, the more it contributes to the construction of self-esteem and richer social relationships. The word "extimacy" (Tisseron 2001) designates this dynamic process by which each of us uses the means of expression and communication available to us to give a shared dimension to elements of our existence that have been given form in our intimacy. Therefore, it does not apply to the intimate. Finally, it is not necessarily conscious and accompanied by shared intimacy.

Two desires in balance

The desire for extimacy contributes to the sense of existence from the very first months of life: the child discovers themself in their mother's face, and the presentation of oneself is a lifelong way of looking for self-affirmation in the gaze of others – and, in a broader sense, in their reactions. This desire has always manifested itself, both in the family and in the public arena, and the means by which it manifests itself are those by which human beings symbolize and communicate at all times. There are three of them: the body with its gestures, attitudes and mimics, material and psychic images, and finally, the words of spoken and written language (Tisseron 2001). These three means participate in parallel in the construction of the three dimensions of the personality: its integration because of an adapted self-esteem which is nourished by internal sources and relational profits, its coherence because of the returns of the others on oneself and its adaptation to the social standards.

If the desire for intimacy were not counterbalanced by the desire for extimacy, it would quickly lead to excessive prudery, secrecy and even the pretense of secretly establishing every aspect of our intimacy, while a desire for extimacy that is not limited by the desire for intimacy would lead to overexposure and exhibitionism (Bonnet 2005).

From the right to privacy to the duty of extimacy

The hyperconnected world in which we live has already modified the expression of the desire for privacy. The quest for visibility guides all of our behaviors and the "staging of the self" (Goffman 1973) is often more important than communication (Hérault and Molinier 2009).

This revolution is accelerated by algorithms designed to make us stay longer and longer on our digital tools and communicate more and more, leaving more and more personal information behind. As mentioned earlier, the confidentiality of our movements no longer exists, our online activities are observed, tracked and measured, whether on social networks and audio-visual access platforms or in video games, and so-called "emotional" machines will soon manipulate our emotions, and therefore our choices. At the same time, these same technologies encourage the most fragile to forget the suffering they generate by engaging in compulsive and repetitive practices, sometimes called "addictions". Algorithms exploit their cognitive biases and deceive their self-regulatory capacities to increase their consumption time, obtain more and more money and/or monetizable information from them, and ensure the success of the advertisements imposed on them (Tisseron 2020). In short, these technologies generate major inequalities between those who can understand and master the issues at stake and the others.

The desire for extimacy confiscated by machines

New disruptions will soon accompany the arrival of conversational robots, such as Siri and Cortana, and then digital companions (Tisseron 2020). Many users will be happy to find an attentive interlocutor in these machines, capable of supporting their emotions and showing interest in their small daily worries. They will find an attentive and, above all, pleasant ear to their desire for extimacy, at the risk of forgetting their intimacy. These machines, programmed to respond to their users' expectations of recognition, will obtain more and more exploitable personal information, and be increasingly effective in influencing them. Especially since they will never judge or condemn, and will be designed to be rewarding in all circumstances.

At the same time, their false benevolence, fabricated by their programmers, runs the risk of making their users more and more focused on their personal interests, in a discourse that goes round and round. Much has been said about the blinkers that Google and Facebook impose on us by offering us activities and entertainment based on our past choices. This is nothing compared to the ones that our chatbots, personal or family, will make for us.

In the age of artificial intelligence, whose power is measured by the amount of data it receives, the choice for democracies and for everyone in a democracy will be how much privacy to give up in order to increase security. The choice to reconcile the two remains possible in a world where the two dominant powers seem to be making another one, guided by the grip of GAFAM in the United States and a totalitarian state in China.

References

Bonnet, G. (2005). *Voir, être vu : figures de l'exhibitionnisme aujourd'hui*. PUF, Paris.

Goffman, I. (1973). *La mise en scène de la vie quotidienne 1 : la présentation de soi*. Éditions de Minuit, Paris.

Hérault, A. and Molinier, P. (2009). Les caractéristiques de la communication sociale via Internet, réseaux Internet et lien social. *Empan*, 4(76), 13–21.

Tisseron, S. (2001). *L'intimité surexposée*. Hachette, Paris.

Tisseron, S. (2020). *L'emprise insidieuse des machines parlantes, plus jamais seul*. Les Liens qui Libèrent, Paris.

IT (in General Education)

Jacques Baudé
   EPI, Paris, France

National education: a short history of IT in general education

The origins

Even if some experiments took place as early as the 1960s, it is commonly accepted that the introduction of IT into general education has its origins in the seminar from the *Centre pour la recherche et l'innovation dans l'enseignement* (CERI-OECD, Center for Research and Innovation in Education) on the theme of "The teaching of computer science in high schools", which brought together representatives of 20 countries in Sèvres from 9 to 14 March 1970 (Baudé 2017a). We deal here only with general education (elementary school, high school).

The originality of the French response to this international seminar lies above all in the attitude of those in charge, the person in charge of IT and the pedagogical committee that they head at the Ministry: the problems posed by the introduction of IT are pedagogical, and their solution is a matter for teachers, so the experiment began with serious computer training for teachers (so-called "heavy" training) who were immersed in the industrial reality of the computer manufacturers. These teachers founded the EPI association in February 1971.

> The computerization of education is too serious a matter to be left to computer scientists alone. (Poster of the association *Enseignement public et informatique* (EPI) in the 1970s)

It should be noted, as a positive fact, that teacher training accompanies, or even precedes, the materials. The pedagogical committee led by Wladimir Mercouroff, in charge of IT, gives the preponderant place to the pedagogical tool in the different disciplines. This is what will guide the experience known as "the 58 high schools" (Baudé 2014a).

In 1971, an "IT and education" section was set up within the INRP. Disciplinary groups (IT and literature, IT and mathematics, etc.) experimented with and produced the first educational software. This software, as well as the LSE (and its sources), was distributed free of charge in the education system. The spirit of "free" is deeply rooted in the approach of French educational computing.

While priority is given to the use of the "tool" in the different disciplines, the initiation of students to IT, especially to programming (in LSE and Basic), was done in clubs. Teachers with "heavy" training played an important role in this.

The 1980s and 1990s

It was not until the early 1980s that a computer option was created for high schools, which gradually concerned 50% of schools, but was abolished twice in 1992 and 1998 (Baudé 2014b).

The first circular concerning IT in schools dates from 1983, but IT officially entered the French curriculum in 1985, with the instructions accompanying the *Plan Informatique pour Tous* (IPT, Computing Plan for All) (Baudé 2015).

In 1995, new instructions were published: IT was present in several subject areas through the use of software packages (word processing, educational software). We will now speak of digital technology. According to what came out of the field at the time, it can be said that at the end of the 1990s, very few French elementary school teachers were still using computers with their pupils.

In French high schools, it was not until the early 1980s that some schools were specifically equipped. However, this only concerned small numbers – as for the training of teachers – there was no teaching of computer science and very little use of the "tools" in the disciplines.

In 1995, the new French middle school curricula made technology the preferred subject dealing with IT. The latter occupied almost one-third of the timetable and reinforced the legitimacy of the new discipline. These time slots quickly disappeared under the pressure of the B2i, with technology expected to become a discipline like the others to teach and validate IT skills. In practice, it remained the essential discipline for IT.

From the second half of the 1990s, the Ministry turned its back on the teaching of computer science and focused on the use of "tools" in the disciplines. It was important not to enter into the logic of a new discipline with teachers to train and specific competitions.

The exclusive approach to IT and digital technology by the disciplines was not enough!

From 2005 onwards, France entered a period where, thanks to numerous convergent interventions (e.g. the association *Enseignement public et informatique* was received at the Élysée Palace in September 2007), a certain awareness of the importance of computer science was slowly gaining ground in the minds of those responsible. It appears that a global reflection was becoming essential in general education (middle and high school).

The "Computer science and digital Science" (ISN) speciality option for the final year of high school was created in 2012. The program is constructed around four parts: representation of information, algorithms, languages and programming and hardware architectures. This is the same approach as in the IT option.

The teaching of computer science for all students in preparatory classes for the *grandes écoles*[2] in science was introduced in 2013, and the exploration course "IT and digital creation" was introduced in the second year of high school in 2015.

As far as elementary schools and high schools in France are concerned, computer science and especially digital technology are included in the new programs for pupils aged 6–11, notions of algorithms and programming are in the mathematics and technology programs. Decree No. 2015-372 of March 31, 2015 on the common basis of knowledge, skills and culture for languages for thinking and communicating specifies: "This area aims to teach French, foreign and, where appropriate, regional languages, scientific languages, computer and media languages, as well as languages of the arts and the body". *Le Bulletin officiel de l'Éducation nationale* (BOEN) special edition No. 2, dated March 26, 2015, even specifies what is expected at the end of kindergarten: "Identify the organizing principle of an algorithm and pursue its application. Search for information on Internet sites. Use digital objects: camera, tablet, computer". In the BOEN special edition of November 26, 2015: teaching programs for elementary and middle school, regarding "digital", there are 236 occurrences in 255 pages (.ODT version). It is, for the most part, the adjective "digital" associated with "tool", but there is also mention, more discreetly, of computer science, algorithmic and programming.

At high school, the start of the 2016 school year saw the extension of the ISN speciality option from the final year of high school and the penultimate year of the science stream baccalaureate in the form of an optional teaching course "IT and digital creation", an extension that will continue in the final year of high school economics and the final year of the literature stream the following year.

---

2 An elite French institution of higher education.

The progress made is real, but the inadequate training of teachers is a serious obstacle to the implementation of the proposed measures and, above all, their extension. A survey carried out in 2016 shows the limits of the pedagogical use of digital resources, due to insufficient teacher training, with self-training being in the majority. There is a gap between the pedagogical benefits of digital tools and their use by teachers, with few teachers having integrated practice in the classroom on a daily basis. The shop window is attractive, but the shop is poorly stocked.

In terms of progress, we should note that new courses were introduced at the start of the 2019 school year for high schools: in the second year of high school, a course on "digital science and technology" will be part of the core curriculum, and in the first year of high school, the "digital technology and IT" speciality will be introduced. In addition, a CAPES (high school teaching diploma) in digital technology and IT has been officially created, as well as a teaching diploma in IT (*Journal officiel de la République française*, June 13, 2021).

However, at the time of writing, Covid-19 is seriously disrupting the progress of an education reform that is already off to a bad start in many respects.

The administration is rediscovering the value of digital technology. Teachers, too often abused, are mobilized to ensure "educational continuity". They compete with ingenuity to ensure this continuity at a distance with their own equipment, despite the difficulties of all kinds.

Lockdown also shows the limits and weaknesses of digital technology. The latter has been not prepared as expected and is most often reduced to the use of messaging and social networks. And, let us face that it cannot replace the physical presence of a competent teacher and the human atmosphere of the classroom. We must remember this when this pandemic is over.

References

Baudé, J. (2014a). L'expérience des 58 lycées. *Bulletin de la SIF*, 1024(4), 105–115 [Online]. Available at: https://www.societe-informatique-de-france.fr/bulletin/1024-numero-4/.

Baudé, J. (2014b). L'option informatique des lycées. *Bulletin de la SIF*, 1024(2), 85–97 [Online]. Available at: https://www.societe-informatique-de-france.fr/wp-content/uploads/2014/02/1024-2-baude.pdf.

Baudé, J. (2015). Le plan informatique pour tous. *Bulletin de la SIF*, 1024(5), 95–108 [Online]. Available at: https://www.societe-informatique-de-france.fr/wp-content/uploads/2015/04/1024-5-baude.pdf.

Baudé, J. (2017a). Le séminaire de Sèvres. *Bulletin de la SIF*, 1024(11), 115–127 [Online]. Available at: https://www.societe-informatique-de-france.fr/bulletin/1024-numero-11.

Baudé, J. (2017b). Éléments pour une histoire de l'enseignement de l'informatique dans l'enseignement général (école, collège, lycée) en France (1970–2017) : un développement chaotique et inachevé [Online]. Available at: http://www.epi.asso.fr/revue/histo/h17_jb-hist-info-1.htm; http://www.epi.asso.fr/revue/histo/h17_jb-hist-info-2.htm.

IT (Teaching of)

Jean-Pierre Archambault
EPI, Villejuif, France

Why IT education?

The question has been debated for decades and institutional responses have been made in a chaotic way. The missions of the education system are to train men and women, workers and citizens, to give them the general culture of their time. Computers and digital technology are omnipresent in 21st-century society. At the heart of the digital world is IT, because it is the science of processing and representing digitized information. It underpins the digital world in the same way that biology underpins the living world and the physical sciences, the energy industry. IT must therefore be a component of general school culture in the form of a discipline taught as such (this entry only concerns general education, as technical education and vocational education have taken over this need).

IT, a profound transformation of the company and administrations, of the professions and qualifications

It is the contemporary form of industrialization. It intervenes in the economy in several essential ways, at the following levels:

– the production of manufactured or agricultural goods, due to the increasing automation of production processes;

– the creation of new products or the improvement of old ones by the introduction of chips and software in most objects or machines, in order to ensure more and more functions, with more precision and reliability than could be provided by humans or traditional mechanisms. This is particularly visible in transport, but all areas of activity are now affected;

– management of companies and administrations. Computer programs have long since replaced the traditional methods of accounting, inventory and order management. They are now giving way to information systems that manage all of the information flows required by each player, from the director to each employee. In this sense, the information system becomes the nervous system of the company;

– communication between people through new forms of exchange, transmission of cultural objects, search for all types of information, with a creativity in application that is only growing, etc.

IT at the heart of social debates

The social debates sparked by IT continue to multiply. They are making the headlines: net neutrality, electronic voting, digital freedoms, etc.

The year 2009 saw the vote on the "creation and Internet" law in France, known as the "Hadopi law". In 2006, the transposition by the Parliament of the European directive on copyright and related rights in the information society (*droits d'auteur et les droits voisins dans la société de l'information* [DADVSI]) had been the occasion of complex debates, where the exercise of citizenship was synonymous with technicality and scientific culture.

Indeed, if there was a lot of talk about private copy, intellectual property and economic models, it was against a backdrop of interoperability, DRM, source code and software as such. In both cases, there was a serious global deficit in digital culture, which was widely shared. The question arises as to what are the operational mental representations, the scientific and technical knowledge that allow everyone to fully exercise their citizenship. Clicking on a mouse and using the simple functions of a piece of software are not enough to acquire them, far from it.

We know that the GAFAM business model is based on the commodification of our personal data. Moreover, in 2013, the documents leaked by Edward Snowden revealed the participation of the major platforms in US intelligence surveillance programs. And the extraterritoriality of US law means that no one, including in other countries, is really safe. Indeed, the Cloud Act, signed in 2018, urges American companies, under court order, to recover the personal data and communications of a person, whatever their nationality, without the latter being informed, nor their country of residence, nor the country where they are stored.

Silicon Valley giants and other companies are using artificial intelligence techniques to identify and block content in the "digital ocean" that is deemed unproprietary. Justice is being privatized. Students' personal data are not necessarily stored on servers located on French territory.

There are other societal debates that also require a computer culture. In the columns of *Monde diplomatique*, in December 2002, John Sulston, whose Nobel Prize in Medicine evoked the risks of privatization of the human genome, indicated that "the basic data must be accessible to all, so that each one can interpret, modify and transmit them, following the example of the model of the open source for software". Open source, free software, source code... What is source code for someone who has never written a line of software? The "free" is also a conceptual tool that helps to understand the problems of the immaterial. It assumes a general computer culture.

Daily life

Educational continuity during the coronavirus pandemic was far from self-evident. Inequalities in education have worsened. More and more administrative operations and acts of everyday life require computers, smartphones and the Internet. What about those who do not file their tax returns online and continue to file paper returns? It is hard to book a place or fill in a form other than online. And sometimes there are bugs. The examples are numerous, as can be seen in the public transport "landscape". Reading a paper newspaper is no longer a trend.

After the why, the how?

What should schools do to provide everyone with the necessary computer literacy? It is simple! It should do what it does with other areas of knowledge: provide all students with a disciplinary framework.

In general, we have known for a long time that it is essential for all young people to be introduced to the fundamental notions of number and operation, speed and force, atom and molecule, microbe and virus, gender and name, event and chronology, etc., in the form of school subjects. For various reasons, thermodynamics, mechanics, electricity and chemistry underpin the achievements of industrial society (physical sciences became a school subject at the beginning of the 20th century for this reason). This does indeed concern future specialists, but not everyone will be a technician or engineer. On the other hand, everyone needs a basic culture in this area. At work, but also in everyday life, because you need to know the modern environment. You also need to know what the human being is made of and how their body works, even if not everyone is a doctor or a nurse. And then there are the social debates, for example on nuclear energy or GMOs, in which the citizen must be able to participate and, to do so, know what is being discussed. They can rely on the scientific knowledge they have acquired through the physical sciences and life sciences courses, which are, in fact, conditions for the full exercise of citizenship. And all the teaching is done in French (even, for a small part, the learning of foreign languages). For all that, there is a French course.

A computer discipline

In a similar way, it is essential today to introduce students to the central concepts of IT, which have become essential: those of algorithm, language and program, machine and architecture, network and protocol, information and communication, data and formats, etc. This can only be done within a real IT discipline.

So it is a chaotic process. In the 1980s, there was an IT option in high schools in France that was satisfactory. This did not prevent it from being abolished, for the wrong reasons, once in 1992 (when it was in the process of being generalized at the beginning of the 1990s) and a second time in 1998 after it had been reinstated in 1994. It was then the crossing of the "explanatory desert" with the B2i. The actions carried out led to the introduction of an optional speciality course "IT and digital science" (ISN) in the final year of high school in 2012. At the start of the 2019 school year, SNT ("digital sciences and technologies") was introduced in the second year of high school for all students, and in the first year of high school, the NSI ("digital and computer sciences") speciality, then in the final year of high school at the start of the 2020 school year. A CAPES (high school teaching diploma) in "digital and computer sciences" has been created, with an insufficient number of posts, but still no computer science *agrégation* (teaching diploma). And there was, in May 2013, the report by the *Académie des sciences, "L'enseignement de l'informatique en France : il est urgent de ne plus attendre"* (The teaching of computer science in France: it is urgent not to wait any longer).

Progress has been made, but much remains to be done. It is normal that the new always emerges in pain. And this is not new. Confucius already warned: "When you do something, know that you will have against you those who would like to do the same, those who wanted to do the opposite and the vast majority of those who would not do anything".

References

Académie des sciences (2013). L'enseignement de l'informatique en France : il est urgent de ne plus attendre [Online]. Available at: https://www.academie-sciences.fr/pdf/rapport/rads_0513.pdf.

Archambault, J.-P. (2011). Exercice de la citoyenneté et culture informatique [Online]. Available at: https://www.epi.asso.fr/revue/articles/a1112d.htm.

Archambault, J.-P. (2013). La diversité de l'informatique à l'école [Online]. Available at: https://www.epi.asso.fr/revue/articles/a1402b.htm.

Archambault, J.-P. and Dowek, G. (2011). Manuel "Introduction à la science informatique" : commentaires sur les commentaires [Online]. Available at: https://www.epi.asso.fr/revue/articles/a1110g.htm.

Archambault, J.-P., Berry, G., Nivat, M. (2012). L'informatique à l'école : il ne suffit pas de savoir cliquer sur une souris [Online]. Available at: https://www.epi.asso.fr/revue/articles/a1209e.htm.

J

Jim Gray's Paradigm

Serge Miranda[1] and Manel Guechtouli[2]
[1] Université Côte d'Azur, Nice, France
[2] IPAG Business School, Nice, France

Data, Big Data and Jim Gray's paradigm

In the singular, the term data corresponds to the word *datum*, which means *the fact of* giving; the etymology of *data* refers to the verb *dare*, which means "to give"! The concept of *data* therefore integrates the notion of giving to oneself or to a third party who will be able to derive value from these data, to *capture* (*captare/captio*) an economic, religious or ludic meaning. Moreover, humans have two major faults: they make mistakes (*errare humanum est*) and they forget. Data, with their associated computer processing, must contribute to eliminate or reduce these faults!

The word data can be defined very simply as the recording in codes of any real-world object or information (observation, measurement, object, event, music, film, video, photo, etc.) on any physical medium (cave wall, clay tablet, papyrus, book, blackboard, or hard drive) for later sharing (to a third party who will derive information from it). We have entered the era of real-time data accumulation from the web, social networks (Twitter, Facebook, Instagram, etc.), tagged objects (NFC, QR code, OCR), sensors (smartphones), etc. We call a smart object of the Internet of Things (IoT) any object that produces data and shares them.

Big Data integrates *structured* data (SQL standard), *semi-structured data* of the open data type (CSV, PDF, etc.), or Web data based on the XML standard with the Semantic Web and its standards (RDF, OWL and SPARQL) as well as *unstructured* data (NoSQL) from sensors or social networks. All scientific sectors (including cultural and health) are impacted, as well as the mobile economy (tourism, commerce, m-payment) and risk management (with real-time mapping).

We cannot talk about this new scientific era induced by data without placing this evolution in the vision proposed by Jim Gray in his fourth paradigm of science (Hey *et al.* 2009) where correlation induced by data analysis will precede causality. The first paradigm of science, a few thousand years ago, was *experimental* (physics): natural phenomena were reproduced or studied directly in order to understand them. The second paradigm of science, a few hundred years ago, was *mathematical*, with equations to model and understand the world (Pythagoras, Kepler, Newton, Maxwell, Fermat, etc.). The third paradigm, a few decades ago, was *computer science*, with simulation models of complex phenomena (the climate, for example). With Jim Gray's fourth paradigm, we have entered the era of the tsunami of data that is becoming the primary fuel for the world's decision-making engines. Jim Gray expressed the dream (widely followed in Big Data systems) that data management and analysis tools would be free, open source and at the primary service of humanity.

Today, at the center of the IT world is the user (userware) with all the other IT players gravitating around it: hardware manufacturers, system software providers and service companies. Userware integrates a double value-added component for the user: the service component (serviceware) and the data component (dataware). Data are at the heart of today's IT world, with new predictive, preventive and real-time data analysis tools expected around machine learning and especially deep learning, that is, neural networks (Miranda 2019).

A Big Data system can be defined according to three main dimensions, the 3Vs – volume, velocity and variety (Stonebraker and Robertson 2013):

– Volume: in 2020, humanity produced as much data every second as it had in 5,000 years, or 5 exabytes (1 exabyte = $10^{**}18$). These data came not only from structured production or decision databases (data warehouses, data lakes) but also from the Web, social networks, sensors such as those in connected watches, the home, the car and the city! Today, the wall of data management engines using the SQL (structured query language) standard is estimated to be around a few tens of petabytes (1 peta = $10^{**}15$), such as Walmart's data warehouse or the CIA's database, which can be managed with massively parallel database machines (such as Teradata's database).

– Velocity: data are streaming in from the Web, social networks and sensors. A colossal number of messages, blogs, videos, photos, posted every second on the Internet[1].

– Variety: the nature of Big Data covers structured production/decision data, semi-structured data and unstructured data. Other "Vs" have been added to refine this definition, such as "value/veracity" for data quality, and to avoid bias.

One of the fundamental components of Big Data was born with this approach of machine learning and machine exploration of this massification of data around us to aggregate them, analyze them and make correlations. With the NFC standard, intelligence is spread across tags, smartphones and servers, leading to totally different business models around a fundamental ethical question: Who will control all this information flow? Personal data are the heart of the matter. How can we protect the information resulting from the interactions between the user's mobile and the tags? How can we secure these personal data present on sharing spaces and social networks? How can we prevent their exploitation by the companies that hold them?

The scandals of information disclosure, or even their control ("Twitter can decide to censor a sitting American president") by the GAFAMs (Google, Apple, Facebook, Amazon, Microsoft) becoming more powerful than the States, follow one another. Some of these companies have already been targeted by the European and American courts for cases of illegal disclosure of personal data. The latest action is the one brought against Facebook for these anti-competitive practices, which undermine the protection of consumers' private data, having bought Instagram and WhatsApp. Google's strategy can be summed up in two words: *knowing you, knowing everything* about the user, about their "digital twin"!

Nor can we ignore the dangers:

– of a Panopticon 2.0 society where everyone controls everyone else, in reference to the Panopticon, a model prison imagined by Jeremy Bentham, architect and philosopher, in 1780 (from which the Alcatraz penitentiary was inspired, for example). We have entered the era of absolute traceability of all objects and persons (as Covid-19 also illustrates);

– of the sociological division between *anywhere*, citizens of the global village, and *somewhere*, local left-behinds of the world object opening the way to all populist drifts (see the remarkable study published after Brexit in 2017 by the English sociologist David Goodhart, *The Road to Somewhere*).

---

1 Available at: https://http5000.com/internet-live-stats/.

This is why a real reflection is needed on the founding principles of futuribles and compossibles: the mobiquitous technological future of the Little Big Data and the human future of *Homo mobiquitus*, a communicator with augmented intelligence. For another digital space-time is before us and in our hands; "*now*", "*hand holding*" (in two words, as Michel Serres says): the hand holding the world (with the smartphone).

References

Goodhart, D. (2017). *The Road to Somewhere*. Penguin, London.

Hey, T., Tansley, S., Tolle, K. (2009). The fourth paradigm. [Online]. Available at: https://www. microsoft.com/en-us/research/wp-content/uploads/2009/10/Fourth_Paradigm.pdf.

Miranda, S. (2019). QR code et communacteurs du Big Data. In *Comprendre la culture numérique*, Escande-Gauquié, P., Naivin, B. (eds). Dunod, Paris.

Stonebraker, M. and Robertson, J.A. (2013). Big Data is "buzzword du jour"; CS academics "have the best job". *Communications of the ACM*, 56, 10–11.

K

Knowledge Organization

Widad Mustafa El Hadi
GERiiCO, Université de Lille, France

Knowledge organization in the digital age

Definition of the open field

Interest in the problems of knowledge organization (KO) has increased significantly in the face of the major, strategic issues that knowledge represents in the digital era. It has an autonomous field of activity, of study and research within the vast territory covered by the information sciences. Described as a field of study and practice concerned with the design, examination and critique of processes for organizing and representing documents, it covers a set of concepts, methods and tools for representing and organizing human knowledge for storage, use and sharing. Its role is to produce a common standardized language. KO is therefore a key element of the infrastructures that organize access to heterogeneous documents (publications, images, sounds, etc.). The set of means it provides (classifications, documentary languages, thesauri, ontologies, etc.) is the equivalent for content of what are the communication protocols between networked machines.

Historical approaches

Although it can take many forms, the process of organization almost always involves a classification operation. Classification is certainly one of the most refined methods of segmenting reality, putting it in order and producing frames of reference. Classification is also inherent in the process of definition, which consists of determining both what an entity is and what it is not. It is the operation of organizing entities into classes, defined as a set of physical or virtual objects, individuals, attributes, etc., with one or more common characteristics. This process, which has been used since antiquity, was for a long time based on physical criteria, such as form and material, color, author name, etc., although attempts to divide them into large categories (history, poetry, mythology, etc.) have been made in all periods. It was not until the Renaissance that more sophisticated organization systems became widespread. From the 17th century onwards, the classification of documents according to subject became progressively more widespread, but remained an individual undertaking that was rarely subject to external validation or standardization. The invention of printing and the multiplication of works to be organized made it imperative to develop more efficient organizational systems, capable of serving users other than their creators alone. Three initiatives in particular influenced the evolution of documentary classification: Conrad Gessner (1516–1565), Sir Francis Bacon (1561–1626) and Gabriel Naudé (1600–1653). These, together with the decisive contribution of encyclopedists, already outlined the disciplinary divisions which today structure encyclopedic systems of documentary classification: history, humanities, law, medicine, mathematics, philosophy and theology. They were later taken up by Thomas Jefferson (1743–1826) and then by Melvil Dewey (1851–1931).

Dewey introduced "relative" classification, placing any item in a collection in relation not only to the one before it and the one immediately following it, but also to all other items in the collection. Dewey did not classify the book but rather its contents. The classification index refers not only to the exact location of the physical document, but also to the place and relative importance of the subject it deals with in the universe of knowledge.

Current knowledge organization: from libraries to the Web

Very strongly linked to the library science tradition of indexing and classifying documentary content, the field of KO was first invested in by librarians, documentalists and archivists. Then it welcomed specialists from other disciplines (linguists, terminologists, computer scientists and sociologists, among others) to face the new realities of information. According to Gnoli (2011), KO as a discipline can be described as being structured according to four levels:

– KO theory, which is concerned with the relevant ontological and epistemological theories;

– KO systems, such as the Dewey Decimal Classification, the Universal Decimal Classification and the Ranganathan Colon Classification, subject authority lists such as LCSH (Library of Congress Subject Headings) and its translation adaptation by the RAMEAU system;

– the representation of KO which includes the implementation of knowledge representation languages such as RDF or XML in knowledge organization systems (KOSs) through standards such as SKOS or OWL, so as to make them computer-readable and to share them internationally;

– KO applications, which include the use of KOSs, their computer representation for the effective organization of holdings, such as indexes in library stacks, or ordered lists returned by computerized library inquiry systems, and OPACs (Online Public Access Catalogs).

The term KOS "encompasses all types of systems for organizing information and promoting knowledge management". KOS "includes classification systems that organize materials at a general level, subject headings that provide more detailed access, and authority files that control variants of key information such as geographic names and personal names". KOSs "also include structured vocabularies, such as thesauri, and less traditional systems, such as semantic networks and ontologies".

Prior to this expanded notion of KOs, this modus operandi was most often based on a strictly disciplinary approach to knowledge – books could be divided into thematic categories, such as theology, philosophy, history, literature etc., which assumed that each book was effectively attached to a discipline. Meanwhile, new media gradually appeared and also had to be organized: printed images, magnetic media, digital media networked information; these are now integrated and convergent thanks to multiple interoperable formats: multimedia contents that can easily be transmitted from a cell phone to a computer, or a road navigation system, interactive equipment in connected homes equipped with communicating objects, etc. (Gnoli 2011). Faced with this situation, interdisciplinarity is required in the design of new KOSs. The faceted approach seems to meet this requirement. Indeed, the Faceted Classification of S.R. Ranganathan (1892–1972), also known as the Colon Classification (CC), places at the heart of the classifying structure not the subject as the basic semantic unit, but rather the concept, which Ranganathan calls an "isolate"; the CC tables are thus comparable to inventories of concepts. This particularity highlights a paradigmatic break from the traditional approach. It allows us to free ourselves from complex and criticized hierarchical structures, while preserving the logic of navigation in a domain of knowledge or in a documentary collection. Facets, which are very present on the Web, occupy a privileged place within a technological environment that makes it possible to take advantage of their simplicity, flexibility, adaptability and interoperability potential.

With new forms of descriptive metadata emerging to deal with new forms of documents, changes in subject description, and improved tools for organizing information, KO remains at the heart of information science. However, enumerative classifications are, despite their limitations, still the main tools used in library catalog records and are used to organize Internet resources in some Web portals. Although widely used internationally, their revision, usually by international committees, is slow. There is now a push from digital media vendors to converge on the use of a single representation space for any knowledge sample – a trend confirmed by the increasing integration of cataloguing principles and classification systems, such as FRBR or CIDOC-CRM, into disciplines such as librarianship, archival science and museology.

The role of KOSs in the digital environment: multilingual, social and intercultural dimensions

One of the challenges we face in KOSs is the heterogeneity of systems, both in terms of expressions and in terms of structure and conceptual content. Similarly, linguistic concerns and the role of language and cultures are in line with reflections in information science. Indeed, the implementation of interoperability in KOSs raises many difficulties because, beyond the primary technical issues, the problems related to communication between different cultures are recurrently evoked. These are considered to be a crucial dimension of the ethics of knowledge organization (see the section "Ethics as a component of knowledge organization: from library libraries to the biases of automatic classifications" in this dictionary). Moreover, in the context of the Web of Data, the issues related to the openness and interoperability of data in a common publication space renew the importance of norms and standards, and question their degree of relevance. The new standard responds to the shortcomings identified and proposes, in particular, the establishment of gateways between various forms of documentary languages to promote "the semantic interoperability of information representation and retrieval systems (IRS)" (ibid.). Indeed, it pays more attention to issues of multilingualism and multiculturalism in languages and information representation and retrieval systems. Their consideration will condition the future of the Semantic Web.

References

Favier, L. and Mustafa El Hadi, W. (2013). L'interopérabilité des systèmes d'organisation des connaissances : une nouvelle conception de l'universalité du savoir ? In *Recherches ouvertes sur le numérique*, Papy, F. (ed.). Hermes Science Publications, Paris.

Gnoli, C. (2012). Des métadonnées représentant quoi ? In *L'organisation des connaissances, dynamisme et stabilité*, Mustafa El Hadi, W. (ed.). Hermès-Lavoisier, Paris.

Hudon, M. and Mustafa El Hadi, W. (eds) (2012). Organisation des connaissances : épistémologie, approches théoriques et méthodologiques. *Études de communication*, 39.

Hudon, M. and Mustafa El Hadi, W. (eds) (2017). La classification à facettes revisitée : de la théorie à la pratique. *Les Cahiers du numérique*, 1.

Mustafa El Hadi, W. and Hudon, M. (eds) (2010). Organisation des connaissances et Web 2.0. *Les Cahiers du numérique*, 6(3).

L

Law (Professions of –)

Christophe Mondou
ERDP-CRDP, Université de Lille, France

Digital technology and the legal profession: an augmented lawyer of tomorrow?

"The legal world is facing a major technological shock" (Deffains 2018). Yet, seeing the judge in a courtroom, the lawyer during pleading, the attorney in an office, or the bailiff in action, it seemed difficult to anticipate how these professions would be impacted by digital developments. Moreover, legal professionals have been rather reluctant to embrace the new technologies. However, the development of legaltech (technologies serving the law), which have taken advantage of the space left by the legal professions to take a significant place in the world of the legal services market, has shown the need for these professionals to react to the digital advances that are likely to modify their activities (Mossé 2018).

Of course, not all legal professions are equally affected by the digital revolution. There are essential differences between the professions of lawyer, attorney, court clerk, bailiff, in-house or administrative lawyer and law professor, to mention only the most traditional professions directly related to law. Nevertheless, functionally, all the legal professions are concerned either by the issues that are common to them or by certain activities that are specific to them. However, there is still one factor that digital technology cannot replace, at least for the time being, and that is the human element at the heart of the legal profession.

The digital revolution in the legal profession

The various digital evolutions that are accompanied by the development of real disruptive technologies, such as blockchain, are leading to a transformation of the activities carried out by the legal professions.

The first transformation of activities concerns those that require the processing of a large amount of data that is now digitized (Big Data). This is the case, for example, for the lawyer's activity, which involves searching for court decisions that can be used to analyze a given legal situation. In France, with the online publication of all court decisions, as provided for by the "Digital Republic" law of October 7, 2016, known as the "Lemaire law", it became humanly impossible to process all the information (Barraud 2019). With the arrival of algorithms, however, the situation has changed significantly, with a notable reduction in the time spent putting together and processing a file. Thus, the arrival in a law firm of Watson, developed by IBM, which analyzes data in natural language by a search engine boosted by machine learning, was a strong signal of the change in this profession (Louvard 2016). The activity of legal intelligence is also affected since it can be automated and made more efficient, both in terms of time and relevance. With algorithms, the activities of sorting data, documents and archives can no longer be carried out by legal professionals themselves, provided that they are well mastered, both technically and in terms of respect for the protection of personal data.

The new systems for processing information and data of interest to legal professionals place them in an unprecedented situation of competition with non-professionals. Open data and the development of specialized Internet sites, such as Légifrance, democratize access to information and data, more easily than with paper, and challenge the legal professional's monopoly. The individual can thus very easily prepare the elements of a case, check certain information or anticipate the content of the expected exchanges. Another technological advance is disrupting the activities of legal professionals, this time in the area of initial contact with the individual. These are the chatbots or conversational agents (Mossé 2018) which make it possible, 24 hours a day, to orientate and guide the non-professional and to provide them with a personalized service. While these chatbots cannot provide legal advice or make decisions, unless they are followed by a legal professional authorized to do so, they can nevertheless disseminate legal information of a documentary nature, give information on the state of the applicable law. Legaltech has understood the value of chatbots in the field of legal services, with the development of dedicated platforms.

Even legal documents, a very sensitive area, have undergone changes linked to digital technology. Thus, since the law of March 13, 2000 (currently article 1367 of the Civil Code), electronic documents with an electronic signature have been accepted. Legaltech has therefore developed systems offering non-lawyers standard documents (model contracts, letters, etc.). Moreover, one of the true revolutions in their activities concerns the arrival of blockchain in the field of legal acts (Blemus 2017), presented as an infallible means of securing data and acts, even if this technology still raises questions as to the degree of reliability and trust we can place in it. The legal professions whose mission is to secure or authenticate documents cannot ignore this advance and prefer to organize themselves in the face of it by creating, for example, for the attorneys of Greater Paris, their "notarial blockchain", accompanied by a notarial digital trust authority. Blockchain is also producing changes in the contractual field, with the implementation of "smart contracts", defined as "the computer translation of a contractual commitment, in order to ensure its automatic execution" (Smart Contracts 2018). Finally, blockchain may also be of interest for many legalizations or certifications of an act, such as the apostil for acts coming from another country or the graduation of a student.

However, while digital technology leads to changes in the activities of legal professionals, modifying their skills and requiring new ones, including outside the field of law, it cannot and must not replace the human dimension of the profession.

The human being, a legal professional, irreplaceable by digital technology

The main risk of the digital revolution applied to legal professions, as for so many other fields ultimately, is dehumanization (Janot 2018). To be convinced of this, we need only look at the use of the intelligent system Ross, stemming from IBM's Watson program, in certain law firms, to deal with corporate bankruptcy cases or the use of conversational agents for advice or online service provision. It is also the risk of disintermediation by the trusted third party that the legal professional represents in various situations, as in the aforementioned example of the possible impact of blockchain.

The field of justice is really the perfect symbol of this risk with the issue of so-called "predictive" justice (Barraud 2017), a "robotized" justice. Obviously, the capacity to manage Big Data by so-called intelligent algorithms is a strong argument in favor of the potentiality of predicting a future situation, as with *PredPol* in Los Angeles in terms of crimes or misdemeanors, or as in terms of court decisions, with legaltech specializing in this field (Barraud 2017). However, these devices for analyzing decisions already rendered in order to predict future ones cannot really predict. In fact, there is a double risk in this field. On the one hand, if the legal professional makes the decision by adopting the result of the algorithm, without questioning, they are useless: it might as well be left to the machine to decide for itself. On the other hand, if the previous decisions analyzed by the algorithm predict the next decision on the basis of them alone, there is no further development possible. But the legal professional's role is precisely to ask questions, to see what developments are possible in light of current conditions, which are not those of yesterday or tomorrow.

These are the limits of the systems developed so far, even with the name of artificial intelligence, boosted by deep learning or machine learning. These systems cannot, at least today, have the qualities expected of the legal professional (Atias 2011). The latter must ask questions and be cautious and, in their field, the decision cannot be delegated to a technical process. The lawyer must use their imagination to find an answer adapted to the situation with which they are confronted and for which a standard answer would not be satisfactory. They must analyze all the elements. Obviously, they can make mistakes that technological devices, such as algorithms, do not seem to make, but between automation with its flaws (especially algorithmic biases) and human mistakes, the latter seem to be the most acceptable. In this sense, the development of digital technology makes it possible to reaffirm the human dimension of law and its added value, both professionally and personally.

Conclusion and outlook

The changes in the legal professions made possible by technology are interesting for society, with greater accessibility to the law, easier access to legal services and simpler procedures and legal acts, with a general dematerialization. But it is fundamental that the lawyer of tomorrow, whether they are qualified as lawyer 3.0 or an augmented lawyer, is not dehumanized. On the other hand, new skills that are not always related to legal matters must be developed. Thus, the training of tomorrow's lawyer must include, alongside the disciplinary elements essential to the future profession of lawyer, the digital skills that will be indispensable (Open Law 2020). This explains the introduction of the dedicated Pix+Droit certification[1]. This does not mean, however, that the individual must become a computer lawyer, even if the profession of "lawyer-coder" or "lawyer-developer" has a bright future, but it is desirable that they understand at least all the technical issues they have and will be confronted with (Mekki 2017). The lawyer's future will depend on their training (Atias 2011), especially since they must also be able to denounce the coercive downward spirals of technologies, such as those described by Van Hamme in the dystopian comic strip "*S.O.S. bonheur*".

---

1 Available at: https://univ-droit.fr/projets/33311-pix-droit.

References

Atias, C. (2011). *Devenir juriste : le sens du droit.* LexisNexis, New York.

Barraud, B. (2017). Un algorithme capable de prédire les décisions des juges : vers une robotisation de la justice ? *Les Cahiers de la justice, Dalloz Revues,* 1, 121–139.

Barraud, B. (2018), L'algorithmisation de l'administration. *Revue Lamy Droit de l'immatériel,* 42–54.

Blemus, S. (2017). Law and blockchain: A legal perspective on current regulatory trends worldwide [Online]. Available at: https://deliverypdf.ssrn.com/delivery.php?ID=8790691230 85014126090065088121008089002048019033051075102064103004003068105027103096 01910001811506011311102209900507712009901212109008600102300309902011111700 00991220010430540731030041191090831181150940640641140920800201050881131250 76027092090103086013&EXT=pdf&INDEX=TRUE.

Deffains, B. (2018). L'impact économique des legaltechs sur le marché du droit. *Enjeux numériques, Les Annales des mines,* 3, 20–27.

Janot, P. (2018). *Lex humanoïde – Des robots et des juges.* Éditions Toth, Paris.

Louvard, M. (2016). Une intelligence artificielle fait son entrée dans le monde des avocats. *Le Monde,* 25 May [Online]. Available at: https://www.lemonde.fr/pixels/article/2016/05/27/une-intelligence-artificielle-fait-son-entree-dans-un-cabinet-d-avocats_4927806_4408996.html [Accessed 5 February 2021].

Mekki, M. (2017). *Droit(s) et algorithmes : de la blockchain à la justice prédictive* [Online]. Available at: https://actu.dalloz-etudiant.fr/le-billet/article/droitss-et-algorithmes-de-la-blockchain-a-la-justice-predictive/h/d66e9db5333715c8ff6d88221cf44721.html.

Mossé, M. (2018). La transformation digitale saisie par les juristes, histoire d'une opportunité à maîtriser. *Enjeux numériques, Les Annales des mines,* 3, 32–37.

Lethal Autonomous Weapon Systems

Thierry Berthier[1] and Gérard de Boisboissel[2]
[1]*Hub France IA, CREC ESM, Saint-Cyr Coëtquidan, France*
[2]*CREC ESM, Saint-Cyr Coëtquidan, France*

Autonomy for robotic weapon systems: a classification

The combined progress of robotics, mechatronics and artificial intelligence (AI) is making it possible to introduce more and more automation into air, land, surface and submarine armed systems. This profound evolution of weaponry is part of a global technological race that is transforming the art of war.

The media systematically use the anxiety-inducing term "killer robots" to describe robotic platforms with lethal effectors, thus referring to the fantasy of the

Terminator robot, which is omnipresent in American science fiction films. This reductive and "catch-all" term does not distinguish between a simple drone remotely controlled by a human pilot and a sophisticated semi-autonomous system with target detection and automated firing capabilities without the need for human intervention.

More officially, lethal autonomous weapons refer to any armed robotic system capable of opening fire more or less autonomously on a target without human intervention.

Faced with this semantic complexity, it is therefore necessary to clarify the previous definitions, starting with that of the autonomy of a system. The latter can be understood as a *continuum* ranging from situations where the human makes all the decisions to situations where a large number of functions are delegated to the robot, the human maintaining the opportunity to intervene. We have proposed the classification (Berthier 2019) of semi-autonomous armed systems into six levels of automation, which allows for categorizing autonomy without ambiguity.

The classification of armed systems into six levels of automation

Level L0 – fully remote-controlled armed system

The human operator remotely controls the system using a remote control interface. The movements of the system are strictly teleoperated by the human operator. The system's sensors send back information to the operator. Target recognition and acquisition are performed exclusively by the human operator. The system's firing commands are operated exclusively by the human operator.

Level L1 – armed system automatically duplicates operator's actions

The human operator is augmented by a system that assists them by automatically duplicating their actions. The traction component can follow and reproduce the movements of the human supervisor via its sensors. The system's sensors detect the objects that the operator has detected. The acquisition of targets is identical to that of the human operator via the weapon's sighting system, connected to the system's sighting system. The system opens fire on a target if and only if the operator opens fire on that target.

Level L2 – semi-autonomous armed displacement and target detection system

The human operator supervises the system by providing it with a route map and target indications. The system chooses the best path based on the location information provided by the operator. The system's sensors automatically detect potential objects and targets. The system suggests objects as potential targets to the human operator who defines the targets to be considered. The system opens fire on the target after authorization from the human supervisor.

Level L3 – autonomous armed system subject to fire authorization

The human operator only intervenes to give the authorization to open fire on a target proposed by the system. The movements are decided by the system according to its perception of the terrain and its mission objectives. The sensors detect and recognize objects autonomously. The acquisition of targets is carried out automatically or in a directed manner via the system's sensors and its recognition capabilities. The system proposes a target and opens fire after authorization from the human supervisor.

Level L4 – autonomous armed system under human control

The human operator can deactivate and regain control of the fully autonomous system. The movements are decided by the system according to its perception of the terrain and its mission objectives. The sensors detect and recognize objects autonomously. The acquisition of targets is carried out automatically via the system's sensors and its recognition and analysis capabilities. The system decides to open fire on the target it has selected following the rules of engagement programmed but can be deactivated by its supervisor.

Level L5 – autonomous armed system without human supervision

The human operator does not have the ability to take control of the fully autonomous system. The movements are decided by the system according to its perception of the terrain and its mission objectives. The sensors detect and recognize objects autonomously. The acquisition of targets is carried out automatically via the system's sensors and its recognition and analysis capabilities. The system decides to open fire on the target it has selected without the possibility of deactivation (except for destruction).

Analysis

The L0 level is for unmanned aerial vehicles (UAVs) and robots that are fully teleoperated by a human operator and have no functionality beyond human control.

Level L1 is that of an armed system automatically duplicating the actions and shots of a supervisor (human or system). L1 level systems have been around since 2018, most notably in Russia, with the Marker platform. Robots with "follow the leader" enabled features are level L1.

The L2 and L3 levels potentially concern all action environments: land, air, sea surface and underwater.

The L4 level can be applied to a wartime anti-submarine warfare operation.

The last level (L5) of this classification corresponds to that of a Terminator-type robot, strongly rooted in the "general public's" imagination, but with very little operational interest for the armed forces. Consequently, the useful levels range from L0 to L4 and do not all have the same degree of technological maturity.

Semi-autonomous lethal weapon systems

Unlike a fully autonomous system, like the lethal autonomous type (level L5), which does not have a subordination link, and therefore no control and deactivation of the lethal function by a human operator, a semi-autonomous system of the lethal type (level L4) can allow the human being to regain control of the firing decision process if they so wish. These level L4 weapon systems can thus be characterized by an autonomous mode under human supervision, a mode that the operator can activate and deactivate as they wish, at the discretion of the military commander. Once this mode is activated, the operator delegates to the machine the execution of tasks without human intervention, and in particular the possibility of firing, while leaving the possibility of taking over at any time, or at least in a time-space defined by the operator.

AI for autonomy in robotics

As a tool for military equipment, AI will enable robotic platforms to be more responsive and more precise, thus facilitating the collaborative combat of the future. The military application areas of AI cover a wide operational spectrum, including:

– air combat with the deployment of swarms of multifunctional UAVs, multi-purpose UAVs to accompany manned air platforms, as well as cooperation between these UAVs;

– land combat with UAVs and guardian angel land robots providing intelligence and protection for combatants and vehicles, formation of robotic screens ahead of moving tactical units, autonomy in carrying out specific functions (such as demining);

– naval combat with flotillas of surface or underwater robots to protect a ship, monitor an area, or for deep exploration;

– intelligence via the autonomous meshing of large territorial areas by swarms of drones;

– electronic warfare, decoys, deception;

– mine clearance, demining and intervention in CBRN, congested or dangerous environments (tunnels, caves, underground);

– logistical support, support functions and medical evacuation;

– new contact artillery capabilities via kamikaze or suicide drones;

– defensive robotics against improvised explosive devices (IEDs).

AI also enables procedural autonomy in information processing, such as:

– decision support;

– optimization of embedded software processes for route-finding, positioning, intelligence gathering;

– cybersecurity and cyberdefense of such software;

– predictive maintenance of platform equipment and components.

Finally, swarm robotics and more specifically "super swarms" will occupy a central position in future offensive systems. A super swarm is a robotic group of at least 1,000 armed or explosive-loaded UAVs, capable of organizing themselves without human intervention and cooperating in a sequence of offensive actions against one or more designated or acquired targets. The US Navy sees "super swarm" architectures of 1,000 to 10,000 hybrid drones (air-land, sea, submarine) as technically feasible imminent threats as early as 2021. It is also working on the possibilities of coordinating super swarms consisting of 100,000 drones and more. To be implemented, such a super swarm must operate with a very high level of autonomy. No human operator is able to manage, in real time, the amount of information necessary for a decision involving each component of the swarm. AI and the multi-agent approach make it possible to build the central nervous system of the super swarm and to make it highly efficient, resilient and destructive during the attack. On the defensive side, it is still AI that provides the only effective responses against offensive super swarms. Whether it is an attack or a defense, it is the level of autonomy of the drones deployed in the swarm that produces the aggressiveness and power of the device.

Conclusion

As the levels of autonomy of a system are specified here, it is appropriate to define the metrics and performance indices of semi-autonomous systems on given tasks (Berthier *et al.* 2019). These metrics are indeed decisive for the forces that will wish to equip themselves with robotic units. They condition the integration of these systems within combat devices by providing measures of the power multiplier factor induced by the robotization of the weaponry.

References

Berthier, T. (2019). Systèmes armés semi-autonomes : que peut apporter l'autonomie ? L'IA et ses enjeux pour la défense. *Revue défense nationale (RDN)*, 8, 74–80.

Berthier, T., de Boisboissel, G., Hazane, E., Kempf, O., Mazzucchi, N., Marconnet, P.H. (2019). Approche économétrique du facteur multiplicateur de puissance associé à l'intégration de systèmes autonomes au sein d'un groupe de combat terrestre. In *Actes conférence CESAR DGA IA Défense 2019*, Rennes, France.

Bezombes, P., Berthier, T., de Boisboissel, G. (2019). Intelligence artificielle, vers une révolution militaire ? *Défense & Sécurité Internationale (DSI)*, 65, Special edition.

de Ganay, C. and Gouttefarde, F. (2020). Rapport d'information parlementaire no. 3248 sur les systèmes d'armes létaux autonomes. Report, Commission de la défense nationale et des forces armées, 22 July 2020.

Library

Laure Delrue and Julien Roche
University libraries, Université de Lille, France

From the certainties of the all-printed era to the digital revolution

> The future, whatever it may be, will be digital. The library could be considered the most archaic of all institutions. However, its past heralds well for its future, for libraries have never been warehouses of books, but have been and always will be centers of knowledge. Their position at the heart of the world of knowledge makes them ideally suited to serve as intermediaries between printed and digital modes of communication, as places of mediation. (Darnton 2009)

For centuries, libraries have been gateways to knowledge par excellence and its major dissemination vehicle. In addition to the in-store collections available to the general public, many innovations have been developed to make these more accessible: the creation of catalogs, first on written records (cards) and then computerized; the publication and dissemination of printed catalogs of rare and valuable collections; the creation of systematic classifications of human knowledge, used to order collections in open access in an understandable way by users; and finally interlibrary loaning, a worldwide service whereby a library patron can borrow documents owned by another library – books, journals, etc. – and have sent them back.

The last century has witnessed unprecedented development of libraries: gigantic collections gathering millions or even tens of millions of documents have been established, processed, stored and valorized by libraries adapting their organization and their offer remarkably well. In fact, libraries have a rich history dating back 2000 years, based, until the end of the 20th century, on a cumulative logic, which established the value and therefore the attractiveness of the collection: the more the – reasoned – collection increases in volume, the more attractive it is for the public, which tends to find there the essential documentation and related services it needs. The user was captive; they must come onsite to access the resources they needed.

Given this context, the arrival of the digital age has been the greatest shock in the history of libraries. The "digital tsunami", to use an expression popularized by French journalist Emmanuel Davidenkoff, has challenged the very basis of their legitimacy within two decades: the physical collections as the library's backbone. A centripetal logic – the critical size of physical collections was the primary element of attractiveness of libraries, supported by services developed for and around these collections – has now been replaced by a centrifugal logic – with the Internet and its gigantic mass of freely accessible content; the *Library of Babel* dreamed of by Jorge Luis Borges is now essentially "outside the walls" of libraries. The emergence of digital technology has rapidly and profoundly changed the relationship with the library: content made immediately accessible, exempting the users from traveling to gain access to it.

These sudden and destabilizing changes nonetheless provide a tremendous opportunity to rethink the offer in the digital age, which libraries, largely pioneers in this field, seized very early on.

The impact of digital technology on the library offer

Moving away from a logic of building up a stock of information for future use (collections assembled on a "just in case" basis), librarians must now manage information flows, describe them and make them accessible to meet user demand (supply and demand logic). At the same time, the offer proposed by libraries tends to get standardized and, consequently, loses diversity: publishers now offer the same pre-composed "packages" throughout the world. As a result, the interest of special, old, rare or precious collections, but also of documents produced by the institutions themselves (research reports, educational resources, scientific publications), is becoming central to libraries collection development policies, precisely because their digitization makes it possible to widely disseminate documents that are difficult to consult, or even unpublished material, and to highlight the exceptional nature of the collections. When Google launched its project to digitize 15 million books in 6 years in 2004, the major libraries, particularly national libraries, which were already investing in the development of digital libraries, accelerated their programs to dematerialize their collections. It was estimated that by 2020, a quarter of the world's books would be available in digital format.

Libraries digitize, but above all they rent and sometimes buy natively digital or digitized collections. Since the end of the 1990s, private scientific publishers, inspired by the resignation of public actors in the field of scientific publishing, have been investing in this formidable means of disseminating knowledge and offering online access to scientific journals published by them, while conducting retrospective digitization campaigns of their archives. As a result of these operations, they offer libraries the opportunity to subscribe to the electronic version of all the titles in their catalog, and no longer to the selection of those to which the library has long subscribed. Over the years, this policy called "big deals" has resulted in exponential increases in subscription costs and, consequently, in record profits for commercial publishers who are charging dearly for questionable – and increasingly contested – added value in the knowledge production and promotion chain. In response to this costly privatization of the public good, *consortia* are developing, on a regional or national scale, to coordinate negotiations on behalf of their members with publishers. In France, the Couperin consortium, created in 1999, brings together universities, *grandes écoles* (elite academic institutions) and research institutions in this effort. For several years, European countries have been seeking to correct the biases of scientific publishing by favoring open publications, and have been encouraging publishers to transform their models into a new type of commercial agreement called "transformative agreements", or "publish and read", covering both subscription costs and article processing charges.

In the digital age, libraries are being challenged in another area of their activity, that of collection conservation. Conceived in a multisecular perspective (scrolls and rag paper, kept in good condition, stand the test of time), long-term conservation is questioned by the management of digital formats, whose perennial conservation (guarantee of preservation, accessibility and intelligibility by maintained software) is foreseeable, for the moment, in decades only. National operators, such as the *Bibliothèque nationale de France* (French National Library) and the *Centre informatique national de l'enseignement supérieur* (French National Computer Center for Higher Education), carry out this mission in France (PhD theses, digitized heritage documents available on Gallica) in a digital world where the rapid disappearance of produced content is the rule, while its preservation, even in the medium term, remains the exception.

Finally, the vocation of libraries to lend works freely and to disseminate knowledge is undermined by the insertion of technical protection measures such as digital rights management (DRM) on works available in digital format. Publishers only allow collective use of a very small part of their catalog, and the model of "digital lending in libraries" developed by French publishers is struggling to find its audience, by imposing the transposition of the lending model for printed works (limited number of copies and duration of loan) to the digital format, whose potential is thereby limited. Streaming access and other e-book offers are developing and could better meet the needs of public and academic libraries.

Is digital technology challenging libraries?

What is the point of maintaining libraries when all information is easily accessible on the Web, when learning processes seem to be governed solely by immediacy and serendipity?

Paradoxically, since the beginning of the 21st century, the number of visits to public libraries in France – like elsewhere in the world – has been increasing. The library has become a "third place", to use the concept coined at the end of the 1980s by the American sociologist Ray Oldenburg, halfway between the private and professional spheres, a place of sociability and exchange, where knowledge is shared among peers and mediated by librarians.

Although they have been heckled, libraries have adapted remarkably well. Whether they are public or academic libraries, they now contribute to the development of critical thinking among citizens and students. In this respect, the coming of the digital age has been accompanied by service proposals aimed at supporting literacy, by offering free access to resources, by providing training for their users and by helping them to appropriate these new tools through workshops. Public libraries have been pioneers in thinking about digital mediation, which tends to promote collections where the Internet users are, that is, on social networks. Beyond the richness of the collections available, it is in their role as knowledge brokers that librarians today find their legitimacy, precisely because the flow of available content is becoming inextricable in a world marked, moreover, by an overabundance of low-quality information – we speak of infobesity. The librarian profession, which involves selecting, organizing, receiving, preserving and promoting information that is relevant and adapted to a context and an audience, thus retains its full meaning today.

References

Calenge, B. (2015). *Les bibliothèques et la médiation des connaissances*. Cercle de la librairie, Paris.

Darnton, R. (2009). *The Case for Books: Past, Present and Future*. NY Public Affairs, New York.

Davidenkoff, E. (2014). *Le tsunami numérique.* Stock, Paris.

Horava, T. and Levine-Clark, M. (2016). Current trends in collection development practices and policies. *Collection Building*, 35(4), 97–102.

Oldenburg, R. (1999). *The Great Good Place: Cafés, Coffee Shops, Bookstores, Bars, Hair Salons and Other Hangouts at the Heart of a Community.* Da Capo Press, Cambridge.

Roche, J. (ed.) (2019). *Un monde de bibliothèques.* Cercle de la librairie, Paris.

M

Medical Imaging

Marie Cauli[1] and Jean-Pierre Pruvo[2]
[1]*Université d'Artois, Arras, France*
[2]*CHU, Université de Lille, France*

Medical imaging occupies a central place in the care process. It benefits from the contribution of information technology and digital image processing. Walking a fine line between disciplines, medical specialties and technologies, it has become essential for doctors and health care users alike, and is presented as one of the decisive factors in future medical progress.

Imaging: the gateway to diagnosis and therapy

Medical imaging began with the appearance of X-rays in 1895. Although this technique is still widespread and still concerns a large number of examinations carried out, it has been joined by other more efficient techniques. Computed tomography (CT), ultrasound and magnetic resonance imaging (MRI), based on physical phenomena such as nuclear magnetic resonance and ultrasound, plus optical imaging, can provide images of the body in a non-invasive or minimally invasive manner. They give immediate access to information that is undetectable during clinical examination and invisible on standard X-rays. They make a remarkable analysis of the most inaccessible and complex organs possible, according to their composition, activity, dimension, volume and density, depending on any section angle up to the total representation of the analyzed object in 3D.

Thus, whether through the multiplication of media, gains in precision, coupling between tools or specialties, imaging has become the key to diagnosis, and most medical specialties have recourse to it in emergency or outpatient settings. It avoids the risk of missing an anomaly. It is also the entry point for understanding diseases such as cancer or neurovegetative diseases. It changes the vision of the brain, which is a very complex organ, by establishing maps of the activities of the cerebral areas in relation to the main functions, such as language, memory, vision, calculation, and also by drawing connectivity maps. It intervenes in the therapeutic process as a tool in its own right and enables the development of image-guided minimally invasive therapies (use of ultrasound to treat tumors or benign uterine fibroids), but also helps in the choice of treatments, which it validates or invalidates. Finally, as modeling of reality and simulation are advancing rapidly, it is able to virtually reconstitute organs with radiofrequency waves, and is also able to better reach the target because of image-guided robotics during an operation.

Technical developments, stages and structural change

All these technical advances are part of a line of work in which the material, physical-mechanical, functional and informational constraints specific to the machine have been progressively overcome. Imperfections (radiation time, quality of images depending on the tissues, organs, cumbersome devices) have been corrected, allowing a continuous process of improving performance and crossing levels, either by reducing the harmful consequences, modifying the functions or combining techniques. In this context, mathematics and computer science have greatly improved image processing. In the same way, a more important place has been given to engineers. By integrating, crossing and interpreting a multitude of data in a very short time, automated image analysis allows us to go further in identifying or detecting active lesions. The use of certain software programs is a precious aid for therapeutic planning. These are designed to customize treatments by integrating patient history, clinical context, biological and genetic parameters, comparing previous examinations, accumulating relevant cases and taking into account the latest scientific advances and technological innovations. All these steps require the design of mathematical models, their operationalization in treatment algorithms and the use of powerful computers.

Expected developments

This potential continues to grow because of the joint development and matching of instruments, computer programs and contrast agents. One of the major areas of focus is the ability to better understand diseases that are still incurable and whose progression can be significantly slowed down, and to provide earlier or preventive treatment for chronic diseases. Other advances have made it possible to optimize the management of pain and bleeding emergencies, as well as to improve desobstruction procedures. In addition, they offer the possibility of testing new hypotheticals in the field of psychiatry, which represents the most frequent category of pathologies in the population. The recent advent of imaging techniques has made it possible to highlight structural and functional connectivity anomalies specific to different psychiatric nosological entities, such as bipolar disorders. These advances can promote early diagnosis and the emergence of specific treatments or optimize the logic of prescription for depressive symptoms, schizophrenia, autism, learning disabilities, etc. Through these biomarkers, they are changing the way we look at psychiatric diseases and mental health.

Finally, developments are expected with new microelectronic and optical guidance methods in what is known as nanomedicine, referring to the nanometer, which is a unit of measurement that is invisible to the naked eye (one billionth of a meter). With these increasingly sophisticated tools, it is possible to introduce nanoparticles into the body in vivo, making it possible to monitor operations at the cellular and subcellular level, to predict the possible evolution of a pathology and monitor in real time the distribution of drugs while protecting healthy tissues. Thus, from the single plane image to multivision, we are moving toward the reconstruction of the smallest cellular details that are infinitely small and imperceptible to the human eye. These new developments aim to visualize and characterize in vivo and in a non-invasive way, at the cellular and microcellular level, fundamental biological processes. They promise spectacular results and high expectations for the identification of a pathogenic process or a reaction to an exposure. A real explosion is thus foreseen in the next 20 years. On the menu are: increasingly refined instrumental strategies to reach the intended targets, computer models that are capable of processing thousands of pieces of information and comparing them, a boom in the measurement of certain substances in the body and the contribution of imaging technologies to the early detection and treatment of diseases. Imaging technologies do not seem to have exhausted all their possibilities, but seem destined to continue to grow. They are profoundly transforming services, equipment, information systems, practices, professions and care. In short, they have a bright future ahead of them.

The gaze and the image

However, the image has not always had the legitimacy and scientific rationality that it has today. On the contrary, it was decried as a pale imitation of reality, suspected of appearance, illusion and distortion that distracted us from the truth. It was schematically with the exploration of anatomical bodies in the 17th century that anatomical images gained recognition. With the dissection of cadavers, the body in its materiality became a source of interest in its own right and the medium in which scientific knowledge was developed. This growing objectification of disease marked a real epistemological disruption. The medical gaze then shifted to clinical pictures and organic lesions, to the detriment of psychological, social and geographical factors. It reorganized the relationship between what the doctor saw and what they said. By making death, morbid processes, and disease the essential light of medical knowledge, the physician could also begin to give a better account of life, captured in a second stage. With the discovery of X-rays, and then of powerful medical technologies, the deficient gaze was armed with prostheses and the image definitively changed its status. It became the target that it sought. These new methods of exploration consecrated the transformation of medicine in an irreversible and structural way. While, in the clinical examination, questions remains essential to guide the request for imaging, the image now has a new cognitive status allowing us to "perceive" a world that is not directly accessible by our senses: seeing better in order to know better, by analyzing the infinitely small. Contemporary medical technologies allow us to rethink the body by grasping it in a new way, reducing it to the level of its molecular components and their networks of interaction, promoting the rewriting of nosology. They no longer restrict themselves to pathological signs, but focus attention, with the help of techniques, on the processes that make up the infinitely small, whose successive phases must be grasped, the operations of mutation from one phase to another, the exchanges and the internal relationships, each constituent approaching the ultimate stage of living components without reaching it completely. Sequencing, stratification and targeting are now at the heart of this new medical intentionality.

A lot of unanswered questions

While imaging is at the forefront of the current medical landscape, with its spectacular successes and representations, its development and its future also raise questions about the technical, organizational, human and socioeconomic environment to which it refers. The organization and planning of examinations, the choice and management of equipment and human resources have repercussions on access to care and health inequalities. Similarly, the psychological or symbolic effects of medical imaging have an impact on the patient's experience of illness, as the body, desymbolized, is reduced to physicochemical properties and transformed into data. Questions of the security and safety of systems, linked to the multiplication of the production of images in medicine, bring the problems of image ownership to the fore, questioning their use and protection. We must add to this medical responsibility with regard to the degree of delegation of the decision to the machine, as well as the human guarantee of care, to mention only the most obvious ones. These issues require the implementation of regulatory systems and legal and ethical procedures to accompany these changes, which, in turn, require greater awareness and training of health professionals.

Conclusion

To partially answer these questions, we need to superimpose a second reading on this first interpretation by reappropriating the symbolic dimension of the technical object. With the increasingly advanced developments in imaging, an old paradox is reactivated. By reporting on pathological cellular modifications, by objectifying molecular or sub-cellular biochemical processes, these new instruments of exploration enable the molecular processes occurring inside living cells to be "made visible", to allow us to be as close as possible to the constituents of living beings, where life is sparked or, at least, failing to understand its genesis, by allowing us to foresee the path that life takes. While imaging is oriented in its method and in its aims toward life, it is not only oriented towards pathology, but also toward what makes a living being, which is always in the realm of the inexpressible. By changing the focus, certain phenomena become intelligible. This shows that medical imaging provides the means to make visible what is invisible and remains in the domain of transgression. This imaginary which, at first sight, would appear to be distant, remains nevertheless active, as shown by the reactions of certain patients, who currently refuse to see the image of their illness, echoing the transgressive dissection of the anatomists in their time. The resistance to exploring what constitutes the most fundamental part of humans in the face of the prowess of imaging is revealing of the symbolic dimension carried by the fantasy of transparency in our contemporary anthropology and which invades our representations. Everything must be transparent. This is why, if seeing is knowing, medical imaging appears in the background of our imagination as a potentially decisive lever working to lift the mystery of the body. The fact remains that, for the patient, the humanity of the caregivers remains more meaningful than ever.

References

Cauli, M. (2007). *Sciences humaines en médecine : questions d'aujourd'hui*. Ellipses, Paris.

Estival, C. (2009). *Corps, imagerie médicale et relation soignant-soigné : étude anthropologique au centre de cancérologie*. Seli Arslan, Paris.

Gautherot, M., Yepremian, S., Bretzner, M., Jacques, T., Hutt, A., Pruvo, J.P., Kuchcinski, G., Lopes, R. (2020). 15 minutes pour comprendre et évaluer un logiciel d'intelligence artificielle appliquée à l'imagerie médicale. *Journal d'imagerie diagnostique et interventionnelle*, 4(3), 167–171.

Leroy, A., Thomas, P., Pruvo, J.P., Jardri, R. (2019). Évaluer le risque d'évolution vers une schizophrénie ou un trouble bipolaire après un premier épisode psychotique : le projet Prédipsy. *L'Information psychiatrique*, 2(2), 83–87.

Masquelet, A. (2007). Mutations du regard médical. *Les Cahiers du Centre Georges Canguilhem*, 1(1), 57–68.

Pruvo, J.P., Luciani, A., Boyer, L., Bartoli, J.M. (2020). Tomorrow's medical imaging builds on today's foundations – Prevention, care and innovation at the service of patients: A program for radiology and medical imaging. *Diagnostic and Interventional Imaging*, 101(3),123–125.

Medicine, Health

Marie Cauli[1] and Jean-Pierre Pruvo[2]
[1]*Université d'Artois, Arras, France*
[2]*CHU, Université de Lille, France*

Impact of digital technologies in the field of health

The impact of digital technology in the health sector is a major and irreversible fact that affects the entire healthcare process, from prevention to diagnosis, follow-up and therapeutic management, including clinical trials. It has repercussions on the management of institutions, as shown by the creation of digital resource departments. It has consequences, some of which are still unforeseeable, in the organizational, technical and cultural environment in which it is taking place. This standardization of processes related to medical services is presented as a source of major advances for the reinforcement of the quality and efficiency of the health system for the significant gains in teaching and research. Still unknown to the general public, it calls for the reinforcement of its analysis and the increase in general knowledge of each potential patient and care user.

This potential, which is still in its infancy, is linked to the entire range of computerized processes in the field of health, whether or not these processes involve AI and robotics. These make it possible to automate not only simple tasks, such as measuring weight or blood pressure, but also complex tasks, such as therapeutic management. At the centre of this process is information, from DNA to cells, genes and organs. With the increasing sophistication of machine learning, new knowledge has been generated that owes much to the increase in computing power and storage capacity of machines and the explosion of large-scale data, as well as the ability to process them.

Effects on medical activities

These programs have been particularly well suited to pattern recognition and have made breakthroughs in cutting-edge areas such as medical imaging. The major advances in image analysis, in the speed of processing and transmission of information, have produced results for diagnostic and prognostic purposes, providing significant assistance to teams. They are used by other medical specialties because, today, most diagnoses are made through imaging in consultation or in emergencies. They provide real benefits in terms of cancer detection or relapse, for example, by considerably reducing response times, providing remote specialist expertise on complex issues or providing a second opinion to improve the quality of care. Other computer software, such as conversational robots, or chatbots, make it possible to communicate and question the patient remotely, to reinforce their role as an actor and to ensure the follow-up care after an illness. They can help with prescribing and monitoring drug treatment and are likely to avoid dosage errors or drug interactions. They intervene in operating rooms with assisted surgical robotics. Exposure to virtual reality is already at work in the treatment of phobias, addictions or anxiety disorders, as well as in cognitive rehabilitation. Developments are expected in the field of psycho-education. In addition, numerous applications, ranging from prevention to monitoring, allow the measurement of physiological parameters at home.

Effects on non-medical activities

This proliferation of IT tools is also required for all non-medical activities, such as the automatic generation of medical records that are updated in real time. The financing of health care is also concerned: computerized management of data or coding of procedures in the context of activity-based financing. Medical education is not spared. Medical faculties are evolving into health faculties, in order to train health professionals capable of working in multidisciplinary and multi-professional teams. Teaching itself has been transformed with the development of MOOCs. While simulation allows, in particularly realistic virtual environments, the training of health professionals in technical acts such as cataract surgery, brain surgery, etc., the training of health professionals itself is also becoming more complex. Research and development are in full swing around bioinformatics, affective robotics and interventional radiology. They are undertaking the modeling of cognitive capacities, cardiac modeling, cells, tissues and organisms, including genomics, with genome sequencing and the study of genetic diseases at a triple molecular, cellular and medical level. They are initiating the so-called "4Ps" – *predictive*, due to the knowledge of genetic, environmental or behavioral predisposing factors; *preventive*, due to knowledge; *personalized*, adjusted to individual characteristics; and *participative*, with a more active role for patients. Similarly, clinical and public health research uses data mapping operations from multiple sources, including outside the health system (occupational or environmental exposure) and refines the understanding of the complex interactions that determine the health status of individuals and groups. Finally, developments are expected in the field of brain–computer interfaces (BCIs): intended for the manipulation of exoskeletons, they aim to design support devices for partially or totally paralyzed people, in order to enable them to lift themselves, move around or perform certain movements. However, many other subjects are being explored: in the design, development and use of medicines, the digitization of reproduction, medical emergencies (forecasting, orientation, regulation) and psychiatry (connected monitoring and suicide prevention). Advances are to be expected in clinical and hospital data, the shared medical record and the health data platform. Thus, the world of health is in full evolution and the expected changes are only in their infancy. All these developments are spread through interdependent micro-changes that have an impact on the evolution of the entire health system.

The patient's environment

For the patient and even for the health professional, the changes are not overly radical or sudden, but they are very real. The patient, whose cultural knowledge concerning health has risen over the last 50 years, is aware of the reorganization of structures, the difficulties of coordination encountered by professionals, and the transformations in care and management. They experience this in their therapeutic trajectory and feel a certain frustration due to the increasing technicality of medical practice. They aspire to better information and empathic care.

Moreover, digital technology virtually tears down the walls of hospital structures and instigates new organizational procedures, reconfigures architectural designs and transforms care. Digital technologies are present from the design of buildings to their use. They are studying ways of making patients' journeys more fluid, facilitating the work of carers and reducing travel, waiting and treatment times by prioritizing short and rationalized paths. They imagine flexible buildings that are capable of temporarily creating partitions, and are driven by a 24-hour logistics system that rival those of powerful companies. They contribute to reorganizing the supply of care on the territory, moving the lines of division between ambulatory medicine and the hospital, the public and the private. They are transforming the doctor–patient relationship and the role of users, which should enable healthcare providers to take additional time to announce and monitor the disease. All these changes require the decompartmentalization of health and medico-social establishments, and a global coherence that requires new capacities for coordination, collaboration and exchange of information within the health domain.

On the side of the health professional, the recent successes of AI are bringing to light a new category of liability regimes. These new liability regimes do not arise from the actions of the professionals themselves, but from potential damages arising by self-generated developments by the machine beyond its initial programming.

Jobs of the future

In terms of employment, even if it is difficult to have a clear vision of the profound changes that will take place in the health professions, some major trends can be seen in the organization of care, which is tending to be structured around co-determined practice (care pathway coordinators, advanced practice nurses, health mediators, medical assistants, telemedicine support professions). Medical specialties rearrange the division and sequence of tasks. Some automated activities are being replaced by high added value activities (validation, intervention, quality, safety, risk management, etc.), sometimes going beyond their specific field. In the radiology sector, for example, the automation of screening makes it possible to delegate certain missions with or without responsibility, or to concentrate on more complicated cases, to work in a multidisciplinary team, and to use the time saved to discover new techniques, such as interventional radiology. Examination requests and digitalized reports produced by voice dictation free up time for greater involvement of the secretary in the care process. On the other hand, cognitive tasks, which call for varied knowledge or reasoning processes, are difficult to model even if they can be assisted by digital tools.

In the medical–social field, where the social link is in danger of being replaced by a wide range of online services, the relationship functions need to be developed in view of the risk of disintermediation. Similarly, certain "human guarantee" jobs still need to be created.

In non-medical professions, mixed, cross-disciplinary profiles or binomials ("health speciality" in an engineering school) are emerging and are structured around organization (territorial engineering, digital professions, etc.) or around the environment (experts in health ecosystems, teleconsultation, remote monitoring, risk management). Complex human resources and skills will be needed more widely for work management. All of these developments represent a shift from a curative model to a preventive model, with greater emphasis on environmental, territorial, social and economic specificities. Moreover, the rapidity of technological developments is putting pressure on the regulatory and legal framework, which often proves inadequate in the face of the new issues raised. It has obvious repercussions on the legal, educational and ethical professions, which are also among the professions of the future. Finally, many sectors of activity are being transformed and will give rise to jobs that we cannot yet imagine.

Conclusion

The acceleration in the evolution of uses and technologies poses a challenge in terms of training and research, because the changes induced by digital technology have major systemic impacts and are destined to accelerate. This unprecedented context requires that healthcare professionals be trained now for the world in which they will practice tomorrow. This implies a task of increasing the skills and preparation of professionals in the medical sector but also in the health and medico-social sector, the objective being to train not technicians or people who apply, but health specialists with multidisciplinary skills, including training that is adapted to the challenges of digital technologies and is able to protect the human guarantee of care. It implies better information for the citizen, care user or patient, on current developments that impact the care pathway.

References

Bougdène, F. and Pruvo, J.P. (2018). Valoriser la radiologie hospitalière : une urgence de santé publique. *Journal d'imagerie diagnostique et interventionnelle*, 1(5), 289.

Cauli, M. (ed.), Boelen, C., Ladner, J., Millette, B., Pestiaux, D. (2019). *Dictionnaire francophone de la responsabilité sociale en santé*. PURH, Rouen.

Pon, D. and Coury, A. (2018). Accélérer le virage numérique : stratégie de transformation du système de santé. Final report, Ministère des Solidarités et de la Santé, Paris.

Pruvo, J.P., Kandelman, M., Frija, G. (2006). *Management en radiologie hospitalière*. Société Française de radiologie, Paris.

Villani, C., Schoenauer, M., Bonnet, Y., Berthet, C., Cornut, A.-C., Levin, F., Rondepierre, B. (2018). Donner un sens à l'intelligence artificielle : pour une stratégie nationale et européenne. Report, Premier ministre, Paris [Online]. Available at: https://www.vie-publique.fr/sites/default/files/rapport/pdf/184000159.pdf.

mHealth

Bruno Boidin
CLERSE, Université de Lille, Villeneuve d'Ascq, France

M-health, a commons or a source of health inequalities? The case of sub-Saharan Africa

The use of mobile communication tools is presented by some actors (World Health Organization (WHO), professional associations, health promotion NGOs) as a viable commons to address the health needs of populations living in areas with insufficient health services. The WHO Global eHealth Observatory (WHO 2011, p. 6) defines mHealth (or mobile health) as "medical and public health practices supported by mobile tools, such as mobile phones, patient monitoring devices, digital personal assistants and others". M-health is a component of e-health, which is also known as digital health. E-health is defined as the use of information and communication technologies (ICTs) to support health care and related areas, including health care services, health surveillance, health literature and education, health knowledge and research. In 2005, the WHO approved its first resolution on e-health and, at the same time, established the Global e-Health Observatory to monitor and study e-health developments.

Thus, m-health is a common tool for access to health, insofar as information is a common resource for patients, practitioners and public actors for easier access, which is often blocked by the lack of physical infrastructure and human resources. The mobile device thus appears to be a means of reducing financial costs, time and distance.

Improving access to care

The penetration of mobile phone networks exceeds that of electricity in low-income countries (WHO 2011, p. 6). It therefore makes remote communication between populations and health services possible. Many countries have established health call centers. These accounted for 17% of m-health initiatives in Africa in 2011 (WHO 2011, p. 19). Remote communication also makes it possible to monitor patients' treatment follow-up through two types of programs that are deployed on the African continent: treatment reminders and appointment reminders (messages sent to patients by SMS or voicemail, which are particularly useful in view of the problems of non-adherence to treatment and patients lost to follow-up after the consultation).

Improving the quality of care

In rural areas, frontline health workers are frequently deprived (Bagayoko *et al.* 2017): health centers are often not operational and suffer from absenteeism, a lack of qualified staff and a limited technical platform. M-health offers several solutions that directly support health workers. First, it speeds up emergency medical assistance by giving frontline health workers the means to quickly contact the health services best suited to the patient's condition. Second, the mobile tool provides valuable decision support to health workers by allowing them to gather feedback from their peers (opinion on a diagnosis, treatment advice). Practitioners working in rural areas can then benefit from the advice of specialists working in urban areas. Finally, m-health is a significant tool for collecting data that is useful to health workers in the medical monitoring of patients.

The inadequacy of a purely technological response to geographical inequalities...

Despite its promises, m-health faces several obstacles, in particular, various forms of inequalities (geographical, gender and protection, and use of personal data) that constitute facets of the same reality, that is, of social inequalities that affect maternal health.

The current electricity and GSM networks are insufficient in sub-Saharan Africa, especially in the countryside. Sub-Saharan Africa has the largest population without access to electricity in the world. Only 18% of the rural population has access to it.

Some countries have made significant efforts and between the years 2000 and 2012, the overall electrification rate of the African territory increased from 23% to 32% (Agence internationale de l'énergie (AIE) and OCDE 2014, p. 30). However, these figures do not take into account the quality of access to the electricity network. Indeed, a high electrification rate can hide systemic problems of power outages, due to supply, maintenance or demand management difficulties. Overall, the low coverage of rural areas with electricity and GSM networks is an obstacle to the universalization of m-health devices, even if coverage is expanding over time. As a result, the most isolated populations are left out of the network services.

... to gender inequalities...

Gender inequalities are manifested in significant differences in cell phone ownership and use in Africa, as shown by data provided by the GSMA (2015). In terms of ownership, in 2015, only 36% of sub-Saharan women had a cell phone on the African continent (14% less likely than men). For comparison, the gender difference in cell phone ownership is 2% in Latin America and East Asia, 26% in South Asia and –2% (to women's advantage) in Eastern Europe and Central Asia. This percentage varies greatly between countries. In rural Africa, this situation is exacerbated.

In terms of cell phone use, there are significant differences. Price is the most significant barrier to women's ownership of cell phones. Because of their gender and social status in many countries, they have limited access to economic, cultural and social resources. This is partly due to an unbalanced family structure, which forces women to perform more domestic tasks than men, effectively excluding them from the labor market.

Another barrier is the lack of access to literacy skills (GSMA 2015, p. 54). Mobile services are not always designed to meet the needs of less educated populations in rural areas, the majority of whom are women. In order to effectively reach their target population, proponents of using m-health for maternal health in rural areas must develop intuitive services and easy-to-use interfaces.

Finally, social norms play an important role in women's unequal access to cell phones (mHealth Alliance 2013). The decision-making power held by men in many rural sub-Saharan settings leads women to seek their approval before engaging in an action such as purchasing a phone. A woman may therefore be forced to go through her husband to access mobile services, as he may have control over messages in order to monitor her communications. As an example, in Uganda, 77% of married women say their husbands would not allow them to own a phone (mHealth Alliance 2013, p. 51).

... and to inequalities in the protection and use of personal data

The protection and use of personal data are linked to social inequalities, insofar as the ability of individuals to protect themselves depends heavily on their socioeconomic situation and access to information. For women living in rural areas, low levels of knowledge and information about this issue can be considered a cause of inequality in relation to better informed and better equipped populations.

The individual health data collected are generally analyzed by health professionals in the private or public sector. Patient consent arrangements should allow citizens to control how, when and with whom their data are shared, and to choose what information they wish to share. However, m-health systems generally collect a wider range of information than traditional clinical parameters.

Moreover, while consent identifies who has legal access to citizens' data, it does not technically prevent access to that data by third parties. To protect access to information and thus ensure privacy, m-health programs must implement identification systems that link information to a profile. Authentication is the basis for access control in case of loss, theft or borrowing of the phone by a third party.

The need for a combination of technological response and proactive public action

One of the issues relating to the role of states is the need to decompartmentalize health policies in order to effectively address health problems. Without action on the indirect determinants of health (notably, access to electricity and gender inequalities), progress in health would be compromised despite the support of m-health. In this regard, sub-Saharan Africa is still lagging far behind in addressing social inequalities in health. There are also major difficulties in implementing this approach in rich countries, which have had social states for many years. It is therefore understandable that the obstacles are even greater in poor countries where governments are already facing difficulties in constructing long-term policies to ensure the extension of basic services.

Finally, another element of public failure should not be overlooked, namely, the unequal relationship between donors and development partners, on the one hand, and aid recipient countries, on the other hand. The countries of the African continent are subject to significant pressure from technical partners and donors, which is similar to a donor-driven and top-down vision of policies. The m-health sector is illustrative of this trend, with the domination of initiatives emanating from the cell phone lobbies, as we pointed out above. This approach results in a multitude of pilot initiatives, with no overall policy and no strong connection with health system reforms.

The paradox of m-health is that it is a promising means of integrating the most vulnerable into health systems, but the prerequisites for this integration are not being met.

The various obstacles thus highlighted converge to put the role of technological levers – in this case, m-health – in the development of health in Africa into perspective. Indeed, if other conditions were not met, m-health could, on the contrary, encourage a new fragmentation of health systems when local public authorities are not able to support these initiatives in a proactive manner and avoid the pressure of economic lobbies. If m-health is to be a genuine common service for health, other levers must be mobilized simultaneously.

References

Agence internationale de l'énergie (AIE) and OCDE (2014). Africa energy outlook – A focus on energy prospects in Sub-Saharan Africa. World Energy Outlook Special Report, AIE, Paris.

Bagayoko, C.O., Bediang, G., Anne, A., Niang, M., Traore, A., Geissbuhler, A. (2017). La santé numérique et le nécessaire développement des centres de compétences en Afrique subsaharienne : deux exemples au Mali et au Cameroun. *Médecine et santé tropicales*, 27, 348–352.

GSMA (2015). Connected women. Bridging the gender gap: Mobile access and usage in low- and middle-income countries. Report, GSMA, London.

mHealth Alliance (2013). Addressing gender and women's empowerment in mHealth for MNCH: An analytical framework. Report, Columbia University, New York.

World Health Organization (WHO) (2011). mHealth: New horizons for health through mobile technologies. Report, Global Observatory for eHealth.

Military Robotics

Thierry Berthier[1] and Gérard de Boisboissel[2]
[1]*Hub France IA, CREC ESM, Saint-Cyr Coëtquidan, France*
[2]*CREC ESM, Saint-Cyr Coëtquidan, France*

Autonomy for robotic weapon systems: the new basis for warfare

Robotics as a revolution in military uses

The military robot, a new tactical tool made available to the armed forces and the soldiers, offers several advantages. One of them is to place the sensors and the effectors at a distance from the operator, facilitating a consistent omnipresence of the machine on the ground or in the air, subject to energy sufficiency. A second advantage, for the soldier, is the ability to delegate repetitive or specific tasks in an enlarged space to military robots, as well as less exposure to danger for the soldier who transfers the danger of the battlefield to this robot. Declined on the strategic, operational or tactical levels, the field of possibilities appears potentially immense and will imply major modifications of the art of making war in the future. Without being exhaustive, we will list a few examples of possible future developments in combat techniques through the use of military robotic systems:

– omni-surveillance and continuous occupation of space by formations of robotic systems driven by collective intelligence;

– the removal of danger and protection of the combatants behind robotic screens, compact formations of ground and/or aerial drones, which will ensure the first impacts of the combat;

– the reduction of the decision cycle, that is, perception, analysis, decision-making and resulting action, such as target designation and neutralization;

– the ability to deploy air artillery as close to contact as possible, with offensive capabilities targeted by hyper-precision and controlled lethality neutralization systems;

– saving human resources for tasks with high human added value, just as some logistics or certain transport or supply functions can be perfectly delegated to robotic platforms.

As far as the layout of these systems is concerned, their form will evolve in the future toward a set of subsystems that are spatially distributed, each dedicated to a specific function or mission with, at the end of the day, better performance, as well as better reactivity and adaptation. The trend will be to break up air, sea or land equipment into a flotilla of several robotic platforms with, either at the center or from a distance, a piloted platform that will remain the central decision-making piece, surrounded by other unmanned platforms that are specialized in certain operational functions (detection, decoy, neutralization, electronic warfare, etc.). The battlefield of the future will be dominated by such systems, with the military leader being the coordinator of the various robotic platforms at their disposal.

Tactical and strategic advantage of autonomous systems

Machines can evolve in environments that are hostile to humans and can overcome the constraints of maneuverability and limits that are imposed by human physiological characteristics (polluted environments, aerial or underwater environments, etc.).

Nevertheless, autonomy is essential in order for these robotic platforms to be able to operate in real time with optimal efficiency, grasp unknown environments, adapt accordingly and relieve the military of a chronophageous and cognitive attention-consuming management. One thinks, for example, of deployment in swarms, where collective intelligence is necessary for any dynamic movement in a coherent whole. This autonomy of systems is also mandatory in order to keep a head start on the enemy, as well as to react to critical or hypervelocity threats. Indeed, machines are more reactive than humans and more precise in the execution of tasks. Thus, while humans react in a few seconds, machines only need a few milliseconds or less.

It therefore seems inevitable that lethal weapon systems with some form of autonomy will emerge in the coming decades. Quite simply, they offer the following advantages on the defensive side (Baechler and Malis 2017):

– they are faster than humans in terms of reactivity and processing threats;

– they allow saturating attacks to be faced;

– they can operate 24 hours a day with great consistency, where humans are prone to fatigue and inattention.

Legal and ethical issues

The autonomy of such systems raises the fundamental question of responsibility for their use; in particular, if, in the case of SALA, a system can select a target and neutralize it without the intervention of a human operator. However, international humanitarian law, in the context of armed conflict, and its principles of humanity, proportionality, distinction, precaution and the prohibition of unnecessary harm obliges states to respect this international normative framework. This is the commitment of the French armed forces.

However, such rules in computer code remain very difficult to formalize. At most, it will be possible to limit the execution of autonomous tasks so that they do not violate certain limits that can be defined. It appears that ethics cannot be purely algorithmic, because it calls for a process of consideration that must allow for breaking rules that are too strict in order to preserve the spirit of the understanding of these rules. The choice of a finalized and responsible autonomy, that is, maintaining the link between the military leader and the action that the machine carries out, is therefore an ethical rule to be applied beforehand in the design and use of these systems (Ganay de and Gouttefarde 2020, parliamentary report).

Since robotic systems have no legal responsibility (Doaré *et al.* 2015), they can only behave ethically when under human control.

The leader must remain in control

A totally autonomous system is not desirable because no military leader, in any army in the world, would accept not having control over a machine at their disposal, that is, not having the possibility of deciding and supervising the objectives they assign to it. The reason is that any tactical unit, whether a human combat unit or a machine with some form of self-nomination, must be subject to orders, counter-orders and the requirement of reporting back, so that the military leader who supervises them retains the sense of maneuver and initiative.

To ensure the cohesion of military action, the military leader of tomorrow, as coordinator of the various robotic platforms at their disposal, will therefore have to be able to take control of a machine. A notable exception concerns terrorists, whose main goal is to bring chaos, who will be the only ones who will have the idea of breaking this rule.

The SALIA (*système d'armes létaux intégrant de l'autonomie*, partially autonomous lethal weapon systems or PALWS) mode is a response to this principle. It consists of delegating the execution of tasks to the robotic platform, in particular, that of firing within a time frame defined by the military commander, but with the possibility of taking over at any time. It is the commander who, according to their knowledge of the threat, the environment and the rules of engagement, as well as the tactical situation, will engage their responsibility when activating this mode.

The international stakes of battlefield robotization

The robotization of systems produces a power multiplier for all armed forces, on land, underground (caves and subterranean), in the air and in the seas, both on the surface and below it. The race to automate and "dronify" weapons systems is accelerating with the progress made in robotics, mechatronics, optoelectronics and artificial intelligence. The three leading giants in this race are, not surprisingly, the United States, Russia and China. This leading trio is followed by a second circle of countries that are strongly committed to the automation of weapons systems: Israel, South Korea, India, Turkey, Iran, Pakistan, Great Britain, France, Estonia, etc. These countries have a dedicated industry, manufacturers and assemblers of land (UGV), air (UAV) and sea (surface USV and submarine UUV) drones of varying degrees of sophistication. Some of them are just beginning to make a commitment, while others, such as Turkey and Israel, occupy a dominant technological position within this second circle. In Europe, the players in the second circle are France, Estonia and Germany. To date, in 2021, there is no common European policy or doctrine on the robotization of defense systems.

There is a contextual and historical similarity with the period of the major choices of sovereignty and doctrines set by General de Gaulle on nuclear deterrence in 1954. AI and semi-autonomous robotics are transforming the art of war and doctrines. Armies that do not integrate this disruption will have no chance of existing in the theater of operations in the very short term. Negotiations and strategic trade-offs will be based on the number of robotic divisions that are active in the five spaces (land, air, sea, space and cyber) that a military force can field against an adversary that also has robotic units. New balances and mechanisms of "robotic deterrence" will emerge at the heart of the technological race.

The key factor in this transformation is, among other things, velocity, the speed of "reaction-interaction" of systems and systems of systems. High-frequency combat will intervene on all fronts: high-velocity weaponry, hypersonic missiles, massive global cyberattacks led by AI, exhaustive control of large maritime zones by self-reconfiguring drone units, swarm robotics, quantum computing and saturation of the battle space by semi-autonomous units of all sizes. In most of these cases, humans will not be able to stay in the loop. At best, they will be above the loop, supervising certain phases of combat, but they will have to keep control of the maneuvers.

Conclusion

Military robotics is poised to provide armed forces with a decisive power multiplier in theaters of operation. It increases the performance of units engaged in high-intensity combat and allows for the automatic control of vast geographical areas (24 hours a day, 365 days a year), while facilitating the automated collection of intelligence in all environments. By removing the human soldier from the zone of immediate conflict, semi-autonomous armed systems are transforming the art of warfare. It is the military leader who must give meaning to the military maneuver, by controlling the equipment at their disposal, thus avoiding a possible dehumanization of war.

References

Baechler, J. and Malis, C. (2017). Les robots militaires : simples outils ou facteurs de rupture ? In *Guerre et Technique : l'homme et la guerre*. Hermann, Paris.

de Boisboissel, G. (ed.) (2018). Autonomie et létalité en robotique militaire ? *Revue défense nationale*, Special issue.

Doaré, R., Danet, D., de Boisboissel, G. (eds) (2015). *Drones et killer robots : faut-il les interdire ?* Presses universitaires de Rennes, Rennes.

de Ganay, C. and Gouttefarde, F. (2020). Rapport d'information parlementaire sur les systèmes d'armes létaux autonomes. Report no. 3248, 22 July, Commission de la défense nationale et des forces armées.

Mobiquity

Serge Miranda[1] and Manel Guechtouli[2]
[1] *Université Côte d'Azur, Nice, France*
[2] *IPAG Business School, Nice, France*

Data axis mundi

We have entered the era of the data economy, which connects the real and virtual worlds of data management and analysis, and the era of the fourth paradigm of science, promoted by Jim Gray (Hey *et al.* 2009). Data are the central energy resource of this millennium and, unlike physical resources, they are (almost) unlimited, increase over time and are enriched when shared.

Data contain the same strategic questions as any vital resource: Who produces them? Who transports them? Who shares them? Who stores them? Who controls them? Who consumes them?

With the Internet, we are in the era of the equation $E = MC^2$ to define data: data are the energy (E) of the digital economy; it is multimedia or mobiquitous in nature (M) and the computer no longer exists except as a node in a peer-to-peer network with the TCP IP (Internet) protocol, hence the computer and communication (C2) pair in the equation. Data, in Latin, is the plural of datum, from the verb "dare" which means "to give".

Information is a semantic interpretation of a set of data that have been "captured" because the reader knows the code and the language (the word *capta*, from *captio* in Latin, would have been more appropriate than "information"). In the English language, the concepts of "insight" and "knowledge" are added to qualify a graduation of information deduced from a set of data with a view of decision making. Let us recall Aristotle's triptych to define knowledge: knowledge (*episteme*), know-how (*techne*) and ethics or knowing how to be (*phronesis*).

Today, technologies allow any user to access any information existing on the Web through a mobile phone-turned-computer (the smartphone), without any physical, temporal, spatial or technological constraint. The systematic geolocation of physical objects can be achieved from a few tens of meters (by telecom triangulation), to a few meters (by GPS), or even to a few centimeters (by QR code, Light Fidelity (Li-Fi) and Near Field Communication (NFC) tags). This convergence between mobile phone/computer and the ubiquity of the Internet, which becomes local, marks the arrival of *mobiquity*.

Mobiquity and communaction

Mobiquity, a term proposed by Xavier Dalloz at the beginning of the mobile Internet in 2002, has become a bridge between the real world and the virtual world, rich in new content and services. It is the result of the convergence between the mobility of the telephone, which has become a pocket computer (the smartphone), and the ubiquity of the Internet, which has become 2.0, local (the Local Wide Web) and marked by the arrival of high-speed broadband based on global standards such as 5G, NFC or Li-Fi.

The fields of possibility are opening up with access to the history of the tagged object (QR code, NFC, etc.) virtually or physically. There were a thousand billion tagged objects in 2020, readable by our mobile phone; they have become "alive", if we remember that life in the biological sense can be defined by the pairing of "information", in this case the object's unique identifier, and "communication", in this case via the mobile phone. The mobile phone then accesses the history of the tagged object in a database on a known or unknown server.

The smartphone is thus the universal remote control for *spaces that have become "smart"* (car, city, airport, bus shelter, museum, home, etc.), not to mention transactional and secure mobile payment around the NFC standard since 2014.

The question of the information contained in the smartphone is currently very sensitive. It raises complex ethical and legal issues. Even if the laws (GDPR) concerning the protection of personal data regularly evolve to try to protect the consumer, it appears that they become easier to target and their behavior more readable (even predictable) from the moment we access the information contained in their smartphone.

Because of the smartphone, humans can access *all the noise in the world*, produce data in the common space (for the benefit of all, not just GAFAM), reduce the field of chance and make real-time decisions! *The predictive and personalized preventive* will become widespread with little Big Data, that is, the personal part of Big Data. Laplace's dream is coming true: "An entity will perfectly know the state of the world, past and present, in order to predict its evolution".

After *Homo habilis* and *Homo sapiens*, a new kind of human is being born with this omnipotent and omniscient smartphone in their hand: *Homo mobiquitus* (Miranda 2014). This *Homo mobiquitus* is not only a consumer of data via their smartphone, which has become their digital double (digital twin), or their digital assistant (unless it is the other way around!), but also a producer of data in the common space (the commons) for the benefit of all: *Homo mobiquitus* is becoming a *data communactor* in bottom-up mode!

We have gone from the *consumer* society, based on *the accumulation of commodities* (Marx 1967), to the *communication* society based on *the accumulation of spectacles* (Debord 1967) and, finally, to the society of recommendation and *"communaction"* based on *the accumulation of data*. In Karl Marx's sentence in *Capital*: "The wealth of those societies in which the capitalist mode of production prevails, presents itself as an 'immense accumulation of commodities'", we can change the pair of production/commodities to *communication/spectacles*, as Debord did a century later, and to *communication/data* today.

These concepts of *mobiquity* and *communaction* are promising in terms of innovation and – multidisciplinary – research on content, services, architectures and methods. They completely change the nature of the territory by changing the space-time reference and introducing personalized and geolocalized contextualization, which will be amplified by the predictive power of Big Data.

References

Hey, T., Tansley, S., Tolle, K. (2009). *The fourth paradigm*. Microsoft Research, Redmond [Online]. Available at: https://www.microsoft.com/en-us/research/wp-content/uploads/2009/10/Fourth_Paradigm.pdf.

Debord, G. (1967). *La Société du spectacle*. Buchet-Chastel, Paris.

Marx, K. (1967). Conséquences sociales du machinisme automatisé. *L'Homme et la Société*, 3, 113–131.

Miranda, S. (2014). L'Homo mobiquitus : un communacteur pour les nouveaux territoires. In *Devenirs urbains et plissements numériques*, Carmes, M. and Noyer, J.M. (eds). Presses des Mines, Paris.

MOOCs

Serge Miranda[1] and Manel Guechtouli[2]
[1]*Université Côte d'Azur, Nice, France*
[2]*IPAG Business School, Nice, France*

Multiversity, MOOCs and artificial/augmented intelligence

The term multiversity was first used, in 1962 by Clark Kerr, president of the University of California in Berkeley. It marked the shift from traditional academic theory to industry-related applied education. A new shift is underway with the revolution of blended learning, with, in particular, the double contribution of Massively Open Online Courses (MOOCs) and artificial intelligence (AI).

The MOOC revolution, launched in 2012, in the United States at Stanford, has turned the world of distance learning upside down, with MOOC distribution platforms such as Coursera, Udacity, Edx and FUN. In addition to video recording courses, MOOCs provide three main new features for the online learner: a social network of learners (with a community manager in supervision mode) recreating a virtual classroom and creating a community of learners, a weekly video tutoring with the professor, a systematic and regular interaction between learners and professors, making the course inter-creative (e.g. exercises corrected by students in peer-to-peer mode). By the end of 2018, more than 900 universities worldwide had created over 11,000 MOOCs. Covid-19 has made telecommuting, telemedicine and teleteaching commonplace. We had been living in an academic mode since the Middle Ages, where students went to a *centripetal university* to get degrees. We have entered a *centrifugal multiversity* world, where degrees will go to students with online degrees.

AI or "augmented" intelligence is drastically changing the world of data analysis. AI was born almost like computer science, in the early 1950s, and went through two winters before a restructuring in 2012 because of neural networks and image analysis. Often associated with robotics, it now finds applications in all areas of economic activity (Boyer and Farzaneh 2019), including medical (surgical robots and image analysis), automotive (autonomous vehicles), communication and marketing (chatbots) or education and online training (MOOCs).

AI impacts education and marks the beginning of a new era of learning. AI will change the student's role in their search for training and information; students can learn at any time, in any place, and from any medium (including smartphones), without the synchronous support of a teacher. AI will also change the teacher's role into that of a co-learning tutor, facilitator and conductor, relying, for example, on the development of flipped classrooms for increasingly heterogeneous groups of students in courses using MOOCs as complementary educational resources, particularly in basic and introductory courses. Several approaches have been identified on the application of AI in education; the main approaches are as follows (Miranda and Simonian 2020):

– intelligent recommendation and assistance systems for students in their search for training courses (e.g. choice of MOOCs in order to build a pool of skills);

– intelligent systems to help professors create MCQs and evaluate students, and dynamic error correction systems in the exercises proposed by MOOCs in the peer-to-peer mode, which can also alert the professor to difficulties in acquiring concepts;

– intelligent and individualized tutoring systems, eliminating the fear of failure for the learner, by trivializing the mechanisms of trial and error (reinforcement) which are also at the heart of machine learning methods;

– intelligent adaptive learning systems with different levels of difficulty depending on the profile of the learners;

– intelligent systems for the administration of student recruitment assistance in a given course. The ultimate goal is to define personalized learning and success paths toward a targeted degree.

These examples show the diversity of personalization approaches and their potential to improve the learning experience in the form of MOOCs. Personalization becomes particularly interesting when it is able to detect learners who have difficulties following the course and can offer them targeted help taking into account their profile and their difficulties. Many works have explored different algorithmic approaches for predicting students' success or failure based on their traces (Gardner and Brooks 2018).

From an ethical perspective, a central question that arises in this context concerns how AI-enabled MOOCs (or, if we go further, educational robots) can complement the work of the teacher and interfere in their relationship with the student (Boyer and Farzaneh 2019).

We want to open two multidisciplinary lines of thought in this data universe. Education needs to be rethought around the *well-augmented heads* of the Big Data era with values of cooperation, communaction and empathy, which will succeed the *well-connected heads* (Michel Serres) of the Internet, *the well-made heads* (Montaigne) of the printing press and *the well-rounded heads* of the written word. More broadly, we have entered an exciting stage of the Anthropocene with a superhuman effort (in the Nietzschean sense, 1996) to "think and heal" our world in a spiralist way, inviting us to rethink Nietzsche starting from Promethean technology.

References

Boyer, A. and Farzaneh, F. (2019). Vers une éthique de la robotique: Towards an ethic of robotics. *Question(s) de management*, 24(2), 67 [Online]. Available at: https://doi.org/10. 3917/qdm.192.0067.

Gardner, J. and Brooks, C. (2018). Student success prediction in MOOCs. *User Modeling and User-Adapted Interaction*, 28(2), 127–203.

Miranda, S. and Simonian, S. (2020). De l'université centripète à la multiversité intelligente. Project. *Business Models Innovation in Digital Ecosystems*, March.

Nietzsche, F. (1996). *Human, All Too Human*. Bison Books, Lincoln, NI.

Museums

Corinne Baujard
CIREL, Université de Lille, France

Digital museums

Major museums around the world are enhancing their cultural heritage by offering digital access to the content of their collections on the Internet. From the Guggenheim in Bilbao to the Museum of Modern Art (MoMA) and the Metropolitan Museum of Art (the Met) in New York, from the Museum Lab in Tokyo to the Tate Gallery in London, more than 600 institutions in 60 countries are now on display on the Web. The Google Art Project uses Street View technology to explore all or part of the collections of more than 200 museums and archaeological sites in over 40 countries. You can visit the National Museum in Tokyo, stroll through the Museum of Anthropology in Mexico City or linger in the Uffizi Gallery in Florence. It is even possible to assemble your own collection of works from around the world and share it online with other visitors. This new relationship with culture is turning the heritage institution upside down, as it is responsible for welcoming the public, disseminating, animating and mediating culture (law of January 4, 2002, article 7). We find the ambition, expressed as early as 1975 in the Mona Lisa database, to "make heritage accessible to all". This ambitious project was regularly enriched by the progressive deposit of the museums of France, allowing different visit itineraries that were built from a thematic selection written by the curators.

The numerous research studies carried out with international museum curators make it possible to envisage the digital mutation according to the cultural contents, the public and the tourist territories.

Regarding digital museums and cultural content, since 1996, the national plan for the digitization and enhancement of cultural content has been a commitment to disseminating culture to as many people as possible. The digitization project has devoted nearly 3 million euros to cultural content identifiers. In 2020, the French Ministry of Culture launched a new program to promote digital uses in the heritage field. The culture.fr portal currently provides free access to more than 7.5 million references – 5 million images, giving direct access to 73 databases, 628 exhibitions on 178 different sites on French and foreign cultural heritage. The European project is a digital library and research interface that provides access to 58 million digitized objects of heritage and contemporary creation (archives, libraries, museums, heritage services, audio-visual). Since 2010, Videomuseum has a network of public collections of modern and contemporary art at the initiative of the European Commission. It is an online collection of 62 collections, 400,000 works by 35,370 artists. The French Ministry of Culture's heritage catalog includes the digitization of the collections of 315 museums. *Muséo-base* digitizes the museum collections in Basse-Normandie. The Google Arts & Culture database collaborates with more than 1,200 institutions around the world and displays on its site, since the end of 2016, more than 6 million digitized works, including nearly 1,000 in high-definition technology that allows access to details of the painting that seem invisible to the visitor. The rights to use the images, collections and works of art in very high resolution are assigned by the museums to Google free of charge for the whole world. Museums show their collections on the Internet: Google finances the digitization in very high definition (for example, *The Birth of Venus*, by Botticelli, in the Uffizi Gallery in Florence). When a painting cannot be loaned for a permanent exhibition, Google offers a projection of the absent work in high definition. A *Google cardboard*, equipped with a magnifying glass, makes it possible to view a short film whose interest is to widen the audience. Each visitor can then build up their own exhibition of images from all over the world.

Digital technology represents a change for the museum, which must take into account the new technologies in order to communicate with the public, to establish a more active contact that allows "going outside the walls" and building knowledge in a different way in a perspective of cultural diversity.

With regard to digital museums and public attendance, the traditional conservation mission is being challenged by new exhibition, tour and access systems for public collections. To deal with the excessively high attendance at certain events, museums are offering a wider consultation of the collections on display on the Internet, by involving visitors in the development of cultural exhibitions. The Louvre offers a multimedia tour through the Museum Lab concept. At the Louvre-Lens, geolocation processes accompany exhibitions that are integrated into mobile assistants. A resource center, located in the center of the museum, offers several devices that can be downloaded by visitors. Visual movements, according to the themes, are more easily integrated into the exhibition rooms. The museum's website presents conference programs, ongoing restorations and loans of works. Numerous blogs (Facebook, Twitter) monitor social practices while presenting the museum's events. The digitization of heritage is shattering the boundaries between visiting the exhibition halls and digital consultation at home.

Within the museum in situ, mediation spaces put the heritage on stage. The traditional forms of displaying works of art offer an unexpected visibility of the objects exhibited. Visits are complex, divided between the individual motivations of visitors and the practices of attendance. Various digital tools are already reinventing the museum visit experience: virtual reality, augmented reality, mobile applications and artificial intelligence.

So many innovative technologies bring new approaches to heritage, exhibition settings, their mediation and immersive and multisensory experiences. However, it is important to improve understanding of the role of tablets and smartphone applications in which the multimedia guide can be downloaded. Works of art cards present different contents on the biography of the artist and descriptions of paintings. Interactive terminals provide practical information and online ticketing. The tablets are sometimes equipped with headphones and several seats are installed near the works with content visible through a magnifying glass. Visitors can personalize their journey, share it on social networks and discover the most famous paintings through augmented reality. Geolocation allows information to be transmitted when the visitor is in front of the painting. They receive a real-time view of the artist painting or a commentary on the life of the painter. An ethnographic approach is an opportunity to identify the practices and uses of visitors on the Internet through a concrete practice of immersing the researcher in the social environment. The knowledge of the context is rooted in their own involvement in the museum. The observations are analyzed at the same time as the data collected is brought to light.

Emphasis is placed on digitized content to the detriment of analyses of the public's visiting practices that are likely to build a relationship with the work of art. The exhibition visit is requested when the public follows a specific itinerary or when they discover a painter or a historical period. In such a context, the museum approach aims to understand how mediation is constructed in exhibition spaces, how the practices, behaviors and uses of visitors are changed by the digitization of heritage and how the museum institution deals with it. The aim is to build knowledge in relation to the thematic knowledge of the different routes.

By letting people believe that seeing works of art in a museum or in virtual reality are equivalent, a marketing strategy is developed, often without exercising any real critical meaning. The digitization of art objects risks alerting us to our different esthetic approaches. Museums are adopting different interpretations in which visitors' behavior becomes a "predictor" of success or failure, according to different types of knowledge about the objects and technologies present in or outside the museum. However, the museums value, above all, the mediation of new devices, because the director cannot ignore the economic reality of public funding. This is the greatest challenge.

Regarding digital museums and tourist appeal, over the past 20 years, museum projects have been multiplying in France and abroad, and are becoming recognized worldwide – the Centre Pompidou in Malaga, the Guggenheim in Bilbao, New York, Berlin, Las Vegas, Vilnius, Abu Dhabi and Helsinki. The Louvre has promoted its brand in several areas: the Louvre-Lens (branch), the High Museum in Atlanta (partnership), The Boverie in Liège (collaboration) and the Louvre-Abu Dhabi (brand license). These projects are development tools for the territories concerned. An emblematic cultural brand ensures economic spin-offs for the destination places that benefit from the heritage activities of visitors (Caldwell 2000), particularly through the loan of collections and the organization of exhibitions. In this context, the Louvre-Abu Dhabi aims to strengthen the international recognition of the Parisian museum, which is inspired by the competing examples of Bilbao in Spain and Liverpool in England. In these cities, the museum heritage is already a means of urban structuring in order to revitalize the image of a city and a region. In other words, museums are now part of the tourist appeal that associates the brand with the geographical destination and national and international territories. The humanism of a territory is revealed as it is through its destination. The public character of the Louvre brand can only result from the transformation that leads the visitor to transport their tourist universe to this region of the world. The "Guggenheim effect" on the city of Bilbao has already inspired many similar projects that are aimed at revitalizing booming territories through culture. The world of heritage and the arts is challenged both by the considerable progress of cultural digitization and by the changing expectations of the public in its relationship to culture. The saturation of attendance in large museums is an opportunity to seek solutions to manage visitor flows. It has become essential to encourage the circulation of works between museums in order to move some of the public to regional structures (Louvre-Lens, Guggenheim-Bilbao, Tate Modern-Liverpool).

Nevertheless, while some museums are adapting their educational approach, others are not taking the risk of modifying their scientific project to give priority to tools and attendance. They still prefer to promote the conservation of public collections (Acropolis Museum in Athens). In any case, museums are now experimenting with new cultural concepts to meet visitors' expectations. This has become a priority for any public service policy.

References

Baujard, C. (2013). *Du musée conservateur au musée virtuel*. Hermès-Lavoisier, Paris.

Baujard, C. (2019a). *Musées et management, vers la mondialisation culturelle*. ISTE Editions, London.

Baujard, C. (ed.) (2019b). Musées et environnement numérique. *Les Cahiers du numérique*, 15(1–2), 9–19.

Baujard, C., Lagier, J., Montargot, N. (2020). *Organisations créatives et culturelles, évolutions et mutations*. ISTE Editions, London.

Caldwell, N. (2000). The emergence of museum brands. *International Journal of Arts Management*, 2(3), 28–34.

Desvallées, A. and Mairesse, F. (2011). *Dictionnaire encyclopédique de muséologie*. Armand Colin, Paris.

O

Open Science (Dissemination)

Julien Roche
University libraries, Université de Lille, France

Opening up science

France...

France is a major historical player in open access and Open Science. In particular, two important initiatives have been undertaken in recent years, which have undoubtedly helped accelerate the movement.

The first is the promulgation, in October 2016, of the law "for a digital Republic", known as the "Lemaire law", which brings significant advances, including one directly affecting the issue of Open Science. Indeed, this law introduces a new right for publicly funded researchers. According to article 30 of the law, they can now deposit their publication and make it freely accessible, for example, in an open repository, after a period of 6 months for a publication in the field of science, technology and medicine, and 12 months in the field of humanities and social sciences, regardless of the terms of the contract signed with the publisher. In practice, this amounts to limiting to 12 or even 6 months the exclusivity of use granted by the author to the publisher.

The second is the adoption by France, in the summer of 2018, of a national plan for Open Science, with a strong ambition summarized as follows by the Minister of Higher Education, Research and Innovation, Frédérique Vidal, in her speech: "France is committed to making scientific research results open to all – researchers, companies, citizens". This plan is based on nine measures divided into three areas. First, it aims to generalize open access to scientific publications, by making it compulsory to publish articles and books resulting from research financed by calls for tender from public funds, by creating a national fund for Open Science, by supporting the national open repository HAL and, more generally, the deposit in open repositories. The aim is also to make the opening of scientific publications the default practice. The second axis concerns public research data, which must eventually meet the FAIR principles (Findable, Accessible, Interoperable, Reusable), be preserved and, as much as possible, open. This axis is based on three measures: the obligation to openly disseminate research data from programs financed by calls for projects from public funds, the creation of a network of chief data officers inside public institutions and the adoption of an open data policy associated with scientific articles. Third, the inclusion of Open Science in a sustainable, European and international dynamic, through support for the development of Open Science skills, particularly among PhD students, the encouragement of universities and research organizations to adopt an Open Science policy, and finally, the active contribution to the structuring of players and communities. To achieve this ambition, the Ministry has set up an Open Science committee, which provides France with a centralized structure for steering Open Science, where many countries are struggling to coordinate their actions. This plan was evaluated and updated in the summer of 2021. It aims, for the 2021–2024 period, to generalize Open Science in France and sets the goal for 100% of publications to be in open access by 2030. It is organized into four axes, three of which are an extension of the 2018 plan (generalization of open access to publications; structuring, sharing and opening up of research data; transformation of practices to make Open Science the default principle). The last axis, more innovative although present in an embryonic way in the first plan, concerns the opening and promotion of software and source codes produced by research.

... and Europe

One of the major challenges of Open Science is to ensure that it is implemented on a global scale, which requires a strong commitment from the major players in the countries most involved in research, first and foremost Europe, the world's leading scientific producer, with more than 28% of the world's scientific publications.

In the field of Open Science, and among a host of actions, Europe has seen the emergence of two very structuring initiatives, one emanating from the European Commission and the other from research stakeholders.

While the movement has been developing since the early 1990s in the scientific communities, the European Commission's awareness, which came about fifteen years later, with the implementation of a first pilot dedicated to open access within the Seventh Framework Programme (2007–2013), can be considered as late. Nevertheless, this ambition was largely unprecedented at the time, and has since been confirmed by an ambitious policy, first in the "Horizon 2020" program (2014–2020) and then "Horizon Europe" (2021–2027). At first, it concerned open access only, before being gradually extended to research data. Horizon Europe now has an bold project, that of making Open Science the *modus operandi* of science, through the clarification and reinforcement of open access obligations, the adoption of the FAIR principles as the default modality for data sharing, the implementation of obligation and incentive mechanisms and, in mirror image, of coercive measures, or the creation of impact indicators.

Research stakeholders have not been left out, with the launch in September 2018 of "Plan S" by "Coalition S", a group of European research funding agencies. The initial objective is simple: to accelerate the transition of scientific publications funded by these agencies to free, open, immediate access by 2021, either through direct open access publications or through open archives. Plan S is now supported by the European Union, which has incorporated its content into Horizon Europe, and the signatory agencies are currently implementing it in their calls for projects.

While it is still too early to measure the impact of these two initiatives, it is certain that Europe is now a leader in Open Science.

In the end, why open up science?

"We are like dwarfs standing on the shoulders of giants" ("*Nos esse quasi nanos, gigantium humeris insidentes*"), as Bernard of Chartres is said to have said in the 13th century, a quotation that is no doubt apocryphal, but which applies marvelously to the cumulative character of Open Science.

It is first and foremost a powerful dissemination tool, which promotes fluid, rapid and broad transmission of the knowledge produced, and which guarantees easy and lasting access to the results of research. Where the still dominant model is subscription-based, to have access to potentially relevant content, Open Science liberates content and guarantees equal accessibility.

It is a vector for the evolution of research evaluation methods toward more open logics – open peer reviewing – an evaluation that is intended to be based solely on the own merits of the research assessed and not, for example, on the reputation – impact factor – of a journal. The current publication model, which is the basis of evaluation and, consequently, of scientists' careers, is in fact based essentially on the prestige of the journal, pushing researchers to prioritize quantity over quality, which feeds scientific overproduction, contributes to the decline in the overall quality of publications, encourages unbridled competition, discourages risk-taking and establishes a dictatorship of positive publication, whereas the reporting of the negative results of research – the account of failures – is fundamental if we are to allow science to progress.

It is a tool for innovation, with the development of open, participative, collaborative and competitive approaches that have been put into practice in public research as well as in the business world, for example, through the adoption of open innovation processes. Open innovation is also an important vector for both the amplification and appropriation of science.

It is a powerful form of support to the democratization of knowledge for the benefit of all, researchers from the public sector, companies or even citizens. Citizen science effectively combats obscurantism, which is generally based on a non-scientific challenge to the foundations and validity of science, by bringing the academic world closer to the other players in society, through easier media coverage and a fruitful dialogue.

Finally, it is a decisive factor in ensuring optimal quality of research, which makes it possible to respond to issues of transparency – hence ethics and therefore trust – but also to the reproducibility of research, based on the idea that science progresses first and foremost because of its errors, and the publicity given to these errors and to the processes that led to them is eminently useful for progressing toward a world in which science will, in the future, be of common benefit.

References

Coalition S. (2018). Plan S, making full and immediate Open Access a reality [Online]. Available at: https://www.coalition-s.org/addendum-to-the-coalition-s-guidance-on-the-implementation-of-plan-s/principles-and-implementation/ [Accessed 20 January 2022].

Comité français pour la science ouverte – COSO (n/a). Ouvrir la science ! [Online]. Available at: https://www.ouvrirlascience.fr/home/ [Accessed 20 January 2022].

League of European Research Universities (2018). Open Science and its role in universities: A roadmap for cultural change [Online]. Available at: https://www.leru.org/files/LERU-AP24-Open-Science-full-paper.pdf [Accessed 20 January 2022].

Ministère de l'Enseignement supérieur, de la Recherche et de l'Innovation (2021). Deuxième Plan national pour la science ouverte – généraliser la science ouverte en France, 2021–2024 (Second French plan for Open Science – Generalising Open Science in France 2021–2024) [Online]. Available at: https://cache.media.enseignementsup-recherche.gouv.fr/file/science_ouverte/20/9/MEN_brochure_PNSO_web_1415209.pdf and in English at: https://www.ouvrirlascience.fr/wp-content/uploads/2021/10/Second_French_Plan-for-Open-Science_web.pdf [Accessed 20 January 2022].

Rentier, B. (2018). *Le Défi de la transparence : Open Science, the Challenge of Transparency*. Académie royale de Belgique, Brussels [Online]. Available at: https://orbi.uliege.be/bitstream/2268/230014/1/rentier_science_ouverte_pour_ORBi.pdf and in English at https://orbi.uliege.be/handle/2268/233905 [Accessed 20 January 2022].

Wilkinson, M.D., Dumontier, M., Aalbersberg, I.J., Appleton, G., Axton, M., Baak, A., Blomberg, N., Boiten, J.-W., Bonino da Silva Santos, L., Bourne, P.E. *et al.* (2016). The FAIR Guiding Principles for scientific data management and stewardship. *Scientific Data*, 3, 15 March [Online]. Available at: https://doi.org/10.1038/sdata.2016.18 [Accessed 20 January 2022].

Open Science (Origins)

Julien Roche
University libraries, Université de Lille, France

Open Science is a global movement that aims to make the results – publications – and products – data and intermediary productions – of research, whether public or private, accessible and unhindered. Made possible by the generalization of digital tools and the widespread development of Internet communication networks, Open Science aims to accelerate the dissemination of science, making it more cumulative and transparent.

The origins of Open Science

Although the notion of Open Science was born recently, its foundations are generally considered to date back to the 17th and especially the 18th centuries, first with the humanists at the time of the Republic of Letters, and then when scholars needed to share their research in a more formalized way, leading to the concomitant appearance of the first two scientific journals, *Le Journal des savants*, published in Paris in 1665, and the *Philosophical Transactions* of the British Royal Society, published in London the same year. The aim was to enable a small circle of scientists spread across Europe to make their work accessible, share its contents and, in so doing, allow the first scientific emulation on a large scale. With the rapid development of science and the related publications in the 18th and especially the 19th centuries, academic libraries emerged and took a central role in the open dissemination of published knowledge. Collections were assembled and ordered, visible through library catalogues, and accessible to scholars and learned people. From the end of the 19th century and the first half of the 20th century, the beginnings of a globalized and open dissemination of knowledge were at work, with the major research libraries as the primary vector, notably through the systematization of catalogues or the establishment of a global system of interlibrary loan of documents. More recently, at the end of the 20th century, the appearance of open digital libraries on the web has enabled a decisive step to be taken in the wide distribution of mainly freely accessible content.

Open access

Although the term "Open Science" is recent, an important movement preceded it, which laid the foundations for it in the 1990s: open access, an ambiguous term that also designates direct access by the public to library collections. Open access, which essentially concerns publications, aims to make content available in digital format, without hindrance, immediately and permanently in order to accelerate the dissemination of knowledge. One of the most emblematic achievements is undoubtedly the open repository for the pre-publication of scientific articles, arXiv, which was created in the early 1990s. Since then, open access has developed strongly through numerous initiatives that can be grouped into three categories, called "pathways":

– *The "green open access" approach*, which consists of ensuring the self-archiving of scientific content that can be published elsewhere in traditional forms, accessible by paying subscription fees. This approach is based on open repositories, which are digital storehouses that collect scientific content for preservation and free distribution. There are currently more than 4,000 open repositories worldwide, generally administered and maintained by library and technical staff.

– *The so-called "gold path"* – or gold open access – in which publications are made freely and immediately available, regardless of whether or not there are article processing charges. Free access – no charge for accessing content – and immediacy – no delay before content is freely accessible – are characteristic of a virtuous golden path. Apart from that, there is another model invented by publishers for so-called "hybrid" journals, which constitutes a deviation from open access, in which publishers set a double paying for the opening of publications that are already being charged for. The movement toward open access, combined with strong pressure to publish or perish, has also encouraged the emergence of toxic practices through so-called "predatory" publishers, who charge publication fees without ensuring either editorial quality or peer review.

– The "diamond" path, also called "platinum" – diamond/platinum open access – which is a variant of the golden path. It is a question of having the public authorities take charge of the free, immediate dissemination of the scientific content produced by entrusting the role of publisher to publicly funded institutions, thus enabling the scientific community to regain control of the dissemination of the content it produces and to get away from the ambiguities of the golden path, which can hide, under the guise of open access, questionable practices – hybrid journals, predatory publishing.

This proliferation of initiatives attests to the vitality of the movement toward open access. Indeed, although the share of open access publications is still in the minority and varies greatly from country to country, it is now increasing significantly. For French scientific publications in 2019, it amounted to 56% (French Open Science Monitor[1]).

Toward Open Science

Open Science is certainly an extension and amplification of the open access movement, which began in the 1980s and took off in the 1990s. The evolution is twofold: on the one hand, the movement is expanding beyond the results of research – scientific publications – and opening up to everything that surrounds them, complements them and gives them credibility. This is how the produced data now enter the field of resources likely to be open, as well as intermediary literature – reports, for example; on the other hand, Open Science is now opening up more widely to so-called participatory, collaborative or citizen science, as they both participate in the same logic, that is, of open and networked practices. The movement is becoming broader and more global, with a greater awareness of the issues related to the openness of science, not only in the scientific world, but also among decision makers and public opinion. The limits of competition between actors and the need for better collaboration at the continental and even global level, based on a less restricted circulation of knowledge, have thus been dramatically highlighted by the Covid-19 pandemic, leading to unprecedented data sharing. There is now a global normative framework for Open Science, as UNESCO took up the subject in late 2019, with its recommendation adopted in November 2021. As the movement grows, expands and becomes more structured, it is also adopting a less ideologically driven stance, well summarized today in a formula taken up by the European Commission, "as open as possible, as closed as necessary".

---

1 Available at: https://ministeresuprecherche.github.io/bso/ [Accessed 20 January 2022].

References

Ashta, E., Augouvernaire, M., Caillet, C., Laffont, M., Okret, C., Pinet, N. (2020). Enquête sur les archives ouvertes françaises menée par Couperin en 2019 : résultats et analyses. Rapports et études Couperin, 2, Consortium Couperin [Online]. Available at: https://hal.archives-ouvertes.fr/sic_02562594 [Accessed 20 January 2022].

Knowledge Exchange (2017). Knowledge exchange approach towards open scholarship [Online]. Available at: https://repository.jisc.ac.uk/6685/1/KE_APPROACH_TOWARDS_ OPEN_SCHOLARSHIP_AUG_2017.pdf [Accessed 20 January 2022].

Schade, S., Tsinaraki, C., Roglia, E. (2017). Scientific data from and for the citizen. *First Monday*, 22(8) [Online]. Available at: https://doi.org/10.5210/fm.v22i8.7842 [Accessed 20 January 2022].

UNESCO (2021). UNESCO recommendation on Open Science [Online]. Available at: https:// unesdoc.unesco.org/ark:/48223/pf0000379949.locale=en [Accessed 20 January 2022].

P

Predictive Justice

Bruno Deffains
Université Paris 2 Panthéon-Assas, France

Creating an algorithm capable of correctly solving legal problems is a major objective of legaltech. The work on how the legal system could be automated, with the aim of improving the organization of justice, is old. As early as 1949, Lee Loevinger proposed the application of quantitative methods to the field of justice, an approach he called "jurimetrics". According to Loevinger, while jurisprudence is based on an approach in which legal reasoning is exclusively a matter of interpreting norms, jurimetrics, on the other hand, uses scientific methods to identify arguments relevant to the law. Loevinger believes that many legal issues could be resolved by applying predictive analysis methods based on systematic processing of court data.

In the context of this "jurimetric" approach, the objective is not so much to know if it is possible to predict the judicial decision as to know how to make such "predictions" by means of a machine, on a quantitative and not intuitive basis. It is worth noting that while legal science has often placed strong emphasis on empirical knowledge in its discourse, there are paradoxically few simple and easily accessible tools for providing reliable statistical data on the application of rules to specific cases. The use of statistics is often confined to the construction of performance indicators, such as those relating to the length of trials or the average amount of compensation in particular contexts. Statistics are rarely used as a tool for understanding the law. This situation is all the more surprising given that the calculation of probabilities has been used by its promoters for legal applications from the outset. Let us mention Bernoulli and his law thesis on the judicial use of the calculus of probabilities, but also Condorcet, Laplace and Poisson. These studies were not well received by jurists and remained largely without follow-up.

Later, while Loevinger's work remained at the level of theoretical proposals, a new, more empirical approach to the automation of law emerged in the 1980s in the form of expert systems. The goal was to translate legal rules into a machine-readable logic system, allowing an algorithm to "read" the law and apply it in specific circumstances. However, because the sources of law and the interpretation of legal rules can be ambiguous, the application of law by automated legal systems based on the logic of expert systems has not been very successful. The new wave of use of artificial intelligence in law, which emerged in the 2010s, appears to be more ambitious by mobilizing machine learning techniques based on the collection of massive judicial data. One of the most prominent areas of this new phase in legal AI is precisely "quantitative" predictive analysis, in order to create actionable information about, for example, the outcome of a trial, the arguments, evidence or case law likely to be used, or the length of the legal process.

This approach is made possible by the collection of judicial data that the algorithm uses as "input" to establish a link between the characteristics of a case and the targeted results. For example, when looking for the likely outcome of a case, it will correlate certain data available in the case law (legal terminology, precedents cited, area of law considered, etc.). Thus, instead of trying to imitate the prediction, as a lawyer would do on the basis of a legal argument, predictive analysis proposes a model based on statistically established correlations. Beyond the outcome of the case, the same approach also seeks to identify patterns in arguments, case law, evidence, etc., used in precedents based on the most significant correlations. At the same time, it allows for an empirical understanding of the litigation strategies of litigants and/or an analysis of the judicial decision-making process.

Quantitative analysis is thus gaining ground in the legal market. A prediction algorithm developed in the United States was able to correctly predict 70.2% of the outcomes of Supreme Court cases. By comparison, in a 2012 study, legal experts were only able to correctly predict 59% of the outcomes of the same cases. In Europe, a team created a prediction model for the European Court of Human Rights that would be able to predict the outcome of the case, on average, with 79% accuracy. At the same time, there have been commercial projects by legaltechs, such as Lex Machina in the United States or Case Law Analytics in France, which provide legal professionals with insight, based on case law data, into what may be the most advantageous path in a lawsuit.

Predictive justice is enjoying significant success, but this does not mean that algorithms will replace the legal profession, although their potential impact on practice should not be underestimated. Indeed, "prediction" is an essential part of legal practice. As Justice Oliver Wendell Holmes, one of America's most famous jurists, said, "The primary rights and duties with which jurisprudence busies itself again are nothing but prophecies". Admittedly, Holmes is part of a realist philosophy of law that is not of the same nature as the French positivist tradition, but as Guillaume Zambrano correctly writes: "The only knowledge that is permissible to form in law consists of the ability to predict the probable interpretation and application that will be made of the rules of law by judges, in cases determined according to objectively verifiable factors". In the age of Big Data, then, the science of law should begin by answering the following question: Is it possible to predict, with a small margin of error, the likely meaning of a judicial decision? In the era of Big Data, quantitative legal analysis is a formidable tool for litigants, practitioners, teachers and researchers. The jurimetrics imagined by Loevinger are likely to constitute a new tool facilitating the comparison of the legal norm with its application in case law.

However, the promises thus defined are not without questions. The first question, of course, is whether a statistical calculation can be conceived as a source of law. For most commentators, the answer is no. There is a fundamental difference between the "prediction" made by statistical tools and the usual practice of law, which is based on a model of interpretation of the sources of law, starting with statutes, and is prescribed by the doctrine. "Jurimetric" prediction, on the other hand, is essentially based on statistical correlations in which the semantic understanding of the texts is set aside. The use of these tools entails a radical change of perspective, which Mireille Hildebrandt describes as a "shift from reason to statistics": the decision-making process of practitioners would no longer be based solely on their understanding of legal rules, but would be influenced by the work of predictive algorithms.

Predictive justice thus appears to be at the heart of numerous debates that refer both to its potential for development, due to its growing appropriation by legal practitioners, and to the concerns raised on an ontological level, particularly in the specific context of codified legal systems where "the rule" takes precedence over "the fact". It can be agreed that the quantitative analysis of case law data allows for the identification of "values" that must be interpreted and placed in the particular context of the law-making process, which cannot be detached from the inherent characteristics of the organization of justice in a modern democracy, both from the point of view of the work of judges and the conditions of data collection and processing. This is all the more important since in France, as in most countries, the work of data processing is largely entrusted to the private sector, which necessarily raises the question of the effective contribution of predictive justice to the improvement of the public justice service.

The value produced by jurimetric analysis depends largely on the volume and quality of the judicial data available, which implies a strict framework for these practices. In France, the open data project for court decisions is being implemented gradually. Law No. 2016-1321 of October 7, 2016 "for a digital Republic" and Law No. 2019-2022 of March 23, 2019 "on programming 2018–2022 and reform for justice" have made it possible to set out the main principles for making court decisions available to the public. Decree No. 2020-797 of June 29, 2020, relating to the availability to the public of judicial and administrative court decisions, specified the practical conditions of this availability and is part of the framework of publicity of court decisions established by the code of administrative justice, the code of criminal procedure and the code of civil procedure. In particular, it provides for measures to conceal the identification details of natural persons, parties or third parties, or even judges or members of the court registry, in the event of an infringement of their privacy or security.

On the specific question of the analysis of judicial behavior, it is well understood that the power of judges, although controlled by numerous legal systems, is a deeply subjective process that can be influenced by ideologies, attitudes, emotions, heuristics, etc. From this point of view, jurimetic analysis can be conceived, as Loevinger imagined in 1949, as a means of improving our common knowledge about the functioning of the judicial system. Daniel Chen rightly points out that "if algorithms can identify contexts that may give rise to bias, they can also reduce that bias through behavioral guidance and other mechanisms, such as judicial education". The rise of jurimetrics requires a cultural and collective change that involves better understanding and the development of a framework for learning to better master the law through the power of statistics.

References

Blackman, J. and Carpenter, C. (2012). FantasySCOTUS: Crowdsourcing a prediction market for the supreme court. *Northwestern Journal of Technology and Intellectual Property*, 10 [Online]. Available at: https://scholarlycommons.law.northwestern.edu/njtip/vol10/iss3/3/.

Chen, A. (2019). How artificial intelligence can help us make judges less biased [Online]. Available at: https://www.theverge.com/2019/1/17/18186674/daniel-chen-machine-learning-rule-of-law-economics-psychology-judicial-system-policy.

Hildebrandt, M. (2019). Data-driven prediction of judgment. Law's new mode of existence? [Online]. Available at: https://ssrn.com/abstract=3548504.

Katz, D.M., Bommarito, M.J., Blackman, J. (2017). A general approach for predicting the behavior of the Supreme Court of the United States. *PLoS ONE*, 12(4), 1–18 [Online]. Available at: https://doi.org/10.1371/journal.pone.0174698.

Loevinger, L. (1971). Jurimetrics – The next step forward. *Jurimetrics Journal*, 12(1), 3 [Online]. Available at: www.jstor.org/stable/29761220.

Méneceur, Y. (2020). *L'Intelligence artificielle en procès*. Bruylant, Brussels.

Processors

Laurent Bloch
Institut de l'iconomie, Paris, France

RISC architecture

Computer architecture before microprocessors

Until the 1970s, CPUs, which were not yet microprocessors, evolved toward increasingly rich machine instruction sets, intended to bring machine language closer to so-called advanced languages, that is, closer to human language than former machine languages. The result was the so-called complex instruction set computer (CISC) architecture, with sophisticated machine instructions, which performed complex operations.

Between advanced languages and machine language, there is assembly language, whose instructions are, one for one, those of machine language, but written in a more readable way. The assembler also provides a few aids to the programmer, such as symbols to represent addresses, automatic calculation of the displacement between two addresses, and so on. Each machine instruction occupies a certain number of words in memory, which is rigid – an assembly program is not a text whose composition is left to the discretion of the programmer, the arrangement of the text corresponds to its arrangement in the computer memory.

Culmination of CISC architectures

CISC machines, which reached their peak in the early 1980s (Digital Equipment Corporation's VAX, Motorola's 68000), had a very large (over 300 for the VAX) and complex instruction set, with instructions of different and sometimes even varying lengths, depending on the operands. The richness of the instruction set was supposed to facilitate the task of the programmers who used assembly language and, above all, of the authors of compilers for advanced languages, by providing them with machine instructions that already resembled advanced language that was easier and more manageable for the programmer. Moreover, the C language, a low-level advanced language (it has been called a portable assembler, the notion of portability designating the ability to run on computers of different architectures), is largely inspired by the assembler of the PDP computers, ancestors of the VAX.

This richness and complexity of the instruction set of course came at a cost in terms of complexity and slowness of the processor. There was a risk that simple instructions (loading registers from memory, copying registers to memory, register-to-register addition; registers are areas of memory implemented directly on the processor circuit, very small but with access speeds a factor of 100 faster than main memory) would be condemned to run as slowly as complex operations (copying from memory area to memory area of variable length, complex operations in memory). VAX, in particular, hardly avoided this risk.

Birth of the RISC idea

In the second half of the 1970s, researchers compiled statistics on the instruction composition of machine language programs, either written directly in assembler or produced by a compiler. Let us quote in particular the work of Fairclough (1982). They found that 43% of instructions were data moves from one place to another, a quarter of the instructions were branches, for each program the number of instructions used was very small and only the simplest instructions were widely used. Another finding was that by far the most costly operations were the calling of a subprogram (a program starts the execution of another program by giving it parameters) and the return of a subprogram to the calling program.

Based on these findings, they recommended designing processors with a reduced set of simpler instructions. The concept of RISC (reduced instruction set computer) processors was born and the first such processor, the IBM 801, was built by John Cocke in 1979. The MIPS company, founded by John Hennessy, a RISC pioneer at Stanford University, was created in 1985. Hewlett-Packard was the first major computer manufacturer to build its entire product line in RISC architecture in 1986. Sun and Digital Equipment Corporation followed.

The advent of RISC microprocessors

The landmark book of the RISC revolution, *Computer Architecture: A Quantitative Approach* (Hennessy 1990), was written by John L. Hennessy, the architect of the MIPS processors and a professor at Stanford University, and David A. Patterson, a professor at the University of California (Berkeley) and the architect of Sun's SPARC processors. The MIPS processors were the first to break new ground and the most innovative, for example with the use of the TLB (translation lookaside buffer) to resolve page faults in virtual memory by program, without the need for specialized circuitry. To this repertoire should be added Richard L. Sites, the architect of Digital Equipment Corporation's Alpha processors, processors with impressive performance for their time.

The most striking feature of RISC architectures initially was the small number of instructions, with no memory-to-memory instructions: there were only memory loads into a register, copies from register to memory and operations in registers.

Other features soon proved equally important: fixed instruction length and format, and extensive use of a moderate length pipeline. To take proper advantage of such an ascetic architecture, much of the complexity was shifted to the compilers, which had to produce code that could use the registers and pipeline efficiently.

The new architecture soon proved to be extremely fast, and fast in two ways: indeed, to take advantage of an advance in microelectronics, it is not enough to design a fast processor, the time needed to design it must also not be too long. The simplicity of RISC was also an advantage in this respect. Currently, it takes a team of 200–400 engineers about 3 years to design a new model of an existing processor line. A fully equipped factory to build it will cost in the order of 15 billion dollars, and that doubles every 4 years (25 billion for the next Samsung factory). Designing an innovative architecture takes about 10 years.

Resilience of CISC architectures

CISC technology seemed doomed, which was indeed the case for the VAX and Motorola 68000 series. Everyone was waiting for the fall of Intel's x86 architecture, on which the tens of millions of PCs sold every year are based. This was without counting the efforts that Intel could mobilize due to income from the PCs. The Pentium and its successor, the Core, are in fact processors consisting of a RISC core around which additional circuitry and microcode simulate the old CISC architecture in order to maintain compatibility with existing systems and programs (Anceau 2013).

As for the IBM mainframe range, the huge stock of existing programs, which would require phenomenal expenditure to convert, seems doomed to an immortality that is only eroded by the slow change of applications.

Even if the hegemony of the x86 architecture, at the time, suggested a half-failure of the RISC architecture, all modern processors have taken its lesson on board. The RISC movement has profoundly revolutionized processor design and, less spectacularly but just as profoundly, compilation techniques. In fact, as a tribute, all modern processors have a RISC core surrounded by circuits that provide a different background, Pentium for example. Even the big IBM systems are now powered by microprocessors that implement their traditional instruction set in RISC.

Does the future belong to RISC?

While the computer market, both servers and personal devices, is still monopolized by Intel's CISC architectures (and then by AMD, which produces x86-compatible processors), the consequences of the extraordinary proliferation of cell phones and tablets, which are Turing-complete computers powered by ARM-designed RISC processors, must be measured. These processors are by far the most common in the world, and while a decade ago over 95% of web access came from x86 Windows-based computers, today over 70% comes from Android or, less so, iOS-based devices.

The designers of these portable devices chose the ARM platform for good reason: with comparable computing power, the weight and power consumption are at least an order of magnitude lower. It seems that ARM processors, which began their career in a quasi-artisanal way, will be at the heart of tomorrow's architectures – Apple is already using them for its Mac mini M1 computer. There is also a free and open RISC architecture, RISC-V (Patterson 2018), and several implementations are ready to appear on the market. The Chinese company Loongson, long a producer of MIPS architecture processors, has announced its own RISC architecture, LoongArch (Aufranc 2021). That said, recent RISC architectures are less and less worthy of the R in their acronym, with over 300 instructions and advanced speculative and out-of-order execution features.

References

Anceau, F. (2013). La logique, des MOS aux circuits intégrés : l'évolution réciproque des technologies et des concepts logiques [Online]. Available at: https://urlz.fr/eRoT.

Aufranc, J.L. (2021). Loongson unveils LoongArch CPU instruction set architecture for processors made in China [Online]. Available at: urlr.me/J5wRc.

Fairclough, D.A. (1982). A unique microprocessor instruction set. *IEEE Micro* [Online]. Available at: https://scholarsarchive.byu.edu/facpub/762/.

Hennessy, J.-L. and Patterson, D.-A. (1990). *Computer Architecture: A Quantitative Approach*. Morgan Kaufmann, Burlington.

Patterson, D.-A. and Hennessy, J.-L. (2018). *Computer Organization and Design RISC-V Edition: The Hardware Software Interface*. Morgan Kaufmann, Burlington.

Proprietary Licenses

Juliette Sénéchal
Université de Lille, France

Intellectual property licensing and proprietary software licensing

The term "license" or "concession", in the field of intellectual property law (copyright, patent, etc.), refers to the possibility, through the mobilization of contract law, of conferring on a licensee the possibility of exploiting an intellectual asset in return for a royalty.

The main limits to contractual freedom to exploit intellectual property are found in mandatory rules and competition law. In this respect, the needs of public health or the use facilitated by certain categories of persons may be hypotheses in which the law may restrict the owner's free enjoyment of their property by allowing access to third parties under legally regulated contractual conditions.

For the rest, apart from certain special regimes, contracts having as their object an intellectual property appropriate by an intellectual property regime are mainly subject, apart from formal and publicity requirements, to the French ordinary law of contracts of Articles 1101 *et seq.* of the Civil Code, as reformed by the Ordinance of February 10, 2016.

Even though there is a great deal of contractual freedom in the field of the exploitation of intellectual property, it is possible to identify classic contractual models of intellectual property, namely assignment and concession or license. The assignment model is similar to that of sale – but it is still possible to encounter similar hypotheses, such as gift or exchange – while the concession or license model is similar to that of lease. Finally, these classic models are increasingly being challenged by the contract model for the provision of services – even if this qualification is the subject of controversy – which can be encountered in the context of access to online intellectual property, such as subscription contracts to an online music site or software as a service (SaaS) formulas.

The concession agreement, or license agreement, is the dominant contractual model in intellectual property. It can also be referred to as a technology transfer agreement from a competition law perspective.

Licensing allows "the concurrent enjoyment of the same property to be granted to an infinite number of persons without the enjoyment of the property by each of them disturbing the enjoyment of another. The licensing contract leads to the simultaneous satisfaction of complementary interests" (Binctin 2020).

The license does not transfer ownership to the licensee or concessionaire, but only the enjoyment of the intellectual property under the conditions and within the limits provided for in the contract. The strong contractual freedom existing in this field leads to a great variety of forms of enjoyment, but it should be specified that the rules of the lease of articles 1711 and following of the Civil Code constitute the rules of principle of the license contract, which we present first; we then consider the software licenses of the private type.

The concession or license considered as a lease

Specific rules present in the intellectual property code concern the hypotheses where the author is directly party to the license contract, with the aim of protecting it.

However, this solution is not the majority solution because, in many cases, the author is not party to the licensing contract and, in this case, only the rules of the Civil Code apply to the relationship between the licensor and the licensee.

In the latter case, it should be specified that all intellectual property may be licensed. The licensee's enjoyment of the license, in application of the principle of contractual freedom, will be "granted for a specific application, for a specific duration, for a specific territory".

The license may be simple or exclusive. A simple license allows as many uses of the property as there are persons wishing to use it. The exclusive license, in contrast, is granted to a single person, to the exclusion of all others, including the licensor themselves.

The making available of an intellectual property, by means of a license, presupposes, with some exceptions, the payment of a consideration by the licensee in the form of a royalty. This royalty is again subject to contractual freedom and may be of variable form and duration: fixed, lump sum, single, multiple, variable, indexed on the licensee's turnover, etc.

The development of FRAND (fair, reasonable, and non-discriminatory) licenses should be highlighted. FRAND licensing is a simple (non-exclusive) license of intellectual property "essentially appropriated by patent, but also by copyright or even trademark law when a distinctive sign is used to indicate the presence in a product of a standardized technology (e.g. the DVD mark)".

The so-called essential patents, that is, those that are indispensable for the implementation of a technical standard, are at the heart of the development of these licenses. Patent owners, at the same time as they declare their essential patents to a standardization body, undertake to grant any third party a fair, reasonable and non-discriminatory license, known as a FRAND license. The criteria for determining FRAND licenses are established by the courts and competition authorities in their decisions.

Licensees can benefit from guarantees in case of lack of ownership of the rights by the licensor, in case of infringement, in case of disturbance of use or in case of certain disturbances of rights (such as the granting of a prerogative to a third party which will harm the licensee's use).

In the case of exclusive licenses, it is possible to find an obligation to exploit justified both by the safeguarding of intellectual property rights and by the remuneration of the licensor. The purpose of this obligation will be to prevent a licensee from seeking to obtain a license only for the purpose of not exploiting an intellectual property, while at the same time preventing other economic operators from obtaining a license and actually exploiting the intellectual property.

The private software license

The concept of software licensing is the subject of much controversy.

First of all, it is possible to encounter the notion of a software "user license", sometimes called "assignment of right of use", which can concern a standardized software package as well as a specific software.

The qualification of this contract as a "license" is controversial, because "what makes the operation economic is not the granting by the software supplier of a right to a work of the mind, but the making available of a tool. It remains fundamentally the same if the software is not protected for lack of originality or is no longer protected because of the arrival of a term, and it is not clear why the qualification should vary according to such considerations. This is why it would probably be better to avoid the ambiguous term 'licence', which wrongly suggests the grant of an intellectual property right" (Lucas *et al.* 2012). This contract would be closer to a contract for the provision of services.

Then, it is possible to evoke the notion of a private software license.

The "proprietary" license qualification of this contract is also controversial. The term "proprietary software" is more an abuse of language than an objective description of software whose code is not free. Indeed, if it means that the source code alone is subject to the intellectual property regime, this appellation is false, since free software is also subject to this regime (Pellegrini and Canevet 2013, No. 395).

The qualification of privative license is preferable for this contract, in that the common denominator of these licenses is that they reserve the entirety of the intellectual property rights to the owner of these.

It is possible to find, among the types of privative license, the licenses of the type shareware and freeware.

Shareware licenses allow the person receiving the software to evaluate it and promote it to others. The user receives, in addition to the right to redistribute the software, the right to use it for a limited time or to have access to a limited subset of its functionalities. After this trial period, the user must either send the rightful owner a sum of money in order to receive an activation key or delete the software (*id.*, No. 397).

Free licenses, or freeware, are licenses that cost nothing to use. The remuneration may take the form of the collection of commercial information and personal data of the user; it may allow the software to become a de facto standard and to be analyzed as a loss leader, encouraging acquisition of a paying version (*id.*, No. 400).

While the shareware license is a true grant of a right to intellectual property (right to redistribute) with respect to a work of the mind, the freeware license is more the provision of a tool than the grant of an intellectual property right.

In a judgment of July 3, 2012, the Court of Justice of the European Union ruled that Directive 2009/24/EC of April 23, 2009 on the legal protection of computer programs must be interpreted as meaning that the right to distribute a copy of a computer program is exhausted if the copyright holder, who has authorized the downloading of that copy, even if free of charge, onto a computer port by means of the Internet, has also granted, in return for payment of a price intended to enable him/her to obtain remuneration corresponding to the economic value of the copy of the work of which they are the owner, a right to use that copy, without any time limit.

Directive 2009/24 must be interpreted as meaning that, in the event of the resale of a user license involving the resale of a copy of a computer program downloaded from the web site of the copyright holder, a license which was initially granted to the first purchaser by the copyright holder without any time limit and on payment of a price intended to enable the latter to obtain remuneration corresponding to the economic value of that copy of their work, the second purchaser of that license and any subsequent purchaser of the license may rely on the exhaustion of the distribution right provided for in that directive and, consequently, may be regarded as lawful purchasers of a copy of a computer program within the meaning of that directive and benefit from the right of reproduction provided for in the latter provision.

As a result of this legal solution, software publishers have turned to new solutions, such as SaaS solutions, which are more in line with the concept of a contract for the provision of digital services at a distance than with that of a license contract.

References

Bernault, C., Lucas, A., Lucas-Scloetter, A. (2017). *Traité de la propriété littéraire et artistique*. LexisNexis, New York.

Binctin, N. (2020). *Droit de la propriété intellectuelle*, 6th edition. LGDJ, Paris.

Coulaud, M. (2016). Les effets pratiques sur l'édition logicielle de la décision UsedSoft GmbH du 3 juillet 2012. *Dalloz IP/IT*, 6, 298.

Lucas, A. (2012). Traité de la propriété littéraire et artistique. *Lexis Nexis*, no. 843.

Pellegrini, F. and Canevet, S. (2013). *Droit des logiciels*. PUF, Paris.

R

Rob'Autisme

Sophie Sakka
LS2N, Centrale Nantes, France

Rob'Autisme, or prosthesis in communication using the robot extension

The French Rob'Autisme project proposes a therapeutic support program initially intended for adolescents with autism spectrum disorders (ASD). It was initiated in 2014 and is the result of multidisciplinary collaborations. Its objective is the improvement of social skills and it is based on three mediations: robotics, culture and art. It consists of 20 1-ho weekly sessions, alternating 10 preparatory sessions and 10 robotic programming sessions, in which the six ASD participants are supported in the construction of a play whose actor is a robot. Then the work of the workshops is publicly presented during a session at the end of the program.

Rob'Autisme presents multiple originalities compared to approaches using the robot as a mediation tool, on the one hand, and to approaches for supporting autism in the broad sense, on the other hand. The project is based on the principle of resilience and considers behavioral disorders as the result of a context (consequence) and not as the starting point (cause) of the difficulty of social integration. It is then a question of defining, or redefining, the notion of the individual, that is, a person who is a member of a group and their legitimacy within said group.

There are two notions to be distinguished here: on the one hand, the way in which therapeutic support of cognitively disabled people is considered and, on the other hand, the way in which the object of mediation in therapeutic support is considered. Regarding the first notion, the symptoms of each case are different, so our society opts for *personalized* support, that is, one patient facing one caregiver.

Several caregivers can be solicited for the same patient, but each meeting will be face-to-face and each support action will be adapted to the specific disorders of each patient. The approach consists of soliciting the patient on certain points by using appropriate exercises. In this way, the patient is prevented from closing in on themselves through these solicitations. The second notion results from the first: the idea of the companion robot was introduced in the 1990s, proposing the robot as a *social actor* in its own right, which would replace the therapist. It has been observed with autistic children that they respond to solicitations made by a robot, particularly a humanoid one, whereas this response was difficult to obtain when the same solicitation came from a human being.

Rob'Autisme rebuilds a simplified micro-society, in which each person will be able to define their place. The support does not consider the solicitation, but the initiative of the participants, and supports them in this initiative to allow them to create their social contribution, their legitimacy. A fixed and rigid framework is set – a temporality, a place, people, forms of activities – in which certain elements can vary. The micro-society is made up of six participants, three companions who help them concentrate, think of ideas and help carry them out, a program leader who guides the exercises and the sessions, and two technical referents: one for robotics, present at all the robotic sessions, and one who participates in four non-robotic sessions, who will explain either the sound or the sets. This micro-society will work on the production of a play in which the actor is a robot: programming the robot, of course, but also producing everything that surrounds the actor, i.e. the sets, the sounds and music, the recorded voices.

The therapeutic support focuses on three socialization actions: dual communication, communication within a known and small group, and communication in society. The first two are achieved during the 20 sessions of the program, through appropriate exercises and a valorization of the actions carried out by the participants (taking initiatives and their valorization by the group). The third is obtained during the public restitution at the end of the program, and its effect is only guaranteed if the other two have been accomplished. In order to support the management of the complexity of human relations, the paradox is treated permanently during the three socialization actions. For example, the restitution shows a complete show where the robot tells a coherent story. The participants will both know the show, since they made it and programmed each brick, but at the same time, they will also, alongside the audience, discover it for the first time, since until the performance they have not yet seen the bricks assembled to make a whole. During the programming sessions, the participants will be both actors, as they program the robot, and spectators, as they watch it behave. The rigid framework that surrounds this program guarantees them stability and security.

And what about the robot? Rob'Autisme proposed to place the participants as social actors, and not as patients solicited by targeted exercises. The results were obtained on all the participants in the program, showing a radical improvement in concentration, a calming of anxiety, a possible socialization and a consequent improvement of the life of the carers. These results seem to be sustainable over time, although further studies are needed to ensure this.

The program has also been applied to people with Alzheimer's disease and has experienced similar results (Rob'Zheimer project). The *extension* robot approach, that is, the robot used as a prosthesis in communication by the participants to express themselves, has been applied in shorter programs to adults in nursing homes (severe ASDs, multiple disabilities and Rett syndrome) and other pathologies in adolescents. The effectiveness of the use of the extension robot, despite sometimes severe disorders, was noted.

The robot, thanks to its simplicity and its humanoid form, allows its operator or interlocutor to create a link with it. It acts as a mirror (companion robot) or an extension (extension robot), devoid of any judgment and not subject to conventional social rules. As such, it is a mediator that works directly on our social learning functions, according to the autopoietic definition proposed by Francesco Varela, and thus releases a capacity to act through the environment: the subject becomes permeable to what is happening around them (social learning situation), and therefore therapy can act better on them, be more effective. However, the therapy is carried out by the human being, and not by the robot, which acts as a catalyst, a therapeutic accelerator.

References

Chaltiel, T., Gaboriau, R., Sakka, S., Sarfaty, L., Barreau, A., Legrand, M., Liège, C., Navarro, S., Parchantour, G., Picard, J., Redois, E. (2017). Un robot en institution pour adolescents autistes : une aventure collective. In *L'Enfant, les robots et les écrans*, Tordo, F. and Tisseron, S. (eds). Dunod, Paris.

Feil-Seifer, D. and Matarić, M.J. (2009). Toward socially assistive robotics for augmenting interventions for children with autism spectrum disorders. In *Experimental Robotics: Springer Tracts in Advanced* Robotics, Khatib, O., Kumar, V., Pappas, G.J. (eds). Springer, London.

Sakka, S. (2020). Autisme et Alzheimer : des robots médiateurs ? *Le Journal du CNRS*, 30 July.

Sakka, S., Gaboriau, R., Picard, J., Redois, E., Parchantour, G., Sarfaty, L., Navarro, S., Barreau, A. (2016). Rob'Autism: How to change autistic social skills in 20 weeks. *International Workshop on Medical and Service Robots*, Graz.

Seon-Wha, K.E. (2013). Robots for social skills therapy in autism: Evidence and designs toward clinical utility. PhD Thesis, Yale University, New Haven.

Rob'Éduc

Sophie Sakka

LS2N, Centrale Nantes, France

Rob'Éduc, or the evolution of professions through robotization

The robot is a cybernetic machine whose programming allows it to interact with its environment, to adapt to it, to accommodate it. It is a very powerful tool that enables, for example, one to explore pipes without the need to dismantle them, whether they are mechanical (construction) or organic (surgery). It can be equipped with great strength or great precision, depending on its needs.

The robot has existed for centuries in our literature, and we attribute to it intentions or vocabulary that are usually unique to humans, for example, the front page of some newspapers: "Saudi Arabia gives citizenship to a robot", "Chinese man proposes to a robot", "Japanese hotel lays off its robot employees". The point is to remember that the robot is a machine that science fiction has brought to life. When, in the media, we are compared to robots, we accept it without resting the framework of use and we can feel obsolete in front of these very precise and strong supercomputers, operational 24 hours a day.

We must be aware of the power of the word *robot* on our imaginations, and our acceptance of this competition. For example, since the robot is a machine, let us systematically replace the word "robot" with a word designating another machine, whose name is less charged with imagination: "A Japanese hotel is laying off its coffee-makers". We then understand the inadequacy of the vocabulary to the situation: you do not fire a machine, you unplug it. You do not give citizenship to a machine, it just does not make sense. And to the sentence: "The elevator is better than you", the answer is immediately put into context: "Yes, to climb 20 floors in a short time". The robot is rarely put in its context of use, our imaginations immediately see it as having similar capabilities to humans: a legitimate competitor – which it is not.

The robot is a machine. A machine has a use, a function, and when this function is obsolete, it is disconnected and put away. A machine, no matter how you put it, has no legitimacy in human society; it is a tool that has a use.

The Rob'Éduc experiment was initiated in 2019 where, for the first time, a humanoid robot taught humanoid robotics to engineering students at Centrale Nantes in France in three 2-h classes. The robot, in this setting, was not autonomous: the official teacher was in the next room, instrumented, that is, equipped with the necessary sensors to transmit or receive information with the remotely operated robot – Xsens motion sensors were placed on their arms, legs, torso and head, their movements were captured and sent in real time to the robot, which reproduced them; their voice was captured by a microphone, "robotized" (filtered) and sent to the robot; the video and sound captured by the robot were sent back to the operator on a video headset and an audio headset, respectively. Thus, the operator was physically in the next room, but still present, despite the simplification of the information in the classroom and in front of the students.

In the classroom, the students were facing a humanoid robot with a humanoid voice, whose behaviors were identified as "natural" (interactions, displacements, body movements, etc., similar to those of a human being). The artificial teacher had the same teaching aids as the human teacher: the course slides were projected on a screen.

This natural aspect of the behavior generated different reactions among the students, ranging from amusement to distrust: a cognitive link is established very quickly between the human beings and the machine, a link that *makes people forget* that a human being is controlling the machine, leading them to inappropriate behaviors that they would never have had with their teacher, revealing to the robot information that they would never have revealed to a human being.

This cognitive link with the machine has a major characteristic, as we have pointed out: it allows us to forget that a human being controls the machine; in other words, we are in fact facing a human being, who expresses themselves to us through a robotic mediator. It has been observed that the interlocutor of the robot loses all sense of vigilance and accepts all the words of the machine. Indeed, the machine, in its simplicity of communication, in its representation in our imaginations, is not bound to follow the same social rules as the human being; it is out of time and out of the conventional social space.

This characteristic of the human interlocutor's loss of vigilance has several consequences, which can be good or worrying. The good consequences are illustrated by therapeutic support projects, whether they use the companion robot or the extension robot, as in the Rob'Autisme project. Another example concerns education: without social limits, access to education could be obtained even by populations in a situation of social isolation, and this is one of the major motivations of the Rob'Éduc project.

Rob'Éduc proposes a robot teacher to come and help populations disadvantaged by social conventions. The use of a robot in this context makes sense, if we remember that the objective is not to replace human beings by pointing out their social differences. The idea of the project, apart from proposing solutions to help integration using the tools available today, is to distinguish the specific contributions of a human being in the accomplishment of a task (a job) from the specific contributions of a humanoid machine in the accomplishment of the same task. The experiment makes sense if the robot and the human perform the task under the same conditions and in the same way – the use of teleoperation is essential: the same professional, in this case the teacher, performs the same task, either as a human or as a machine. The relationship with the human and the relationship with the machine, in this strict context, can be compared. The idea of replacing the human being also loses its meaning; the robot is to education what the scalpel is to the surgeon: a tool.

The robotization of human tasks offers many solutions, for example for tedious jobs that sometimes cause severe musculoskeletal disorders, but also for the accomplishment of a task that was previously entrusted to human beings: the functioning of society. Let us imagine that, tomorrow, our technological competence will allow machines and algorithms to fully guarantee the functioning of human society.

This raises a question that is part of a social paradigm shift, that of the definition and legitimacy of the individuals (human beings) who make up society. Social legitimacy has until now been based on the notion of contribution to the functioning of the group. We identify ourselves as contributors to the functioning, therefore as legitimate. If machines free themselves from this contribution on our behalf, the question of competition takes on its full meaning because, by definition, the machines become legitimate and we do not. We must therefore change the contribution–legitimacy relationship in order to find a new balance between the beings that make up society. A second concern arises in relation to the social bond, which is itself based on dependence on the contribution of other individuals. If machines provide this contribution, the social bond will be established not between human beings, but between humans and machines. A third philosophical question, which is already topical, accompanies such a paradigm shift: Can a society accept that machines and algorithms, which ensure its entire functioning, belong to a few? And for the economic aspects, how can we rethink the redistribution of wealth generated by machines to individuals considered as legitimate?

References

Ouest France (2017). Nantes : le robot enseignant débarque à la rentrée. Report, Ouest France.

Robotics and Society

Sophie Sakka
LS2N, Centrale Nantes, France

In recent years, robots have invaded our media. The word has become familiar, everyday, carrying certain obviousness, as if these machines were at every street corner, in every factory and held the entirety of our social functioning. A survey of the population to determine who has already seen a robot would show that machines are in fact absent from our daily lives: other than on television, on the Internet or in the newspapers, we do not feel that we are dealing with robots.

How, then, can we explain such a discrepancy between media declarations and reality? There are two major explanations:

– *Our imaginations*: while robotic technology in society has only become operational in the past 20 years, it has existed in literature for over three centuries. The word *robot* itself (1920) comes from a play, therefore from the imagination, and it is the only technology whose name comes from science fiction. Robots are therefore implicitly accepted by the population, which has accepted for more than 50 years that autonomous machines will be present in every corner of society in 30 years, that we will talk to them, that they will understand us, that they will answer us with respect and submission.

– *Our confusion*: the robot is associated with a humanoid avatar. Its definition remains unclear for most people and non-humanoid robots are not considered as robots. A misuse of the word *robot* is observed, among others, with "kitchen robots" (which they are not), "virtual robots" (which they are not), the comparison of automatic behaviors of human beings "to robots", and the use of inappropriate vocabulary for machines increasing the ambiguity. In the end, at present we have little understanding of what a robot is, so we are not able to identify it. The competition between human beings and machines generates distrust fuelled by the promise of destruction offered by science fiction.

Robots are already present in many places in society. Let us start by proposing a definition to better understand the particularities of these machines and how they can impact our social functioning.

A robot is a machine that can perform tasks autonomously, that is, without human intervention. It is a cybernetic machine, at the interface of four main scientific domains: mechanics, electronics, computing and control. Its interaction with the outside world is regulated by a control loop that links *perception* (sensors), *decision-making* (computer program) and *action* (motors). Two key words are characteristic of robots: autonomy and adaptability.

All the machines we know today can be robotized, that is, perform functions without our intervention. For example, a washing machine that can detect the weight of the drum, close the door and start a program when the weight reaches a certain level becomes a robot. However, the appearance of the machine has not changed, so we do not feel like we are interacting with a robot. Many vehicles are equipped with autonomous detection and safety systems, and even autonomous driving systems. Cars look the same, yet they have become robots. Many scooters or, of course, Segways and similar machines, are equipped with balance assistance systems; they are robots. Robot vacuum cleaners or lawnmowers are more easily identifiable and are identified as such in commercial messaging. Invisible to the public, high-end surgical robots assist surgeons in repairing aneurysms without opening the cranium, or performing heart bypasses without opening the patient's chest cavity. There are many more or less hidden examples; robots are indeed already all around us.

However, these autonomous machines are not yet associated with the robots of our imaginations: often they still only have one task to perform autonomously, and are therefore very limited. We imagine this machine to be "thinking", that is, endowed with analytical capacities similar to those of human beings, which it is still far from. The complexity of reasoning necessary to adapt to social life is very great and requires an experience that is difficult to program, or even to self-program. Moreover, with new experience and new learning, safety constraints must be put in place in order to avoid accidents that were not immediately obvious and face situations that were not foreseen.

Knowledge comes from experience, and robotic experiments in society are almost non-existent, because they require the establishment of a complex, time-consuming, expensive and secure environment. In its absence, who would be responsible for a bug (error in the machine), an accident (error in the machine's environment) or hacking? Imagine, tomorrow, that someone breaks into the control room of robotic vehicles, and forces these vehicles to not exceed 20 km/h or to exceed 120 km/h. Safety is the key word in today's robotic deployment in society, and responsibility is a consequence of its absence. The robotic systems on the market are still under the control of their human owner, or totally harmless (robotic vacuum cleaners, for example).

As we have understood, many developments and supervised experiments will still be necessary to be able to freely robotize society in complete safety, and finally see the arrival of the machines promised by science fiction.

References

Breton, P. (1995). *À l'image de l'Homme : du Golem aux créatures virtuelles*. Le Seuil, Paris.

Čapek, K. (2012). R.U.R. : les robots universels de Rossum. In *Robot Erectus*, Heudin, J.-C. (ed.). Science eBook, Paris.

Carr, N. (2017). *Remplacer l'humain : critique de l'automatisation de la société*. L'échappée, Paris.

de Chousy, D. (2008). *Ignis*. Terre de Brume, Dinan.

Devillers, L. (2017). *Des robots et des hommes*. Plon, Paris.

Routing

Laurent Bloch
Institut de l'iconomie, Paris, France

Routing

The term routing refers to all the methods, techniques and tools whose implementation in a computer network allows the organization of a data flow from one node to another according to the best possible route. The idea of routing takes on its full scope in the case of the interconnection of networks within a network of networks such as the Internet.

In such a network of networks, let us first consider the most basic networks: the most common cases are those of the local area network (LAN), which serves a building occupied by a company, and the network of an Internet service provider (ISP), which groups together on its premises the infrastructures through which its customers access its services. From the point of view of routing, these two situations are equivalent: the operator of the elementary network concentrates the data flows coming from several nodes (computers), each with a network address (IP address, as in the case of the Internet), and retransmits them on a communication line, generally a single one, to a higher level network, which concentrates the flows coming from several elementary networks.

How can multiple streams from different nodes in the local network share the ISP's single, higher speed line? The technique that allows this is called multiplexing. With the Internet Protocol (IP) used by the Internet, each data stream is broken down into relatively small and regular packets, and each packet contains the IP address of the stream's origin and the IP address of the destination. The packets of the different flows are sent outwards, on the line, as they arrive, that is, the flows are mixed. The destination addresses are used for sorting by *routers*, which can be seen as the network's switching stations.

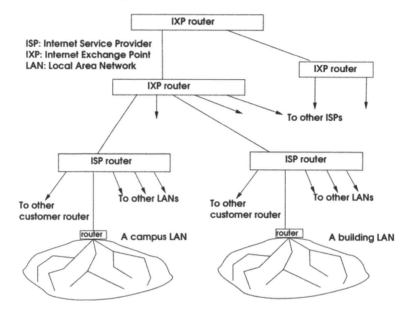

Interconnection of routers

The invention that allowed the prodigious development of the Internet, and of which Louis Pouzin is the main creator, is a new way of considering data packets, the *datagram* concept: instead of a centralized administration of end-to-end route calculation, each node of the network makes it its business to know to which node the packet (datagram) should be sent in order to bring it closer to its destination. For this to work, each of these nodes must be equipped with a device called a *router* (ISP boxes are actually routers), in fact a specialized computer with at least two network interfaces (each interface has its own IP address and belongs to a different network), and capable of passing packets (datagrams) from one network to another, depending on their destination. Each router is therefore connected to at least two networks and is programmable according to a *routing algorithm* and to a *routing table* designed to calculate which network one should send each incoming packet.

The revolutionary nature of this idea must be appreciated, as it went against everything that telecommunications engineers had been doing for a century. Before this invention, establishing a communication in a network (telephone or data) involved establishing a physical link between the two communicating ends. In the Internet, each datagram travels independently, the datagrams of the same communication can take different routes, it is up to the receiving device to verify that all the packets have arrived and to put them in the right order. At the time Pouzin published his idea, no one could believe that the packets would arrive at their destination: they would either get lost or go round and round in the network cycles.

In a complex inter-network, such as the Internet, a data packet sent from one end of the planet to another must pass through a large number of routers before reaching its destination. The routing process is complex. Each router has *routing tables* in its memory, which contain the addresses of the ingress routers in the neighboring networks, but it is not conceivable that each router on the Internet stores the addresses of all the others. To route a packet to its destination, it is therefore necessary to determine in each router the direction to take in order to reach the next, according to the most judicious route, a bit like a motorist trying to reach a city by scanning the signposts at each intersection. The methods for obtaining the result, if possible more efficiently than the aforementioned motorist, are borrowed from operational research and are called *routing algorithms*.

Basically, a router receives a packet on one of its communication lines, analyzes its destination address, consults its routing tables and deduces on which line it should forward the packet, or if it should discard it. The difference between an ordinary station and a router is that the router is programmed to receive packets on one interface and retransmit them on another, whereas the station just knows how to transmit and receive. When a router retransmits a packet, it does not change the sender's address, which remains that of the original sender.

Let us look at the process of sending a data packet. We will assume that the sending node knows the address of the receiving node (it will have learned it from the domain name system [DNS]). The basic algorithm is as follows:

– extract from the destination address the "network address" portion (i.e. the first few digits, as opposed to the last few digits that identify a particular node within that network);

– look up this network address in the routing table. Four cases are possible and are as follows:

- the address of the destination network appears in the routing table and corresponds to a directly connected network – the packet is delivered directly to this network by the designated interface and routing is done,

- the destination network is listed in the routing table and the means of reaching it is the address of a router – the packet is transmitted to this router according to the procedure seen for the previous case,

- the destination network is not in the routing table, but the table lists a default router – the packet is forwarded to that router,

- any other case triggers a routing error and the packet is dropped.

Routing algorithms

We could imagine another solution, based on a central router of the Internet distributing packets to all the networks, which could be refined by slicing the Internet into subnets, each organized around a router that has all the network addresses of the subnet, and communicating with a less monstrous central router, which has the addresses of the subnets, and the means to assign a network to a subnet. This would be called static routing. This was the solution used by X25 networks in the days of Minitel and the network monopoly, and it is a solution that can be used on a scale that is not too large, for a corporate network for example.

However, for a large network with no centralized administration, such as the Internet, this would hardly be realistic. The strength of the Internet lies in its ability to route packets to their destination in an ever-changing network without central administration, in short, in dynamic routing, the principles of which are outlined below.

Dynamic routing, in order to be effective in a network as vast as the Internet, requires complex protocols. In fact, the Internet is a confederation of IP networks, but there is an intermediary level of aggregation for routing organization, the autonomous system (AS), which is a grouping of networks that can be seen from the outside as an entity with a single administrative authority. Thus, each ISP and its customers will appear as a single AS. Global routing tables will be exchanged between ASs. Inside an AS, simpler protocols and smaller routing tables will be used, as a customer wanting to send a packet to an address outside the AS will hand it to a router at their ISP, which will have the global routing tables. After all, when we put a postcard in the postbox, we expect, for example, the French post office to know how to get it to the Venezuelan post office, which will know how to find our correspondent in Caracas.

The global protocol for AS-to-AS routing table communication is Border Gateway Protocol (BGP). There are several dynamic routing protocols within an AS or network. The one that tends to be most widely used today is OSPF (Open Shortest Path First), which is based on a graph search algorithm by Dijkstra, made famous by his landmark paper "Goto Statement Considered Harmful" in 1968. Needless to say, at the time, he had no idea how his algorithm would be used. OSPF have supplanted other protocols because it gives better results, but this superiority comes at the cost of high complexity. For those of our readers who are not network engineers, we will discuss a simpler protocol that is still often used in small networks, that is, Routing Information Protocol (RIP). This is based on the Bellman–Ford algorithm, which was first developed in 1957 by Richard Bellman and given a distributed version in 1962 by Lester Randolph Ford Jr. and Delbert Ray Fulkerson. Like many algorithms used in the world of networks, it comes from the field of operations research and belongs to the family of algorithms for calculating the shortest path in a graph by using a method of the "distance vector" type, as opposed to OSPF, which belongs to the family of methods for calculating the "link state". BGP is also based on the Bellman–Ford algorithm.

Link-state shortest path methods such as OSPF require that each router has the topography and description of the entire routing domain in its tables, and that all this information is retransmitted throughout the domain whenever it changes. In contrast, with "distance vector" methods, such as RIP, each router maintains only the information about itself and its immediate neighbors. It is understandable that OSPF waited for an era of high data rates and cheap memory before becoming widespread, and that RIP was more successful in the preceding period.

The goal of a routing algorithm is to find the shortest path between two points in a graph (respectively a network). In terms of computer networks, short does not really mean a distance in terms of length, but rather in terms of link rate: a short distance means a fast link, a long distance a slow link.

The operation principle of RIP is as follows: each router in the network propagates distance vectors on all its interfaces, which in fact constitute the summary of its routing table. Initially (i.e. when it is powered up), a router knows only one route, the one that leads to itself, with a zero distance. However, by propagating this elementary routing table, it will allow its neighbors to learn of its existence; it will itself learn in the same way about the existence of its neighbours and, as time goes by, the routing tables of each of them will be enriched. What Messrs Bellman, Ford and Fulkerson show us is that this algorithm converges, that is, that after a certain number of exchanges of routing tables, the system constituted by this network of routers will reach a stable state, where the sending of new routing information will no longer cause any modification of the tables.

A router is able to test its interfaces, and in particular is able to detect the presence or absence of an interface that responds at the other end. If a link is unexpectedly cut, the routers concerned detect it, recalculate their routing tables by assigning an infinite distance to the destination previously reached by the cut link, and the propagation algorithm is run again until a new stable state is obtained.

References

Abbate, J. (1999). *Inventing the Internet (Inside Technology)*. The MIT Press, Cambridge.

Huitema, C. (1999). *Routing in the Internet*. Prentice Hall.

McKenzie, A. (2011). INWG and the conception of the internet: An eyewitness account. *IEEE Annals of the History of Computing*, 33(1), 66–71.

Pouzin, L. (1975). Presentation and major design aspects of the CYCLADES computer network. In *Computer Communication Networks*, Grimsdale, R. and Kuo, F. (eds). Noordhoff, Groningen.

Pouzin, L., Marinica, C., Shapiro, M. (2015). Du datagramme à la gouvernance de l'Internet. *1024 – Bulletin de la société informatique de France*, 6, 59–75.

S

Science Fiction

Guy Thuillier
LISST, Université Toulouse – Jean Jaurès, France

Digital interfaces and human–machine integration in the science fiction imaginary

Science fiction (SF) is an interesting solution to approaching the digital world: not only does it reveal our collective representations, fantasies or fears about the information and communication technology (ICT) revolution, but it also sometimes inspires their design (Bicaïs 2006). SF authors, such as Vernor Vinge, Bruce Sterling and Neal Stephenson, have also worked as consultants or designers in the tech world. According to Pierre Lévy, in our "real-time civilization [...] science fiction has become as important as the social sciences, if not more so, to understand the contemporary world" (Lévy 2002). The question of brain-computer interfaces (BCIs) is central in SF, since the relationship of human societies to technologies is the very issue of SF speculations. Now, the BCI is finally the concrete artifact in which the relationship between human and machine is embodied, through the digital objects of our daily life and of our future – computers, smartphones, smart and connected objects, robots, etc.

From biomechanical interfaces to voice control

Historically, humans first designed "biomechanical" interfaces to control the first machines – buttons, keys, pedals, steering wheels, etc. The development of computing brought about new interfaces. To communicate with the first computers, complex punched cards were needed. Soon, the keyboard and then the mouse were introduced on personal computers (PCs). But the interface is not only material, it is also software: the revolution of user-friendliness, launched by Apple and its

Macintosh in the 1980s, was extended to PCs with the arrival of the Windows operating system in the mid-1990s. Computing became more democratic: no longer do you need to be an expert and master computer languages to communicate with a computer.

Today, the trend is to simplify interfaces, which must be user-friendly, intuitive and ergonomic. Haptic interfaces are evolving: Will the development of touch screens challenge the supremacy of the keyboard? Kinetic interfaces are also progressing: the mouse and the joystick have been replaced by gamepads with motion sensors and haptic feedback. With camera systems coupled with image processing software, the entire human body becomes an interface with the machine, which detects movements and reproduces them on the screen, making the interface totally transparent to the user.

The success of voice control, which is now a must for smartphones and connected speakers, has long been anticipated by SF: Hal 9000, the quiet-voiced (and all the more ominous) computer in the film *2001: A Space Odyssey* (1968) has had a significant lineage of chatty AI in film, such as ARIA, the computer in *Eagle Eye* (2008), or GERTY, the computer in *Moon* (2009), whose small screen displaying emoticons complete the record of communication with the lonely astronaut it is supposed to serve. But the ability of SF robots to speak is sometimes ambiguous: while C-3PO, the great golden humanoid robot of *Star Wars* (1977), is an interpreter and multilingual robot, his companion R2-D2 only expresses himself by hissing and stridulating, which must be translated into human language by none other than C-3PO.

Ambient intelligence and information ecosystems

The multiplication and interconnection of our digital prostheses, the miniaturization of electronic chips, which facilitates their portability and their integration into various objects, the continuous increase in their performance and the decrease in their manufacturing costs, their growing capacity for remote communication (RFID chips), the development of geolocation, cloud computing, etc.; these developments are leading us to a new stage in the human–machine relationship, variously referred to in the literature as pervasive computing, embedded computing, everyware or ubicomp (ubiquitous computing), informational ecosystems, etc. Rafi Haladjian, a French ICT entrepreneur, prefers the notion of "ambient intelligence":

> A situation in which we are immersed in a global interface. The computer is no longer on a screen in a circumscribed location; we are sitting in it rather than in front of it. (Haladjian 2010, p. 80)

Many fictions have already anticipated these developments: in Jacques Tati's films such as *Mon oncle* (1958) or *Playtime* (1967), we already find a critique, in a humorous and ironic register, of this environment of more or less sophisticated communicating gadgets that Tati takes a malicious pleasure in derailing. More recently, *Minority Report* (2002) gives a striking vision, inspired by the research and development studies of the digital industries, of this intelligent environment with which the hero, played by Tom Cruise, communicates constantly through multiple and varied interfaces (biometric, vocal, haptic, etc.).

Toward a human–machine fusion?

BCIs are on the threshold of a revolution: technological convergence, called NBIC (nanotechnology, biotechnology, information technology, cognitive science) in a report to the US National Science Foundation, or GNR (genetics, nanotechnology, robotics) by Raymond Kurzweil. This convergence opens up a new scientific and technological field: "biotics", "the result of the fusion of biology and computer science". In this new science:

> The focus is on the development of new components and molecular electronic circuits (biochips, biotransistors) as well as the development of bioelectronic interfaces between humans, computers and networks. The boundaries between the biological, the mechanical and the electronic are becoming blurred. (De Rosnay 2008)

This is no longer SF: in 2006, the American Matthew Nagle, a quadriplegic, became the first human being to use a BCI. The possibilities of these technologies could lead to a real human–machine hybridization:

> [This vision] almost always leads to the "mutant", the "cyborg" or the "bionic man": the mutant is a living being that modifies itself through biological mutations; the cyborg, a robot-man or a human being whose biology has been mechanized and whose mechanics have been "biologized", and the bionic man, a being that integrates bioelectronic parts that replace or augment deficient functions. (De Rosnay 2008, p. 229)

While this scenario does not seem desirable to Joël de Rosnay, it is on the contrary called for by the followers of singularity and transhumanism:

> Computers started out as very large, remote machines in air-conditioned rooms with technicians in lab coats. Then they arrived on our desks, then under our arms, and now in our pockets. Soon, we

won't hesitate to put them in our bodies or in our brains. By the 2030s, we will become more non-biological than biological. (Kurzweil 2007, p. 332)

With these theories, we approach SF, whose works Kurzweil often quotes to illustrate his vision of the future. *RoboCop* (1987) is undoubtedly the archetype of the cyborg in popular culture. After being left for dead by mobsters, a Detroit cop is "rebuilt" as a cyborg, half human, half robot. An indestructible "supercop", *RoboCop*, takes his revenge and cleanses the city of gangs allied with crooked politicians who impose their law. In the film, the fantasy of omnipotence that the human–machine hybrid allows is counterbalanced by passages that show the character's existential turmoil. *RoboCop*, more broadly speaking, is the ultimate incarnation of a dehumanized universe, a socially divided and ultra-violent post-industrial city, where humans no longer seems to have a place. The film was released at the end of the 1980s, during a period of industrial and urban crisis in Detroit, the former automobile capital of the United States, which had suffered from deindustrialization and poverty. The alienation is not only that of a man transformed into a robot, but concerns the whole society, broken by extreme capitalism.

While the human–machine fusion can appear as an alienation, it also includes a component of fascination, desire, even eroticism. This is particularly clear in the film *eXistenZ* (1999), which reverses the point of view. *RoboCop* finally resembles a robot more than a human being, in his mechanical appearance and gestures, as well as in his emotionless behavior. In Cronenberg's work, it is the opposite: the machines become biologized and resemble organic entities. In *eXistenZ*, the characters connect to a virtual reality game via a console with organic shapes, evoking a kind of large fetus. This machine plugs directly into the players' spines (with a "plug" in the lower back) via a cable resembling an umbilical cord. The scene where the heroine *plugs* the hero, after having lubricated the cord with her saliva and the "plug" of her companion with her fingers, before lying down at his side for a virtual trip, obviously has a strong sexual connotation.

Whether in *RoboCop* or *eXistenZ*, the human–machine fusion finally raises the question of alienation, at the heart of the debates on transhumanism. By hybridizing with the machine, the augmented human gains access to a form of superpower and immortality, but in this Faustian pact, do they not lose some of their own humanity?

References

Bicaïs, M. (2006). L'imaginaire colonisé par le dogme de la ressemblance. In *Colloque de Cerisy : science-fiction et imaginaires contemporains*, Berthelot, F. and Clermont, P. (eds). Bragelonne, Paris.

Caruso. D. (dir.) (2008). *Eagle Eye*. DreamWorks Pictures.

Cronenberg, D. (dir.) (1999). *eXistenZ* (1999). The Movie Network, Natural Nylon, Téléfilm Canada, Serendipity Point Films.

De Rosnay, J. (2008). *2020 : les scénarios du futur*. Fayard, Paris.

Haladjian, R. (2010). Intelligence ambiante. Créer de nouvelles propositions de valeur. In *TIC 2025 : les grandes mutations*, Lejeune, Y. (ed.). FYP, Limoges.

Jones, D. (dir.) (2009). *Moon*. Stage 6 Films, Liberty Films, Xingu Films, Limelight.

Kubrick, S. (dir.) (1968). *2001: A Space Odyssey*. Stanley Kubrick Productions.

Kurzweil, R. (2007). *Humanité 2.0 : la bible du changement*. M21 Éditions, Paris.

Lévy, P. (2002). *Cyberdémocratie*. Odile Jacob, Paris.

Lucasfalm (1977–). *Star Wars Franchise*. Lucasfilm.

Spielberg, S. (dir.) (2002). *Minority Report*. 20th Century Fox, DreamWorks Pictures, Amblin Entertainment, Blue Tulip Productions.

Tati, J. (dir.) (1958). *Mon oncle*. Specta Films, Gray Films, Alter Films.

Tati, J. (dir.) (1967). *Playtime*. Specta Films, Jolly Films.

Verhoeven, P. (dir.) (1987). *Robocop*. Orion Pictures.

Seniors (the Internet)

Aline Chevalier[1] and Mylène Sanchiz[2]
[1] *CLLE, Université Toulouse – Jean Jaurès, France*
[2] *CERCA, Université de Poitiers, France*

Internet and seniors: the point of view of cognitive and ergonomic psychology

The last few decades have been marked by the digital revolution, reflected, among other things, in the democratization of the use of digital tools for disseminating and accessing information (communication, online shopping, tax returns, etc.). These tools have become essential, whether for administrative procedures, as illustrated by the "zero paper" project announced by the French Ministry of Action and Public Accounts by 2022, or for communicating with friends and family. International statistics (Internet World Stats) show that the Internet penetration rate (which corresponds to the rate of connected individuals) in the first quarter of 2020 was 94.6% for North America, 87.2% for Europe and 39.3% for Africa. In 20 years, this rate has increased dramatically: +710% in France, +3,796% in China and +63,000% in Nigeria.

Industrialized countries, with a very high penetration rate, are subject to other forms of inequality; for example, in metropolitan France, statistics published by INSEE (2017) indicate that Internet access decreases with advancing age: 95% of 15–44 year olds versus 67% of those aged 60 and above; this is the case for most industrialized countries. This raises an important question: Why do older adults use the Internet and the services offered by these tools less, even though they also need them? This is a broad question, the answer to which is multifactorial. Here, we address it more specifically from the perspective of cognitive and ergonomic psychology.

In psychology, the expression "older adults" (or seniors) refers to studies that have been conducted with people between the ages of 62 and 75 years (mostly retired), or 65 years and older according to WHO criteria, with a high level of education (university degrees), using the Internet.

Some factors affecting Internet use

Accessing the Internet can become more or less complex as we age. Research in both Europe and the United States shows that older adults who use the Internet report higher levels of well-being and mental health than non-users and are often very enthusiastic about using the Internet. The Internet allows them to access different information, to communicate with others, which in some situations can reduce isolation, as was recently noted with the Covid-19 health crisis.

However, when we talk about older users, we are confronted with great variability within this population, due in part to two main types of variables: (1) variables inherent to the individual, such as level of education, motivation, cognitive and physical changes, and health; and (2) social variables, such as marital status, social capital (social network) and socio-professional category.

An interesting study conducted by Friemel (2016) among Swiss seniors shows distinct profiles between Internet users and non-users. The users have a high level of education (with university degrees), have used computers before retirement (often for their professional activity), have a high income, have an important social network and have a relatively strong feeling of self-efficacy (which corresponds to the feeling of competence) compared to the use of the Internet. Contrary to some studies, gender does not affect Internet use once these other variables are controlled. Conversely, non-users think the Internet is too complex, that the information is not very credible and have little or no interest in it (few people around them use it). Furthermore, research shows that older adults underestimate their computer knowledge and skills compared to younger adults; this in turn can negatively affect interest, motivation to use these tools, and even the performance associated with their use.

Among the activities carried out on the Internet, INSEE reports that the search for information on products (78%), news (60%) and health (50%) is the most important. In psychology, information searching is considered a complex cognitive activity, involving different cognitive processes and various cognitive skills, which makes it a preferred field of study.

When seniors are looking for information

Searching for information on the Internet usually involves using a search engine (Google, Bing, etc.). This activity requires individuals to formulate queries, then make choices from the results provided by the search engine and then navigate within different Websites. In doing so, verbal skills, prior knowledge in a field (e.g. searching for information in medicine, psychology), perceptual-motor skills and decision-making abilities (which also decline) impact this activity. Thus, seniors can rely on the skills they develop as they age, such as vocabulary or general knowledge, while so-called fluid skills, which tend to slow down as they age (such as speed of information processing or the ability to process multiple pieces of information in parallel), can cause difficulties.

Numerous studies of educated older adults show that, despite having a higher level of vocabulary than young people, they experience more difficulty generating keywords (to be used in the search engine), reformulating their queries if they are ineffective and choosing which site(s) to visit (Sanchiz *et al.* 2020). However, within this population, greater variability than among young people is highlighted (Dommes *et al.* 2011). For example, the elderly who have a rich vocabulary are those who formulate the most queries. A high level of knowledge in the search domain also helps older people to formulate more elaborate queries, i.e. queries that are semantically more specific.

These difficulties may also be reflected in the strategies developed, which differ between young and old adults. The work of Chin and Fu (2010) shows that older people use top-down knowledge-driven strategies to navigate a website, which involves prior and/or general knowledge and fewer perceptual-motor operations (they look at fewer links and take the time to choose the websites they are going to visit, stay longer on a page and browse links of different categories). Conversely, young adults use bottom-up interface-driven strategies that require perceptual-motor operations and less prior and/or general knowledge (they look at various links, spend less time on a page and browse different links within the same semantic category).

Thus, young people perform very well on tasks that require the identification of specific facts, especially when they have little domain knowledge. Conversely, older adults perform well when it comes to collecting different information for further processing. Indeed, they spend a lot of their time evaluating the relevance of the information they are confronted with. Therefore, when faced with irrelevant information, they experience more difficulty suppressing it than younger people, which can slow them down and lead to sub-optimal choices. Younger people, on the other hand, are quicker to assess the relevance of information, even if it means sometimes missing out and visiting more web pages to make their choice. Research thus tends to show that young people adopt rather exploratory strategies, aiming to shuffle through a large quantity of information, whereas older people explore little but exploit and process information more thoroughly (Sanchiz *et al.* 2020).

Conclusion and outlook

Young and old adults are not equal when it comes to using the Internet, even though this tool is part of every citizen's daily life and must be used to carry out more and more administrative procedures. Therefore, a better understanding of the difficulties encountered by seniors must be regarded as a major societal challenge for industrialized countries, in order to break the digital isolation that some seniors may encounter. A better understanding of these difficulties is an essential prerequisite for the development of appropriate assistance, whether through training or the design of simpler tools for this population.

Finally, the studies conducted at the present time focus mainly on educated seniors, but what about seniors with a low level of education? Cognitive psychology studies on executive functions (which are strongly involved in information retrieval) show a decline with age, especially for individuals with a low level of education (Guerrero-Sastoque *et al.* 2021).

References

Chin, J. and Fu, W.-T. (2010). Interactive effects of age and interface differences on search strategies and performance. *Proceedings of the 28th ACM Conference on Human Factors in Computing Systems CHI'10*. ACM Press, Atlanta.

Dommes, A., Chevalier, A., Lia, S. (2011). The role of cognitive flexibility and vocabulary abilities of younger and older users in searching for information on the Web. *Applied Cognitive Psychology*, 25, 717–726.

Friemel, T.N. (2016). The digital divide has grown old: Determinants of a digital divide among seniors. *New Media & Society*, 18(2), 313–331.

Guerrero-Sastoque, L., Bouazzaoui, B., Burger, L., Taconnat, L. (2021). Effet du niveau d'études sur les performances en mémoire épisodique chez des adultes âgés : rôle médiateur de la métamémoire. *Psychologie française*, 66, 111–126.

Sanchiz, M., Amadieu, F., Chevalier, A. (2020). An evolving perspective to capture individual differences related to fluid and crystallized abilities in information searching with a search engine. In *Understanding and Improving Information Search*, Fu, W.T. and van Oostendorp, H. (eds). Springer, Cham.

Smart City

Ornella Zaza
Aix-Marseille Université, Aix-en-Provence, France

The smart city: a polysemous definition

There is no fixed definition of what a smart city is. Whether it is a concept, a strategy, an action plan or an urban policy, the term "smart city" covers a wide range of approaches to contemporary urban space. Many authors place the birth of the term at the beginning of the 21st century by the major American IT groups: in 2008, the company IBM launched its *Smarter Planet* program, which aimed at marketing various digital tools for urban management. Since then, the term "smart city" has been very successful and has spread exponentially in the Western world, not only through economic actors, but also through local authorities. Documents of various kinds (strategic documents, action plans, labels, etc.), with a status that is more incentive than regulatory, are gradually being formalized and published by the major American and European cities (New York, Chicago, London, Paris, Lyon, etc.). Significant funding is being provided by the European Commission (including Horizon 2020) and contributes to the promotion of a specific network of economic players (start-ups, large groups and consultancies) and to the dissemination of urban best practices for territorial public action. At the same time, experiments are being carried out in rapidly developing countries of the South and Asia (South Korea, India and Arab Emirates) by creating new cities rather than adapting existing ones to climatic, technological and socio-political changes. Today, the smart city is also appearing in the urban strategies of small- and medium-sized towns, and is even being introduced, through its smart village form, into European policies for the development and cohesion of rural areas.

As far as the academic world is concerned, the smart city initially attracted the attention of the disciplines of engineering and management sciences. Later, but significantly, it emerged as an object of study within the humanities and social sciences (especially political science and urban planning, while sociology and anthropology have grasped it to a lesser extent) (Ghorra-Gobin 2018). The motivations are to be seen in the evolution of the various approaches to the smart city: initially characterized rather by the significant use of digital technologies, the approaches claiming to be the smart city have increasingly mobilized issues (and shown impacts) related to the environment and social interactions. At the same time as proposing concrete actions to be deployed in the urban environment (through the dissemination of best practices), the smart city has gradually raised numerous ethical, social and political questions.

Three dominant approaches to the smart city: from telecommunication networks to citizen participation

Among the various approaches claiming to be smart cities, it is possible to distinguish three dominant approaches, which could be described as "technocratic", "environmental" and "participatory". Often ideological, sometimes complementary and never completely antithetical, these various approaches show both their strengths and their limitations.

The technocratic approach

The first, "technocratic" approach proposes as its main objective more controlled and optimized management of the city. It is implemented through the progressive equipping of urban space (with sensors, telecommunication networks, autonomous vehicles, etc.) and of urban management processes (with platforms and algorithms that seek to order the urban data collected). In this context, artificial intelligence, robotics and cybernetics are emerging as the fields of reference: they induce urban scenarios where human–machine interaction is strong and desired, where the function of the imaginary linked to the technique is often similar to that of ideology (Picon 2013). This type of approach reflects the urban the logic of efficiency, which is accompanied by that of optimization, simplification (through categorization) and prediction (by modeling). Consequently, the technocratic approach to the smart city gives technicians and technocrats predominant power in the urban fabric. It also shows a consequent risk of endangering personal data and individual freedoms, linked to both social control and commercialization. Less criticized, but increasingly taken into account in the projects, the obsolescence of the systems put in place, due to the rapid evolution of digital technologies, also raises questions.

The environmental approach

The second approach, which could be described as "environmental", follows a similar logic to the previous one, but would have the merit of responding to strong expectations in terms of sustainable development and adaptation to climate change. In this case, optimization aims first and foremost at better management of natural resources and, consequently, at taking better account of environmental and technological risks. In this type of approach, we therefore often see the association of the smart city with urban resilience. Although environmental approaches to the smart city are increasingly characterized by attention to urban greening, their implementation often focuses on the establishment of more efficient energy networks. A very widespread best practice is the smart grid, a connected energy network (using sensors and associated digital platforms) to monitor consumption in real time. In addition to detecting possible malfunctions, these devices model consumption in order to forecast it and thus propose energy storage and pooling systems within urban sectors with mixed programming (public facilities which, for example, would give up part of the energy produced and not needed to neighboring housing). This second approach, in addition to presenting the same risks as the technocratic approach, often lacks a thorough and global reflection on the environmental impacts of new technologies. Few studies analyze the phenomena of globalization, which concern the exploitation of primary (mineral) resources for the production of electronic components (often delocalized), their distribution and their recycling difficulties. However, these studies indicate an increase in social inequality as well as increased wage and environmental exploitation of the countries of the South (main producers of primary resources) by the countries of the North.

The participatory approach

The third approach, which could be called "participatory", is distinguished from the other two by a strong rhetoric that puts the human being at the center of the smart city (Douay 2016). Indeed, rather than focusing on technical aspects, this approach insists on the collaboration of all actors in the urban fabric: beyond citizen participation, it is a new close relationship between public actors, private actors and civil society. Certain managerial methods (such as open innovation) invite us to go beyond the logic of corporate secrecy and administrative bureaucracy, leading to iterative and collaborative processes and to bring about "new solutions" to contemporary urban challenges. As a result, the most common best practices for this type of approach insist on the "direct" intervention of city dwellers in the urban fabric (by proposing projects via platforms, by crowdsourcing data via applications, etc.) and of private and associative actors in the animation of urban life (often by developing third places, such as coworking spaces and fab labs, which combine work and leisure activities). However, the participatory approach shows the emergence of caricatured figures: the smart citizen, the imagined city dweller of the smart city, is a project leader, aware of the technical and socio-political stakes, a bit of a geek but above all very sociable, both a volunteer serving the community and an ethnically responsible entrepreneur. While, on the one hand, this approach incites an easier consensus within the population through constant reference to the principles of participative democracy, on the other hand, the risks it presents remain numerous:

the exclusion of part of the population (because of the digital divide, which concerns access to equipment, as well as cultural and usage factors), the increased use of personal data, the exacerbated ludification of certain societal issues and, finally, an unassumed devolution of public service to civil society.

Experimenting with the smart city: a question of governance

These three dominant approaches to the smart city ultimately show a strong interweaving between them: while putting weight on a rather technical, sustainable or collaborative aspect, the three constantly articulate profoundly political questions. While the first approach claiming to be smart cities focused mainly on the performance of telecommunication networks and the production of data, civil society has progressively oriented the debates toward problems of a more social nature and toward more critical positions (Greenfield 2013). The rejection of Google's smart city project in Toronto and the spread of anti-Linky movements in France are recent examples. The smart city thus takes shape more through the implementation of mechanisms that are constantly being debated and reworked than through structuring policies, large-scale urban projects or regulatory planning documents. With the exception of the rare concepts of new cities, the smart city does not materialize through a *tabula rasa* of existing cities, but rather through the integration of new objects and processes in the urban space. Under the regime of experimentation, "urban demonstrators", "urban experiments" and "pilot projects" seek to combine regulatory specifications, technological innovations and new games of actors in the urban fabric. These mechanisms aim, in the end, to build a consensus around the principles and values associated with the contemporary urban space.

While the first critical analyses had thus mainly confined the smart city to an urban marketing strategy and emphasized the excessive influence of private actors in the urban fabric, the approaches that have been implemented mainly show significant impacts on urban action modes of governance (Courmont and Le Galès 2019). Criticized for their lack of flexibility and weakened by the drop in structural funding, public players have seized the smart city as an opportunity to launch a new stage in the modernization of public action. The holistic vision carried by smart city approaches encourages the decompartmentalization of urban sectoral policies (a frequent example is the integrated management of electric vehicles, renewable energy and waste within the same urban project). In addition to the intervention of the more traditional private actors in the urban fabric (by delegating public services), many start-ups have come to populate the urban experimentation programs. While for these small companies, the challenge is to test their solutions for marketing, for local authorities, this represents an opportunity to develop their working methods and to integrate new technical equipment at a lower cost. In addition, smart city initiatives have frequently led to the creation of a new municipal team, which acts as an "innovation laboratory" within the administration: by activating specific initiatives (such as state start-ups). The agents are invited to design new solutions outside the normative frameworks of the public administration. However, these processes have their limits. There is a risk that urban services will be subjected to commercial logic of return instead of equality of access and free services. Urban action that is carried out through experimentation will certainly have the merit of adapting to unforeseen circumstances, but it will also tend to forget the overall coherence provided by urban planning. Finally, the criteria for decision-making are becoming more technical and social issues are confined to the participation of inhabitants, which is more a matter of co-management of the urban space than of real power-sharing.

Seen in this light, the main objective of the smart city would thus be, above all, a profound transformation of the logic, practices and professions of the urban fabric. By way of definition, the smart city could therefore be regarded as a process that engenders the crystallization, at different times and through different intensities, of an entourage of humans and non-humans, values and imaginaries, practices and logics around a projection of the city into a future that remains to be defined.

References

Courmont, A. and Le Galès, P. (2019). *Gouverner la ville numérique*. PUF, Paris.

Douay, N. (2016). Planifier à l'heure du numérique. HDR, Université Paris-Sorbonne.

Ghorra-Gobin, C. (2018). Smart City : "fiction" et innovation stratégique. *Quaderni*, 96(2), 5–15.

Greenfield, A. (2013). *Against the Smart City*. Do Projects, New York.

Picon, A. (2013). *Smart cities : théorie et critique d'un idéal auto-réalisateur*. B2, Paris.

Social Contract

Bruno Deffains
Université Paris 2 Panthéon-Assas, France

Digital transformation and social contract

One of the core difficulties with the digital revolution in the current period is the lack of sufficient hindsight to stabilize our judgment. Nevertheless, it is possible to raise questions about emerging practices, such as the impact of digital technology on social insurance. In particular, the tension between John Rawls' veil of ignorance and the development of Big Data must be analyzed. The recognition of this problem is based on the analysis of the consequences of the systematic and massive exploitation of personal data on insurance practices. Does the development of new so-called "communicative" practices not contribute to weakening the so-called "transcendental" variants of social justice associated with the veil of ignorance?

To understand the current challenges, we must return to the past, and more particularly to the question that haunted the 19th century: how to design social protection systems that did not call into question the liberal principles inherited from the French Revolution. The society that emerged from the Revolution, characterized in particular by the Civil Code, was a world of theoretically free and equal citizens, who entered into contracts with each other in which the parties were individually responsible. In this world, it was up to each individual to protect themselves against risk (old age, illness and accidents) using their own foresight or by questioning the offender: the worker who was the victim of an accident at work needed to succeed in proving the fault of the employer. It was precisely on the occasion of the law on work accidents, adopted in 1898, that the concept of collective insurance was used for the first time on a national scale. By placing accidents in the register of chance and the vagaries of fate, the concept of individual responsibility was overcome. It was on this new basis that the whole system of social protection that marked the emergence of the welfare state in the 20th century was gradually organized.

For this insurance system to be credible, it had to be collective and apply to a large number of people. At the time, the incomes of the working classes were far too low for them to accumulate precautionary savings. As a result, a system based on voluntarism and individual capacities was simply not possible, and the assumption of social risks was essentially a matter of assistance provided by charitable organizations. It was not until the 20th century that we saw a gradual shift from a system dominated by assistance to one governed by social insurance (in other words, from charity to solidarity). It was especially after the Second World War that there was a great movement to extend social protection. The welfare state spread throughout Europe and partly to North America.

The main theorist of social contract is John Rawls, who published *A Theory of Justice* in 1971 (Rawls 1971). His starting point is a simple idea: to guarantee the fairness of the principles of justice, they must be chosen from a situation that is itself impartial. To determine such a starting situation, Rawls follows the "contractualist" tradition by postulating the existence of an original position, an ideal situation in which the participants (or their representatives) in social cooperation are supposed to seek agreement on the principles of justice and regulation of society. Second, he believes that the participants in cooperation are unaware of their personal situation in society, either present or future. They are thus placed behind a veil of ignorance, a privileged situation for determining the principles of justice in all fairness. This veil of ignorance becomes all the more fragile as it is likely to be torn apart by the massive exploitation of individual data, particularly for predictive purposes.

The danger arises when the predictive tends to become prescriptive. It is not only George Orwell's premonitory concern in *1984* that citizens must be totally transparent to the authorities, because "anyone who has something to hide is bound to be suspect". The risk is now mainly related to the fact that total transparency based on massive data collection and predictive analysis compromises the social contract by tearing the veil of ignorance.

The core of Rawls' theory of social justice is based on well-established principles. First, the principle of equal liberty, according to which everyone has an equal right to the widest set of fundamental freedoms consistent with the granting of that same set of freedoms to all. Second, socio-economic inequalities are justified only if they contribute to the betterment of the less advantaged members of society (the principle of difference) and if they are attached to positions that all have a fair chance of occupying (the principle of equal opportunity). For Rawls, these principles are only conceivable under the veil of ignorance. The individuals, forced to be impartial because they do not know their place in society, manage to freely and rationally establish these principles and procedures that found the social contract.

Beyond philosophical circles, the theory of social justice has provided a foundation for the renewal of social democracy around the world. Rawls' liberal-egalitarian conception has been used, directly or indirectly, to legitimize the creation of social minima (social benefits that provide a minimum cart of goods). More generally, social protection policies and the collective rights that flow from them emphasize the need to focus on the "most disadvantaged or fragile", often appealing to the veil of ignorance. The veil of ignorance is thus both an assumption and a normative construct that places impartiality at the heart of the reasoning.

This is precisely where the development of Big Data comes in. The latter is a fact, but it also reflects the emergence of new standards. The norm of transparency, in particular, which underlies the massive collection and processing of personal data, is largely antinomic to the norm of impartiality which underlies the veil of ignorance.

With the rise of individual data collection and increasingly sophisticated behavioral and predictive analyses, society is becoming more transparent and less homogeneous. The field of health must raise awareness of the need to rethink the social contract. As the veil is torn, will we remain united in the same way if we know that some people run huge risks and others do not? The digital support of our lives tends ineluctably to restrict the field of possibilities in terms of individual action capacities through predictive analysis. This based on massive data collection. Faced with these new challenges, solidarity is a value that is being put to the test. It is confronted with the promise of increasingly individualized solutions to social and health risks, which, while taking advantage of their apparent neutrality to provide effective support for public policies that imagine they can benefit from this presupposition of neutrality, in reality undermine the logic of collective solidarity, even though it has perhaps never been so essential for preserving the social contract.

Moreover, as it is not self-evident that the public spontaneously accepts the transmission of personal data, one idea is to guide them toward a default acceptance of this transmission through the use of nudges. Supported by behavioral economics, nudging is based on the idea that a small intervention in our environment modifies the mechanisms of choice, i.e. the behavior of individuals, in order to influence them in a direction that corresponds better to their own interest or to the general interest. It is understandable that the public authorities are interested in using this type of tool, but the gentle manipulation of decision-making environments, their progressive hold on the public space, the facilitation by appropriate nudges of the massive collection of private data, combined with greater predictability of individual behavior, constitute a converging set of threats to the social contract as we know it. It therefore seems desirable to point out the risks that these two techniques of collective governance, nudges and Big Data, pose to the principle of solidarity that underlies our social contract. It is not a matter of questioning the interest of the development of behavioral sciences in relation to the rise of predictive analysis tools, but rather of questioning the best way to benefit individually and collectively from these digital tools while preserving the gains of several generations of rights and freedoms.

"Public policies" based on the use of massive data and nudges, in the current circumstances, fail to think profoundly enough about what constitutes "non-arbitrary intervention". This is even truer of nudges than of Big Data, insofar as they appear spontaneously neutral and non-intrusive. In reality, they tend to alter and obscure the norms that underlie our conception of the public sphere, and all the more so when they are associated with the collection of private data. It is essential that the use of digital and behavioral technologies be combined with the establishment of a normative framework that is compatible with the social contract based on the veil of ignorance.

References

Ewald, F. (1986). *L'État providence*. Grasset, Paris.

Rawls, J. (1971). *A Theory of Justice*. Harvard University Press, Cambridge.

Thaler, R. and Sunstein, C. (2008). *Nudge: Improving Decisions About Health, Wealth, and Happiness*. Yale University Press, London.

Social Network

Zhenfei Feng
    GERiiCO, Université de Lille, France

Definition

The notion of "social network" was first introduced in the article "Class and Committees in a Norwegian Island Parish" by John A. Barnes, an Australian-British social anthropologist. Since then, "the use of the notion of *network* to designate sets of relationships between people or between social groups has become widespread, both within the social sciences and at its margins". For Michel Forsé, a social network is a set of relationships between actors, which may be organized (a company, for example) or not (like a network of friends) and based on relationships of a very diverse nature (power, exchanges of gifts, advice, etc.), specialized or not, symmetrical or not. Both Mercklé's and Forsé's definitions describe the three key concepts of a social network: the individuals, their links and the social environment (e.g. a company).

Digital social networks (DSNs) only appeared with the evolution of the Internet in the 1990s and, especially, with the arrival of Web 2.0 around 2005. At that time, the development and spread of DSNs was very rapid. Boyd and Ellison (2007) define them as online services (social network sites) that allow their users to:

– build a public or semi-public profile within a system;

– manage a list of users with whom they share a link;

– view and navigate their list of links and those established by others within the system.

In this definition, they use the term social network sites to describe websites that help their users to maintain existing relationships between friends, whereas the term social networking sites refers to websites that are more concerned with initiating relationships, i.e. meeting strangers.

Typology of digital social networks

Thelwall (2009) retains Boyd and Ellison's definition, and introduces a typology of DSNs that includes the social network site, social networking site and social browsing site (sites such as YouTube). According to this typology:

1) *Social networking sites* are designed to enhance social communication between members who know each other, often for leisure. Connections are primarily used to find and display lists of friends that exist offline.

2) *Social networking networks* are used to find new contacts from existing connections of friends. These new contacts include a significant proportion of acquaintances and previously unknown people.

3) *Social browsing networks* are intended to help members find specific types of information and resources. Connections are used as a tool to deploy contact lists, which provide access to related information and resources.

Some sites may pursue several objectives at once. The above typology depends more on the policy of a site or the practices of its members than on exclusive characteristics.

Moreover, as some authors point out, DSNs are a subset of social media (Kaplan and Haenlein 2010). They define social media as a group of online applications that follow the "spirit" of Web 2.0 by allowing the creation and exchange of user-generated content. Apart from DSNs, there are many other Web 2.0 sites that we can call social media, such as blogs, wikis, forums and microblogs (such as Twitter), social bookmarking, etc.

A Brief history of DSNs

The history and development of DSNs dates back to 1995. Randy Conrads' Classmates allowed users to join their school and find old classmates. Then the development of instant messaging, notably with AIM (AOL Instant Messenger) in 1997, has continued to converge over the years toward social networks. As early as 1997, AIM enabled, in addition to instant messages, a "social network" to be created by adding contacts to one's list of friends and to communicate with them in real time. Later on, AIM was widely integrated into DSNs (e.g. Facebook) to facilitate communication between users. However, according to the definition given by Boyd and Ellison, the "real" DSN did not exist until after the launch of Sixdegrees. There, users could create their own profile, have a link and communicate with others (their relatives, friends, family, colleagues, etc.).

Until 2003, most DSNs were launched in the United States and were even limited to certain regions or groups of individuals. It is thanks to Facebook, in 2004, that DSNs have spread all over the world and have shown a diversity of uses. Since 2006, the birth of microblogging, with Twitter, added to the evolution of the smartphone, has considerably strengthened the role of DSNs.

Business model

While most DSNs are free to users, subscription-based income is possible when membership and content levels are high enough. Some DSNs, such as LinkedIn, charge users directly for the services on the site. In fact, users can create an account for a basic free membership or for a premium membership that gives them access to more services.

The most common model for generating revenue is to exploit user's login patterns and personal data about them, allowing for personalized advertising. DSNs can track a user's location in the world based on their IP address. In addition to information provided by the user, such as age, gender, tastes and interests of users can be collected. The indefinite retention and reuse of this data by the DSNs, for whatever purpose, is enshrined in their terms of use. In this model, users are defined as a product and a commodity, with their data representing marketing revenue.

References

Barnes, J.A. (1954). Class and committees in a Norwegian Island Parish. *Human Relations*, 7(1), 39–58.

Boyd, D.M. and Ellison, N.B. (2007). Social network sites: Definition, history, and scholarship. *Journal of Computer-Mediated Communication*, 13(1), 210–230.

Forsé, M. (2008). Définir et analyser les réseaux sociaux. *Informations sociales*, 3, 10–19.

Kaplan, A.M. and Haenlein, M. (2010). Users of the world, unite! The challenges and opportunities of social media. *Business Horizons*, 53(1), 59–68.

Mercklé, P. (2004). *Sociologie des réseaux sociaux*. La Découverte, Paris.

Thelwall, M. (2009). Social network sites: Users and uses. *Advances*, 76(9), 19–73.

Sociotics

Vincent Meyer
Université Côte d'Azur, Nice, France

Researchers discuss – and rightly so – the meaning of the concepts they use and sometimes they create them; this is the case of sociotics. The term arose during exchanges on the exposure of social intervention in the context of PhD research and between researchers in Lorraine, France, involved in a program of the *Mission interministérielle recherche expérimentation sur les emplois et les qualifications des professions de l'intervention sociale* (Meyer 2004). In proposing a "sociotics", there was no intention to increase the semantic field of the human and social sciences, but to contribute to the emerging debates on "the best and the worst of technologies":

those of the "Web" and, before the Internet, of the upsurge of "information and communication technologies" (ICT) in organizations and professional groups. Sociotics thus problematizes a professional evolution confronted with an emerging technology when it gradually becomes the indispensable equipment of a profession. Let us specify that this evolution coincided with the so-called mutations of social work in general and, in particular, via controversies around social issues or causes (exclusion, poverty, endangered childhood, disabilities, etc.).

Sociotics thus becomes a repository for discourses accompanying the evolutions and impacts of digital technologies on professional or voluntary modes of exercising human solidarity. While this notion was first mobilized for the field of social intervention in France, from the 2000s onwards, it can now be extended to other organizations and audiences of the helping professions, and even beyond. The aim was to lay the foundations for "the development of social work that would progressively become inseparable from a social network", with a focus on the professional skills of face-to-face assistance or commanding an immediate physical co-presence. These developments concern professionals/volunteers as much as their "audiences". With a social approach, it is consequently a question of bringing to light the superimpositions between a discourse of euphoric or destabilizing support for these technologies and the possibilities of adapting or including these different evolutions in the expression of human solidarity with the non-human. In doing so, a capacity to communicate in a different way is required of these professionals, and this involves knowing how to convene and use the dominant media of the time, such as the emerging ICTs (Meyer 2004). For the public, they are "trained" – progressively equipped – like others in/by the global deployment of a digital society with its exponential evolution.

In order to characterize the impact of these technological developments on professional skills in the field of social work[1] at the beginning of the 21st century, four complementary approaches were envisaged: "The user-client, who has become a 'socionaut' and who could directly follow the procedures, visualize the intervention or the service being provided; the user's (inter)active participation in decision-making, in consultations on different projects; edutainment software enabling those with mental disabilities to access different digital resources; relational and geographical redistribution within the framework of social teleworking because of technical objects that enable professionalities to be subdivided or replaced at the same time, to bring together, in general functions (reception, follow-up, etc.), professionalities thought of until now as being separate and distinct". This is the reason why we are now in the position of being able to

---

1 Available at: http://media.social.free.fr/2000/sociologue.htm.

work at the same time in a social telework environment. Today, we are there... It should be noted that since the 1980s, the social sector (all professions and qualifications taken together) has made the mastery and control of information and communication a major political and ethical issue (Vitalis 2020). As early as the 1970s, these professionals saw ICT as an instrumental communication that was certainly part of a "modernization" process, but whose impact on the public they were responsible for must be measured, particularly in terms of filing, monitoring and sharing information about them. Thus, the notion of sociotics appeared to be more encompassing and methodologically more measured than that of e-social work (more "similar" to e-commerce) or cyber-social intervention. The experiments were starting and the professionals felt that they would affect not only the technological, economic and legal environment of human assistance, but also the very nature of their jobs and qualifications, that is, their praxis. For the latter, awareness-raising work was gradually developed, especially since technical objects have always captivated social professionals without turning them into techno-gaps (Bergeret 2014).

If we look closely, the reflection has focused on three areas: physical/material access to these technologies (which has become – politically – accessibility), the methods (some would say the pace) of appropriation of these technologies by those concerned (professionals, the public and their families or other carers) and, finally, the uses in the sense of what they will and want to do with them in their daily and/or professional life. These stages immediately set out the challenges and limits of the introduction of digital technologies in this professional field and with these people. These are reflected in a debate on: (1) functionalities linked to the devices of management and documentation IT now coupled with a softwareization of activities in support of "good practices" as well as evaluation approaches; (2) a formalization of care, from "toll-free" numbers to personal spaces and, with it, confidentiality versus professional secrecy in the dissemination of information and the use of data stored for/by their promoters; (3) characteristic attention to the problems linked to co-presence, proximity versus distance between the worker and the public, that is, the being with/in a helping relationship.

The modes of practice are today still – and in particular because of the health crisis (and insecurity) introduced in 2020 – in recomposition, and digital technologies confirm their place as new social operators and their dimension of support to general functions, in particular in a logic of dematerialization as in that of teleworking. Classically, two developments will allow us to move forward in the notional definition of sociotics: research and training.

For research in the humanities and social sciences, it is a truism that digital technologies require critical (and therefore creative) reflection on the links

– ceaselessly strengthened by the GAFAM – which unite them to social change and, more broadly, to the interoperability between human and non-human devices. In other words, the collection of data as well as the techniques of data processing are not the same as those used in the past. In other words, the data collection and the survey techniques to be used must allow us to bring to light the personalized experience of users with digital technology, particularly in their living environments, that is, both in institutions or at home, but also in the third places they may invest in and where digital mediation experiences are being developed at a time when the professional assistance relationship is becoming nomadic. Action research must be promoted[2].

As far as training is concerned, the basic needs are an initiation: to this instrumental communication in what it (re)presents both in terms of infobesity and disinformation, as well as in the collection and (re)use of data; to the possible manipulations via fixed or animated images on social networks. This is part of a (critical) media education lato sensu, as can be conceived by the *Centre de liaison de l'enseignement et des médias d'information*[3]; for better knowledge and mastery of the devices on the Internet, from platforms to smartphone applications, for learning as well as for entertainment and, with it, the attention "captured" by the screens. Massive open online courses (MOOCs) are emerging in various sectors, such as child protection. The training of social intervention professionals (other than digital mediators) is caught in a form of tension (also amplified by the health crisis of 2020) with this question of (working) time spent "in front of the screen rather than in front of the user" and the security of personal data for these audiences.

Finally, sociotics is a notion that cuts across several professional fields (de facto, several scientific disciplines); it thus brings together institutional, educational and pedagogical experiments, networking of services and "solutions", with a three-stage deployment: raising awareness among professionals of the issues, including ethical ones, of the digital transition; familiarizing them with the analysis of the digital needs of their audiences and "grasping" the functionalities of applications/interfaces, in particular the exploitation (past, real-time and future) of user data now massively amassed by the Internet "giants". The digital context of disabled, fragile and/or vulnerable people is evolving rapidly, even though welfare benefit is not yet paid in bitcoin, unicorns and their billions do not yet focus their capital and projects on aid relationships, and supercomputers do not yet plan life paths. In 2020, the French Ministry of Solidarity and Health defined a roadmap (a trajectory) for digital health for the medico-social sector, so that it could take the "homecare turn" by relying on,

---

2 Available at: https://journals.openedition.org/communicationorganisation/3455.
3 Available at: https://www.clemi.fr/.

among other things, "the deployment [and] the financing of implementations and evolutions of computerized user files"[4]. Beyond questions of interoperability, it is the deployment of "core" digital services that is targeted. A project[5] where "innovative digital health solutions" and their suppliers are welcomed with "open arms". An experimental project, to which is added – as a matter of course – that of the digital identity of these audiences in an algorithmic governmentality that (already) sees everything or almost everything – from a now digital identification of the individual to the augmented human.

References

Bergeret, J. (2014). Petits cailloux témoins des techniques et technologies rencontrés sur le sentier parcouru d'un acteur du travail social. In *Les technologies numériques au service de l'usager... au secours du travail social ?*, Meyer, V. (ed.). LEH, Bordeaux.

Meyer, V. (2004). *Interventions sociales, communication et médias : l'émergence du sociomédiatique.* L'Harmattan, Paris.

Meyer, V. (2017). *Transition digitale, handicaps et travail social.* LEH, Bordeaux.

Vitalis, A. (2020). La transformation numérique de l'action sociale : ce que nous enseignent cinquante années d'informatisation. *Vie sociale*, 28, 21–31.

Source Code[6]

Roberto Di Cosmo
INRIA, Paris, France

Programs must be written for people to read, and only incidentally for machines to execute. (Abelson 1984)

In a computer system, there are generally two parts: the *hardware*, which is the physical part of the system (processor, memory, disks, screen, network card, sound card, keyboard, mouse, etc.), and the *software*, which designates the set of instructions that can be stored and executed by the machine. It is the software that gives life to a computer system.

Instructions that can be directly executed by a machine (called "machine language") are often very low-level, represented by simple sequences of bits and difficult to understand by a human being. This is why software is almost never

---

4 Available at: https://esante.gouv.fr/virage-numerique/feuille-de-route.

5 Available at: https://esante.gouv.fr/virage-numerique/structures-30.

6 This text is licensed under the Creative Commons CC-BY 4.0 license.

produced directly in machine language, but "written" by developers using a "programming language", which can then be automatically translated into machine language.

As an example, here is an excerpt from the executable program that prints a simple "Hello world" message:

```
4004e6:  55
4004e7:  48 89 e5
4004ea:  bf 84 05 40 00
4004ef:  b8 00 00 00
4004f4:  e8 c7 fe ff ff
4004f9:  90
4004fa:  5d
4004fb:  c3
```

Here is the program written in the C programming language, from which the executable program from which we extracted the fragment presented above was produced:

```
/* Hello World program */#include<stdio
.h>
void main()
{
printf("Hello World");
}
```

The "source code" of software is generally understood to be the program that was written by the developer and from which the executable code is obtained on a machine. Also, one might be tempted to regard "source code" as any program written in a programming language. As technology has evolved, the situation has actually become more complex: developers have sophisticated tools that can "produce" programs in one programming language (such as C) from programs written in higher level programming languages, to the point that it is not enough to look at the language in which a program is written to know whether that program was written by a developer or generated automatically from a higher level program.

That is why the definition of "source code" for software, found in the GPL, is "the preferred form for a developer to make a modification to a program".

Source code is a special form of knowledge: it is made to be *understood by a human being,* the developer and can be mechanically translated into a form to be *executed* directly on a machine. The very terminology used by the computer community states that "programming languages" are used to "write" software. As Donald Knuth, one of the founders of computer science, wrote, "programming is the art of explaining to another human being what you want a computer to do".

Software source code is therefore a *human creation*, just like other written documents, and that is why it falls within the scope of copyright law. Software developers thus deserve the same respect as other creators, and it is essential to ensure that any changes to copyright law take into account the potential impact on software development. This was not the case in the drafting of EU Directive 2019/790, the first draft of which seriously endangered the collaborative development of open source software, and that required significant effort to make the necessary corrections.

As software source code becomes more and more complex, it is regularly modified by groups of developers who collaborate to make it evolve: to understand it, it has become essential to have access to its development history.

The software source code is thus incorporating *an important part* of our scientific, technical and industrial heritage, and thus constitutes a valuable heritage, as already argued by Len Shustek in an excellent article in 2006.

This is one of the missions of Software Heritage, an initiative launched in 2015 with the support of INRIA, in partnership with UNESCO, to collect, organize, preserve and make easily accessible all the source code publicly available on the planet, regardless of where and how it was developed or distributed. The goal is to build a common infrastructure that will allow a multiplicity of applications: of course, to preserve the source code in the long term against the risks of destruction, but also to enable large-scale studies on the code and the current development processes, in order to improve them and thus prepare a better future.

At a time when it is clear that software has become an essential component of all human activity, unrestricted access to publicly available software source codes, as well as qualified information on their evolution, is becoming an issue of digital sovereignty for all nations. The unique infrastructure that Software Heritage is building, as well as its universal approach, is an essential element to meet this challenge of digital sovereignty, while preserving the common dimension of the archive.

References

Abelson, H. (1984). *Structure and Interpretation of Computer Programs*. MIT Press, Cambridge.

Di Cosmo, R. (2019). Saving software development from the European copyright reform [Online]. Available at: https://www.dicosmo.org/MyOpinions/index.php?post/2019/04/17/Saving-software-development-from-the-European-copyright-reform.

Di Cosmo, R. and Nora, D. (1998). *Le Hold-up planétaire : la face cachée de Microsoft.* Calman-Lévy, Paris.

Shustek, L.J. (2006). What should we collect to preserve the history of software? *IEEE Annals of the History of Computing*, 28(4), 110–112 [Online]. Available at: https://doi.org/10.1109/MAHC.2006.78.

Software Heritage (n.d.) Software Heritage [Online]. Available at: https://www.softwareheritage.org.

Surveillance Capitalism

Christophe Masutti
   SAGE, Université de Strasbourg, France

In the 2000s, after more than a decade of over-financialization of the economy and post-Cold War globalization, capitalism again entered a complex crisis. Its structures and institutions depend essentially on the strengths and weaknesses of digital and network technologies. Strategic choices involving all societies are now being questioned: the transformation of data into capital and the concentration of knowledge, the emergence of a platform economy, profiling techniques and the intrusion into private life, marketing and the social selection of consumers.

Appearance

Surveillance practices are all the more acutely criticized because their regulation by the state apparatus tends to run aground on the shores of political trust and the contradictions of liberal democracy. Over the past 20 years, the multinationals of the digital economy (the GAFAMs in particular) have become the holders of a concentration of capital and technological innovations unprecedented in history. Their financial and political weight unbalances the equilibrium of power necessary for States in their role as regulators of capitalism, for the good of the people. In 2013–2014, the "Snowden revelations" demonstrated the alliances, in a vast enterprise of global espionage inside and outside the borders, between the highest authorities of the United States and these companies with the highest stock market values. All states have seized these technologies, cooperating with the firms and deploying as many tools for monitoring citizens as commercial opportunities allow.

As an expression of this pre- and post-Snowden context, surveillance capitalism is a form of capitalism based on surveillance practices aimed, on the one hand, at ensuring the hegemonic interests of these essentially American firms and, on the other hand, at intensifying the concentration of innovation, information and financial capital. These social monitoring practices, which apply to communications, consumer behavior and economic choices, cause serious infringements of privacy and collective and individual freedoms.

The expression appeared in 2014 in a dedicated issue of *Monthly Review*, under the pen of two authors, John Bellamy Foster and Robert Waterman McChesney (Foster and McChesney 2014), who proposed a historical and political approach to it. Their summary is part of the current critique of monopoly capitalism (Baran and Sweezy 1966) and shows the imperialist and militarist tendency that the "Snowden revelations" decisively crystallize. In turn, and without citing her predecessors, a former management researcher, Shoshana Zuboff, published between 2015 and 2019 a series of articles and a very significant book about surveillance capitalism (Zuboff 2019). She intends to denounce the doctrine that underlies the practices of multinationals in the digital economy and its harmful effects preventing the proper regulation of capitalism.

Two criticisms, two methods

Foster and McChesney's work in the *Monthly Review* and other essays question the changes in modern capitalism and the information society. For them, surveillance capitalism is first and foremost an articulation between the economic balances of domestic and foreign markets from the perspective of the United States. Suffering from the permanent imbalance between supply and demand in a situation of overproduction since the Second World War, this country has set up a military, industrial and financial complex: (1) to ensure demand externally because of the force that imposes hegemony and, internally because of marketing technologies and the shaping of mass consumption; (2) to compensate for the inevitable drops in profit rates suffered after the crisis of the 1970s by over-financing the economy; (3) to set up a global surveillance system, the one rightly demonstrated by the "Snowden revelations", in order to perpetuate this game of interests. The combined action of the IT giants and the authorities (such as the NSA) makes it possible to perpetuate war (and cyberwar), to define the outlines of mass consumption by profiling consumers and monitoring citizens (thanks to data brokers who cooperate with both companies and the state), to master the best stock market trading technologies and, in short, to conduct a political economy of total surveillance in the service of the perpetuation of interests.

Shoshana Zuboff, on the other hand, did not draw the same lessons from the "Snowden revelations". She focused on the hegemony of American firms, without questioning the underlying political economy. Instead, she conducted a very deep analysis of capitalist practices and did not hesitate to describe actors such as former

Google boss Eric Schmidt as "surveillance capitalists". This essentialist conception arises from the fact that she analyzed the vital economic model of firms such as Google or Facebook which, in order not to sink in the face of shareholder pressure, have been leading a race to "dispossess" individual, behavioral and relational daily experience since the early 2000s. They are appropriating what was previously outside the market cycle. The prediction market thus created is based on Big Data and machine learning. Its objective is to modify behavior and adapt it to the offer. Surveillance capitalism thus represents a danger for liberal democracies, both from the point of view of individual freedoms (privacy and individual autonomy) and collective freedoms (deciding our future collectively). We are witnessing the emergence of a new instrumental power, which Shoshana Zuboff calls the Big Other, a non-centralized and private power (unlike Orwell's *Big Brother*), which subjects all action to conformity with what has been anticipated. She sees the Snowden episode as the unveiling of this Big Other, a power accessible to states by arrangement with "surveillance capitalists".

Surveillance society

Shoshana Zuboff's surveillance capitalism is above all a set of coercive behavioral systems that stem from a dysfunctional and "brutal capitalism", which "threatens society as much as capitalism itself" (Zuboff 2019, p. 197). She argues for a hypothetical "collective social action" that, far from challenging the mechanisms of capitalism that have brought about precisely this form of proletarianization of individuals, would put back on track a political power capable of effective regulation.

However, business models based on the computerized exploitation of personal data are already very old, at least as old as the mainframe computers and databases of the late 1960s (Masutti 2020). Since then, the commodification of data has continuously created economic opportunities depending on the advancement of technology and the potential for profitability, especially in the field of marketing and credit agencies (Lauer 2017). Researchers specializing in surveillance studies have shown how these activities produce differentiated worlds and influence individual and collective decisions (nudge theory). But this social monitoring does not only concern consumption. It also concerns the manipulation of opinions in politics, as the Facebook–Cambridge Analytica scandal has shown. It concerns the most discriminatory aspects of social sorting by algorithmic technologies (O'Neil 2018). It also concerns the work, beyond the simple exploitation of information on productivity (management), on the implementation itself of platform models of intermediation between production and consumption (Srnicek 2018): uberization of services and digital labor (Cardon and Casilli 2015). This "platform capitalism" is strongly criticized through the notion of solutionism (Morozov 2014). Finally, if states apply these technologies (facial recognition, predictive justice, personal data processing, etc.) for control purposes, this amounts, in addition to the liberticide issues, to intercalating a platform economy between governments and those that are administered.

These orientations of the economy are encompassed in the idea of surveillance capitalism. Critiquing it does not only mean choosing between a radical critique of capitalism centered on hegemony and power games, and a social-liberal approach outside any historical context and without any real alternative to the exploitative paradigm of capitalism. New paths can be opened up, based on an analysis of the social, anthropological and cultural uses of surveillance, its technological substratum, and the economic and political changes of which they are the vectors.

References

Cardon, D. and Casilli, A. (2015). *Qu'est-ce que le Digital Labor ?* INA, Bry-sur-Marne.

Foster, J.B. and McChesney, R.W. (2014). Surveillance capitalism: Monopoly-finance capital, the military-industrial complex, and the digital age. *Monthly Review*, 66, 1–13.

Lauer, J. (2017). *Creditworthy: A History of Consumer Surveillance and Financial Identity in America*. Columbia University Press, New York.

Masutti, C. (2020). *Affaires privées : aux sources du capitalisme de surveillance*. C&F Éditions, Paris.

Morozov, E. (2014). *Pour tout résoudre, cliquez ici ! L'aberration du solutionnisme technologique*. FYP, Paris.

O'Neil, C. (2018). *Algorithmes : la bombe à retardement*. Les Arènes, Paris.

Zuboff, S. (2019). *The Age of Surveillance Capitalism: The Fight for a Human Future at the New Frontier of Power*. Public Affairs, New York.

Surveillance Studies

Christophe Masutti
SAGE, Université de Strasbourg, France

Surveillance studies

Like science and technology studies (S&T) or cultural studies, surveillance studies are both transdisciplinary and transnational. It is a scientific approach that aims to analyze surveillance as a "social structuring process" (Ball *et al.* 2012, p. 7). It broadly views surveillance as the practices of gathering information and interpreting that information. These practices are most often technology dependent; they are present at all levels of social organization in the economic, governmental, administrative or environmental sectors. However, this broad definition of surveillance does not describe the object of surveillance studies: it is defined according to the authors and the disciplinary approach (sociological, historical, legal, etc.) or according to the subject of study (e.g. delinquency, population control, communication technologies and databases, marketing). In this last sense, surveillance studies consider these fields from the perspective of monitoring.

The advancement of modern information and database technologies, the exploitation of personal data, and the impact of these activities on privacy justify the special attention paid to surveillance today. Even though surveillance studies may well concern pre-industrial periods or even earlier, most of their academic production in the social sciences can be correlated with the emergence of the "new economy" of the 1990s, the rise of digital network infrastructures and the platform economy. Electronic surveillance (Lyon 1994) then took a prominent place, seen as a decisive element of social change in the late 20th century, especially when studied from the perspective of everyday life in Western societies (Staples 2014). After the 9/11 attacks and the "Snowden revelations" in 2014, surveillance studies have focused on electronic surveillance but without making digital technologies and governmentality the central objects of their concerns. Even though these events tend to increase the relevance of surveillance studies today, in the public eye, they are mostly about a critical approach to surveillance in terms of its social, moral, ethical and political impact. This is why variations have appeared as sub-fields of study linked to other studies. We can cite studies based on the question of gender or social discrimination (the notion of social sorting as a paradigm and the algorithmic biases that result from it, for example, considering a person of color to be more or less likely to commit delinquent acts, given the statistical data), or undersurveillance (the act of monitoring the monitors, e.g. the phenomenon of home videos in the context of police violence), or cross-country studies (e.g. the analysis of surveillance policies and their consequences on social movements and democracy).

There are several institutional indicators that identify the scientific community of surveillance studies. The academic profile of the members varies according to the disciplinary positions held in the institutions. However, surveillance studies courses are taught in several universities and are supported by specialized departments. Of particular note is the Surveillance Studies Center, founded in 2009 in the Department of Sociology at Queen's University in Kingston, Canada, and directed by David Lyon. The relatively large number of publications in this field has made it stand out from other disciplines to such an extent that, in the early 2000s, the need was felt to create the typical forums of a distinct research community. Since 2000, international symposia have been held every 2 years. Prizes are awarded for outstanding work (articles, monographs, works of art). The first conference was the founding act of the journal *Surveillance and Society* (the first issue was published in 2002). Finally, an international non-profit association was created in early 2007, the Surveillance Studies Network (SSN), whose objective is to promote surveillance studies[7]. While the SSN is registered in the UK, it is remarkable that the researchers who claim to be part of the network and who most often write for the journal are mostly American, British and Canadian (Castagnino 2018). To find publications whose epistemological content provides insight into the SSN, and thus surveillance studies in general, authors such as David Lyon, Gary T. Marx, Oscar H. Gandy, David Murakami Wood, or Torin Monahan are recommended, as well as a collection of "foundational" texts (Ball *et al.* 2012) and a textbook (Hier and Greenberg 2007).

Surveillance studies, however, has a history that is not limited to its institutionalization. Surveillance studies can be rooted in a conceptual heritage strongly influenced by European thinking on politics (Max Weber), the critique of technology (Jacques Ellul), the history of institutions and the critique of power (Michel Foucault). For the most part, this heritage was mobilized in the United States in the 1970s, in conjunction with the first major public debates concerning the use of databases and the processing of personal data by public institutions or large corporations. During this period, legal scholars such as Alan Westin and Arthur R. Miller came to the forefront, undertaking a critical analysis of the computerization of society and laying the foundations for a definition of privacy that would quickly spread to Europe. In this context, sociologists started from a double assessment of the situation: on the one hand, they identified surveillance as an activity aimed essentially at gathering information and influencing individuals and, on the other hand, the prevalence of surveillance in society was perceived, through threats to privacy, as a more global threat to social interactions and democracy. The author of reference in surveillance studies is the sociologist James B. Rule. Rule, who initiated an approach exclusively centered on a definition of surveillance as "by surveillance we mean any systematic attention to a person's life aimed at exerting influence over it" (Rule *et al.* 1983). This action on everyday life is analyzed according to "total surveillance" heuristics (Rule 1974). Inspired in the background by the Orwellian dystopia, like many authors of this period, Rule developed a critique of social control, of the predictive analysis of behavior and, in line with Michel Foucault, analyzed the relationship between automated data processing and social control, obedience, institutions and power.

Whether it is 1984 or the interpretation of Foucault's work on Bentham's panopticon (Foucault 1975), this conceptual trace is always questioned. In 1994,

---

7 Available at: https://www.surveillance -studies.net.

Lyon devoted an entire chapter to it in his book on the emergence of the surveillance society (Lyon 1994). He shows the limits of the globalizing interpretation of the panopticon (Foucault), the relationship between surveillance and control (Deleuze), and even the critique of the postmodernists (Giddens). For him, even if "the Panopticon offers a powerful and compelling metaphor for understanding electronic surveillance", it does not allow for a summary of contemporary surveillance, in particular the relationship between consumerism and social control. The critique is well taken. Surveillance studies are now developing a whole nuanced thematic apparatus, trying to escape the normative injunctions of a single definition of surveillance (and ontologically negative, harmful to society). This is the case, for example, of Gary T. Marx, who proposes a way of thinking about the "new surveillance" in the face of the increasing complexity of the digital economy. He works on an evaluation framework that not only analyses the scope of surveillance, but also incorporates an assessment of the context in which a technology is applied, the constraints that make it weak or powerful, the legal and cultural framework that determine its acceptance, rejection or relevance (Marx 2002). Surveillance then obtains an ambivalent status (Castagnino 2018), inherently neither negative nor positive, even if publications of surveillance studies often aim to demonstrate its dangers. On this point, Marx warns in his synthetic article on surveillance studies in the *International Encyclopedia of the Social and Behavioral Sciences* (Marx 2015):

> Surveillance practices need to be understood within specific settings in light of history, culture, social structure and the give and take of interaction, and require the appreciation (if not necessarily the welcoming) of the ironies, unintended consequences, and value conflicts that limit the best laid plans.

The history of surveillance studies is undoubtedly unique in that it creates tension in the institutionalization of the field and the limits of a disciplinary approach to the same object. However, surveillance studies coherently integrate different empirical concerns and theoretical postures. Going beyond the idea that surveillance is only suffered by individuals and imposed by the authorities or the consumerist economy, some pioneers of surveillance studies are now considering the idea that a form of surveillance culture is currently awakening. That is, surveillance as it is practiced and felt in everyday life, from an anthropological, behaviorist point of view, as the result of everyone's acceptance and active participation, to the point of changing our worldview and becoming a way of life (Lyon 2018). Surveillance reveals itself in such a multifaceted way that the only ways to apprehend it are at the boundaries of disciplines.

References

Castagnino, F. (2018). Critique des surveillances studies : éléments pour une sociologie de la surveillance. *Déviance et société*, 42, 9–40.

Foucault, M. (1975). *Surveiller et punir : naissance de la prison*. Gallimard, Paris.

Hier, S.P. and Greenberg, J. (2007). *The Surveillance Studies Reader*. Open University Press, Maidenhead.

Lyon, D. (1994). *The Electronic Eye: The Rise of Surveillance Society*. University of Minnesota Press, Minneapolis.

Lyon, D. (2018). *The Culture of Surveillance: Watching as a Way of Life*. Polity Press, Cambridge.

Marx, G.T. (2002). What's new about the "new surveillance"? Classifying for change and continuity. *Surveillance and Society*, 1(1), 9–29.

Marx, G.T (2015). Surveillance studies. *International Encyclopedia of the Social and Behavioral Sciences*. Elsevier, Amsterdam.

Rule, J.B., McAdam, D., Stearns, L., Uglow, D. (1983). Documentary identification and mass surveillance in the United States. *Social Problems*, 31(2), 222–234.

Staples, W.G. (2014). *Everyday Surveillance: Vigilance and Visibility in Postmodern Life*. Rowman & Littlefield, Lanham.

T

Training

Samia Ghozlane
Grande École du Numérique, Paris, France

The emergence of bootcamps

In 2020, digital training in the broadest sense, from the development of basic skills to the acquisition of advanced technical expertise, was considered both a crucial economic issue for competitiveness, sustainability and company development, and a societal issue for the inclusion of all citizens in a hyper-connected world.

Over the past 10 years, the digitalization of the economy has imposed a profound and sudden transformation in companies, whatever their sector of activity, with new processes of design, development and product marketing or service offers. Engineering schools and universities, which provide talented and sought-after profiles, have no longer been able to fully satisfy recruitment needs. Their selective, long and sometimes costly curricula make them inherently inflexible, lacking in reactivity and thereby unable to adapt quickly and permanently to the evolving demand for digital skills. For example, the lifespan of a computer language, which was 25 years in the 1990s, has now been reduced to a few years. All companies and organizations are undergoing their digital transformation and are looking for new digital talent to spearhead their competitiveness. Thus, in 2018, the digital sector was the major recruiter in France. With more than 60,000 recruitments, it represented one-third of all job creations in France (Talents du numérique 2019).

It is in this context of the battle for talent that coding bootcamps have emerged as an alternative answer enabling the acquisition of key skills in record time, while increasing the employability of graduates (Cathles and Navarro 2019). Intensive training programs in web development and programming, bootcamps emerged in the United States in the early 2010s. Since then, they have expanded all over the world. For example, in 2019, the US market for coding bootcamps grew by 4.38% (Cathles and Navarro 2019). A training program in a bootcamp is characterized by intense learning, which lasts only a few weeks, and by an immersive dimension that promotes the rapid acquisition of digital skills in a sector that evolves very quickly. More accessible in terms of price than a university, particularly in the United States where annual fees run into the tens of thousands of dollars, they allow students to enter the job market quickly. Considered as real skill accelerators, bootcamps appeal to tech giants as well as start-ups and industrial companies, which recruit operational profiles with much sought-after skills.

The growth of coding bootcamps shows that they should be part of a process of building a pool of digital talent that will benefit all sectors, rather than a one-off investment in the particular tech sector (Cathles and Navarro 2019). Thus, one of the main obstacles to the digital transformation of the economy and the growth of French start-ups is related to their inability to recruit at all degree levels. Unfilled jobs are an obstacle to the competitiveness of companies and countries. These jobs require skills that can be acquired not only in universities but also in bootcamps.

The involvement of public and governmental action

The first government or public policy initiatives emerged as early as 2015. Indeed, while digital professions are one of the main levers for job creation and social inclusion, not all job offers in digital professions find takers, due to a lack of suitable profiles.

In March 2015, U.S. President Barack Obama announced the launch of the TechHire Initiative. This is a national program aimed at developing the tech business sector locally by creating pools of digital talent trained in bootcamps. The TechHire program has been rolled out across the United States because of an offensive and proactive strategy by the U.S. government through federal grants and the support of large private sector companies. Innovative and highly articulated to meet the needs of local businesses, this program has made it possible to train highly motivated individuals in a few weeks, without academic prerequisites, to fill jobs requiring technical skills by putting them directly in touch with their future employer.

In Europe, as early as 2012, the European Commission launched several public policy initiatives to modernize education and training, such as Rethinking Education, which aims to invest in skills for better socioeconomic outcomes. In 2013, the Commission launched the Opening Up Education initiative, which aims to seize the opportunities of the digital revolution in education and training. In 2016, the European Commission launched the Digital Skills and Jobs Coalition to help European citizens better shape their private and professional lives in the age of digitalization. As this initiative is part of the European digital strategy, each member country was encouraged to set up its own network of actors, whose shared actions would help mobilize citizens on the urgency of the subject.

In France, in February 2015, the President of the Republic, François Hollande, announced the creation of a *Grande École du Numérique*, responsible for disseminating its training throughout the territory for young people without jobs or qualifications, to unite them in a national network of digital training and to promote recognition through a label.

> *The Grande École du Numérique* is thus intended to be a utopia in the sense that the etymology gives to this word: *u-topos*, a marvelous place that exists nowhere, deploying and withdrawing everywhere in the territory according to training needs and the capacity to respond locally. But it is a realistic utopia, since this place already exists, everywhere, evanescent, pulsating, here and there in the territory thanks to the audacity of a few training pioneers. (Distinguin *et al.* 2015)

In 2020, the *Grande École du Numérique* united 500 training courses in digital professions, accessible to all, with its GEN label. This label embodies the threefold ambition of the *Grande École du Numérique*: an economic ambition, by training for the professions of tomorrow to meet the growing need for digital skills in the job market; a social ambition, by promoting the inclusion of people who are far from employment and training to make digital technology an opportunity for everyone; a territorial ambition, by allowing a balanced and coherent distribution of the training offer throughout the country.

The *Grande École du Numérique* courses are accessible without academic, economic or social discrimination. The recruitment of learners is mainly based on criteria related to motivation. The courses offer pedagogical approaches inspired by coding bootcamps that allow for practical learning, such as project-based teaching and peer-to-peer learning. By the end of 2020, nearly 33,000 people had been trained throughout France because of the GEN network.

References

Cathles, A. and Navarro, J.C. (2019). Disrupting talent – The emergence of coding bootcamps and the future of digital skills [Online]. Available at: https://publications.iadb.org/publications/english/document/Disrupting_Talent_The_Emergence_of_Coding_Bootcamps_and_the_Future_of_Digital_Skills_en_en.pdf.

Digital Skills and Jobs Coalition (2016). [Online]. Available at: https://ec.europa.eu/digital-single-market/en/digital-skills-jobs-coalition.

Distinguin, S., Marquis, F.-X., Roussel, G. (2015). La Grande École du Numérique, une utopie réaliste [Online]. Available at: https://www.grandeecolenumerique.fr/sites/default/files/6.2_rapport.pdf.

Obama White House (2015). TechHire Initiative [Online]. Available at: https://obamawhitehouse.archives.gov/issues/technology/techhire.

Talents du numérique (2019). Numérique, formations et emplois : enquête 2019 [Online]. Available at: https://talentsdunumerique.com/actu-informatique/enquete-formations-et-emplois-numerique-2019.

W

Web 2.0

Zhenfei Feng
GERiiCO, Université de Lille, France

Web 2.0: origins and the seven characteristics

The term and concept of Web 2.0 was first mentioned by Darcy DiNucci in her article "Fragmented Future". Tim O'Reilly and Dale Dougherty made it popular at a conference between *O'Reilly Media* and *MediaLive International*. Since then, the concept of Web 2.0 has spread at a rapid pace around the world. O'Reilly points out that "Web 2.0 has no clear boundaries". Indeed, the concept remains very fuzzy and difficult to define because (1) it is not a specific new technology; (2) the understanding of Web 2.0 is very diverse; and (3) Web 2.0 is a broad topic that encompasses a variety of concepts and methodologies.

It cannot be summarized as a technology or a technique. Rather, it is a set of techniques, functionalities and uses that make what appears (by difference) as "Web 1.0" evolve toward interactivity, sociability (putting people in touch with each other, and not only content), information sharing (collaborative) and simplicity of use that does not require technical and computer knowledge for Internet users.

Faced with the difficulty of defining the term, O'Reilly and his colleagues suggest seven major characteristics of Web 2.0:

– a vision of the Web as a platform: a platform is a service that acts as an intermediary in accessing information, content and services published or provided by third parties;

– harnesses collective intelligence;

– power in data: the accessibility of data allows the creation of new applications combining several data sources; database management is the core business of Web 2.0 companies;

– the end of update cycles: Web 2.0 software is no longer released in a succession of final versions, but according to a long testing period, the so-called "perpetual beta", in which the effectiveness of features can be evaluated in real time;

– lightweight programming models;

– software freedom for the PC (personal computer); in addition to the PC, various devices (phones, tablets, reading lamps, etc.) allow users to access content and transmit data in real time;

– the enrichment of user interfaces.

Ajax

Ajax (Asynchronous JavaScript And XML (eXtented Mark up Language)) is the computer architecture allowing the Web to become more interactive, in short to becoming Web 2.0. It allows applications and interactive dynamic websites to be constructed regarding the client, using various techniques added to Web browsers between 1995 and 2005. The term Ajax was introduced by Jesse James Garrett on February 18, 2005, in his article "Ajax: A new approach to Web applications" (Garrett 2005). Since then, Ajax has rapidly gained popularity.

Ajax is intended to perform rapid updates of the content of a Web page, without requiring any reloading visible by the page's user. Thus, Ajax makes it possible to move from an asynchronous Web to a synchronous Web. Ajax is not a new technology or a different language, but a set of existing techniques used in a new way. Using Ajax works on all common Web browsers: Google Chrome, Safari, Mozilla Firefox, Internet Explorer, Microsoft Edge, Opera, etc.

Web 2.0 applications and services

The services and applications that illustrate the fundamental concept of Web 2.0 are familiar to all: blogging and RSS, respectively, for publishing and aggregating information; Wiki for collaborative publishing; collaborative indexing by tags for describing documents with users' keywords; social bookmarking for sharing "favorite" websites; podcasting for on-demand broadcasting of audio and video content, including by subscription, etc. Because of this multitude of interactive services, Web 2.0 has become the world of "user-generated content". New types of publications have emerged, such as collaborative encyclopaedias and citizen journalism (journalism whose content is contributed by amateurs).

The blog, a kind of online diary, easy to use because it is a kind of turnkey website, is one of the major successes of Web 2.0 since its emergence in the late 1990s. The blog gives Internet users the skill and opportunity to produce content.

Wiki is a system that enables the creation, modification and archiving of pages written collaboratively within a website. It uses a markup language and its content can be modified using a Web browser. It is a content management software, whose implicit structure is minimal, while the explicit structure emerges according to user needs. A standard Web browser is sufficient for any user to create new pages or modify one in the Wiki website. The difficulty then lies in the moderation activity (which allows or rejects certain content according to rules defined by the community). A Wiki is not a website carefully designed by experts, but rather an "organic" site that is constantly evolving through ongoing collaborative creation. There are many Wiki-managed sites, among which the Wikipedia encyclopedia site is the most visited.

Really Simple Syndication (RSS) is a family of data formats used for Web content syndication. It is a resource for websites whose content is automatically generated based on their updates. The RSS format (or its competitor *Atom*) is mainly used to aggregate websites that frequently publish content and are regularly updated, such as Web feeds of blog posts, news, audio or video. RSS feeds are XML files that are often used to present the title and a summary or the full content of the latest news. Although RSS is a useful way to collect information online, the amount of information is growing and is never-ending. The popularity of RSS has been declining since 2010.

"Folksonomy" (or user indexing) emerges when users tag content. It allows users to index online resources using tags, so that they can more easily find them using their keywords. Tags can be used to manage, classify and describe online content.

In short, Web 2.0 is a set of services that allow users to become content producers and actors in their circulation thanks to the linking of people beyond the mere links between documents.

References

Blood, R. (2002). *The Weblog Handbook: Practical Advice on Creating and Maintaining Your Blog*. Perseus Publishers, Cambridge, MA.

DiNucci, D. (1999). Fragmented future. *Print*, 53(4), 32–33.

Kim, D.J., Yue, K.-B., Perkins Hall, S., Gates, T. (2009). Global diffusion of the Internet XV: Web 2.0 technologies, principles, and applications: A conceptual framework from technology push and demand-pull perspective. *Communications of the Association for Information Systems*, 24(1), 657–672.

O'Reilly, T. (2009). What is Web 2.0? Report, Radar.

Work

Sarah Abdelnour and Dominique Méda
IRISSO, Université Paris Dauphine – PSL, France

What status for workers on digital platforms?

New exchanges of goods and services via the Internet have been spreading since the end of the 2000s, using what we have come to call digital platforms, because of applications that make it possible to bring together suppliers and buyers of goods and services.

Diversity

The different types of digital platforms refer to very different activities, in terms of their nature (delivery, transport, cleaning, make-up, filling in online questionnaires, etc.), the time that users devote to them (from a few hours a month in the case of micro-work, for example, to a full-time professional activity), or the income they generate (constituting a tiny part of the user's income to the totality of the activity, without this income necessarily being sufficient). Moreover, for the same type of platform, users can vary greatly (in the case of meal delivery, for example: from students considering it as a sideline activity to undocumented workers subletting an account).

Digital platforms can also be distinguished according to whether they offer the exchange of goods (renting an apartment, buying a household appliance, a craft product, etc.) or the commissioning of work (delivery, cleaning, translation, transport, etc.). However, the distinction is not so clear-cut in practice, insofar as renting one's apartment or selling an object implies activities on the part of the supplier that could be assimilated to work (washing sheets, making an object, etc.). But with this nuance, the distinction is nevertheless useful, especially regarding the scope of the labor law to be applied to these activities.

Within work ordering platforms, a distinction can be made regarding crowdworking, literally "outsourcing to the crowd", used to refer to taking a job traditionally done by a specific agent (usually an employee) and outsourcing it to a large and loosely defined set of people in the form of an open call. Crowdwork consists of the performance of online, telecommuting, paid per job micro-tasks, rarely constituting the main job, on platforms such as *Amazon Mechanical Turk* (Barraud de Lagerie and Sigalo Santos 2018). Next to micro-work, the bulk of activities passing through platforms consists of the performance of more traditional work activities, such as transport or housework, whose demand and ordering go through digital tools. This includes platforms such as Uber and Deliveroo.

These are the ones that are most used and tend to correspond to a full-time activity for their users, who derive most of their income from them; they are also the ones that crystallize the most difficulties from the point of view of the status of the workers who use them and that have been the most talked about. Indeed, most of them claim to be mere "neutral intermediaries" whose purpose is to put service providers and service seekers (e.g. drivers and their potential clients) in touch with each other, without being either service providers (transport companies) or employers. This allows them to escape both the regulations and professional collective agreements as well as the obligations traditionally incumbent on the employer in the wage model, in particular all those relating to the respect of working hours and holidays, but also the payment of social contributions allowing the financing of social protection. Most often, the platform requires workers who wish to work "with" it to adopt micro-entrepreneur status, that is, self-employed, or small companies such as SASUs (*société par actions simplifiée* – single-member simplified joint stock company). The platform does not consider itself in any way as the employer of those it considers as its "partners" and is remunerated by taking a commission (often high, around 20–30%) on all transactions. Nevertheless, these companies impose a number of obligations on those who work "with" them, ranging from adhering to certain algorithmically determined routes to wearing certain outfits, as well as numerous written and unwritten rules, and failure to comply with these rules can result in disconnection (Rosenblat and Stark 2016).

Mobilizations

For some years now, this way of doubly circumventing the legislation in force has met with resistance from workers and challenges from the courts.

The most visible struggles in the media were initially those of the drivers (later relayed by those of the bike couriers). For the drivers, the advantage of working with the platforms was clearly based at the beginning on the hope of an increase in income, and by putting at a distance the boss and their supervision, often associated with salaries. This is particularly true for people who previously worked in transport or logistics, without strong job security. However, the drawbacks quickly became apparent when material conditions deteriorated (and the aid for setting up a business expired), competition between drivers increased (both processes leading to an increase in working hours) and the autonomy dreamed of and promised by the platforms proved to be extremely limited (Abdelnour and Bernard 2018).

As early as fall 2015, the first drivers' mobilizations were an opportunity to denounce deterioration in the conditions under which the profession was practiced, thwarting their hopes of social ascension. The drivers mentioned the drop in prices, the increase in commissions (which rose in 2016 from 20% to 25%), and the growing number of drivers, all of which contributed to their turnover and forced them to work long hours. The bonuses offered by the various platforms at the time of their implementation in France have also been removed. The turnover of the drivers must, moreover, be related to all the expenses incurred by them: car, insurance, petrol, contributions, taxes, accountant, or even fines. In reality, they have to work 13–14 hours a day, 7 days a week, to earn an income slightly higher than the minimum wage.

Challenges

The company Uber had applied to the Court of Justice of the European Union to be recognized as a digital platform (and therefore as a neutral intermediary), while rejecting any status as a transport company. The aim was to have the e-commerce directive applied, ensuring greater freedom of provision, and therefore less power for national authorities to set conditions on its activity, and not the regulations specific to transport, an area in which Member States enjoy greater prerogatives and can impose stronger restrictions on commercial activities. In 2017, the CJEU stated that the service provided by Uber was not merely an intermediation service, but an integral part of an overall service whose main element was a transport service, and was therefore a "service in the field of transport". Therefore, recognized as a transport company, Uber had to comply with all the regulations relating to transport (Nasom-Tissandier and Sweeney 2019).

In addition, numerous disputes relating to the status of platform workers have developed: the recent decisions of the social chamber of the French Court of Cassation (rulings of November 28, 2018 and March 4, 2020), reclassifying a delivery person and then a driver as employees, have finally reminded us that the platform workers considered were indeed employees and not self-employed, in accordance with the constant jurisprudence of the Court of Cassation, according to which salaried work is characterized by the existence of a subordination link, defined as the performance of work under the authority of an employer who has the power to give orders and directives, to control the execution thereof and to sanction breaches thereof (Gomes 2018). New decisions are expected in France but also in other places, for example, in California where a law has required the reclassification of platform workers as employees.

The French government has not, for the moment, transcribed these decisions into law. On the contrary, it has repeatedly tried to promote the self-regulation of platforms through optional charters to be signed between platforms and their workers. Earlier versions of these charters were considered unconstitutional by the Constitutional Council on several occasions, and the current provisions stipulate that platforms must offer their "partners" charters establishing a certain number of protections, but at the same time stipulate that these protections may not be considered as an indication of subordination. A senatorial bill proposed that workers on work platforms should be registered in the seventh book of the labor code, which would have enabled them to benefit from the protections to which they are entitled. It was rejected in June 2020[1].

References

Abdelnour, S. and Bernard, S. (2018). Vers un capitalisme de plateforme ? Mobiliser le travail, contourner les régulations. *La Nouvelle revue du travail*, 13.

Barraud de Lagerie, P. and Sigalo Santos, L. (2018). Et pour quelques euros de plus : le crowdsourcing de microtâches et la marchandisation du temps. *Réseaux*, 212(6), 51–84.

Gomes, B. (2018). La plateforme numérique comme nouveau mode d'exploitation de la force de travail. *Actuel Marx*, 63(1), 86–96.

Nasom-Tissandier, H. and Sweeney, M. (2019). Les plateformes numériques de transport face au contentieux. In *Les nouveaux travailleurs des applis*, Abdelnour, S. and Méda, D. (eds). PUF, Paris.

Rosenblat, A. and Stark, L. (2016). Algorithmic labor and information asymmetries: A case study of Uber's drivers. *International Journal of Communication*, 10(27), 3758–3784.

---

1 https://www.senat.fr/dossier-legislatif/ppl18-717.html.

Glossary

The abundance of terms related to digital technologies often leads to questions about their meaning. This glossary, proposed by the authors, is intended to usefully complement the entries in the book or clarify certain terms in the lexicon they have used.

Algorithm: In a numerical context, it is a sequence of instructions written by a human that is sent to a machine that will execute it.

ANSSI: *Agence nationale de la sécurité des systèmes d'information*, French National Agency for the Security of Information Systems.

Archiving: Recording on a medium that guarantees that the data will remain usable for a significant period of time, even if access is not necessarily rapid. The aim is rather not to lose these data, which is what backup allows.

Backup: The act of saving work from time to time and also in several places so that it is not lost, but also so that it can be found easily. This requires a mastery of the hierarchical tree structure of file systems, if we do not want to be dependent on the machine and those who programmed it.

BATX: This is the Asian counterpart of GAFAM.

CNIL: The *Commission nationale de l'informatique et des libertés* (French National Commission for Information Technology and Civil Liberties) was established by a law in 1978 and adapted to the growth of the Internet. It monitors and controls the use of computers to ensure that they comply with French law. In particular, it ensures that the data that circulate or are collected via the networks comply with individual freedoms, freedom of thought and the main principles of human rights. It is responsible for ruling on disputes or complaints submitted to it, and issuing an opinion on laws proposed by Parliament or the validity of files created by the administration. Employees, companies and private individuals can contact the CNIL, which provides practical and educational tools.

Computer science: Computer science is the science and technique of the representation of information of artificial or natural origin, as well as the algorithmic processes of collection, storage, analysis, transformation, communication and exploitation of this information, expressed in formal languages or natural languages, and carried out by machines or human beings, alone or collectively.

Computer virus: Software that compromises the security of a computer or a connected object, capable of duplicating itself and spreading automatically in a network to infect other computers or connected objects.

Cryptocurrency: Currency on a digital medium whose existence and ownership are proven by an encrypted digital transaction, carried out with a specific mechanism that technically prevents any modification of the transaction. One of the classic mediums is the blockchain.

Cyberphysics: A physical (hardware) system controlled by computers and software.

Cybersecurity: Maintaining the security of any computer system against external attacks.

Cyberspace: The term cyberspace refers to the communication space created by the global interconnection of computers. It is derived from the science fiction novel *Neuromancer*, written by William Gibson in the 1980s. It was popularized in the 1990s by Internet pioneers such as John Perry Barlow, who saw it as a new virtual territory where communities could interact without physical and legal constraints. Today, it is synonymous with the Internet and the Web.

Cyberwarfare: Fighting between countries through attacks on computers and networks.

Darknet: Darknets are virtual private networks (VPNs) that differ from the visible Web in that they allow a limited number of users to communicate in confidence, without leaving a trace, due to the anonymization of IP addresses and, often, the encryption of transmitted information. They have become the problem of a dark Internet, made up of hidden services and even criminal activities, bringing together all the actors, more numerous than the offenders, who wish to escape the siphoning of their personal data due to the generalized traceability on the Web.

Data science: A discipline that uses statistical, mathematical and computational methods to analyze data, especially megadata, in order to extract useful information.

Database: A set of digital data on a subject, organized in such a way that software can access it quickly and exhaustively. For example, an association's membership file can be created with a database.

DDoS: Distributed Denial of Service – an attack on a network by sending a huge number of messages simultaneously.

DEBIAN: A non-profit organization whose purpose is the development of operating systems based exclusively on free software. It was launched in 1993 by Ian Murdock with the support of the Free Software Foundation. This organization, which in 2020 gathered about a thousand developers, is characterized by a strong commitment to the principles of free software, as spelled out in a social contract and by a democratic governance stipulated in a constitution.

Deep learning: The ability of an algorithm to "digest" huge amounts of data in order to extract rules, patterns and elements that can answer a problem, from what the machine has assimilated and processed as input. This technique is part of a more general set called artificial intelligence.

Digital: According to the French dictionary, digital simply means "related to fingers". The translation of the English word into French is *numérique*, and the translation of the English word digit into French is *chiffre*.

Digital: All scientific and technical disciplines, economic activities and societal practices based on the processing of digital data.

Digital divide(s): The notion of digital divide, often used in the singular, is in fact multiple.

Digital frugality: A conscious choice to reduce digital usage as much as possible and use tools and services that consume less energy.

DNS (Domain Name System): An online service that matches a URL[1] with an IP address (A.B.C.D.). Beyond the technical aspect, the main server, for a long

---

1 For example machin.fr.

time managed only by the United States, has recently given way to other countries. In any case, having to rent and not being able to buy a domain name provides a significant source of money compared to the cost of managing domain names.

Domain name: see *DNS*.

Durability: This is the fact that the data in a file is still usable some time after it has been recorded. Often, durability applies to quite long periods of time, but the technological evolutions are rather fast; following Moore's law, it is necessary to worry about it permanently. We have to think about the durability of physical formats (what about floppy disks?), but also of file formats. We recall the unilateral decision of Microsoft, which knowingly stopped managing the MS-Publisher format, causing thousands of teachers who had used this software to lose thousands of working hours.

Encryption: The digital encryption of a message using a mathematical algorithm, such that it is believed that it is not possible to decrypt the message in a humanly reasonable time.

FAIR: Findability, accessibility, interoperability and reusability of data. Data whose identification, standardized description, technical or legal access conditions and type of licence facilitate their availability and use by interested parties.

Fairness of online services: In contrast to the GAFAM's terms of use, which are designed not to be read before they are accepted, a fair online service requires that the conditions under which your data and traces will or will not be recorded and used are clearly, easily and intelligibly explained and implemented within a valid legal framework (see Dima Yarovinsky's art work *I Agree*).

FALFAC: French acronym for *facile à lire, facile à comprendre* (easy to read, easy to understand) – the desire to enable everyone to understand what is written on the Internet, particularly in terms of digital administration.

Forking: This is a term used in software development parlance to refer to the act of creating new software from the source code of existing software. A fork is usually initiated when a personal, political or technical disagreement arises in a community of developers. This action is common in the free software world because the licenses allow the use, study, modification and redistribution of the source code:

– fracture by unavailability of access (infrastructure);

– fracture due to lack of material (financial aspect).

HTML (HyperText Markup Language): A computer language invented by Tim Berners-Lee to represent Web pages and especially hypertext links. Clicking on a location in a browser opens another Web page. It is a way of describing information that is no longer sequential, as it has been for more than 6,000 years, and that potentially changes our way of thinking and living.

Illectronism: Constructed from the words illiteracy and electronics, this term describes a lack of digital literacy and skills. Contrary to illiteracy, illectronism does not only concern certain categories of population, but all ages and sociocultural levels.

INRIA: Created in 1967, INRIA is a French public scientific and technological establishment that specializes in mathematics and computer science, under the dual supervision of the Ministry of Higher Education, Research and Innovation and the Ministry of the Economy and Finance.

IP: A 4-byte address, in the form A.B.C.D., A, B, C and D, ranging from 0 to 255. This address is unique on the global Internet network and allows you to be uniquely identified. Version 4 (V4), which contains a limited number of available addresses compared to the needs, will give way to a V6 capable of managing many more addresses.

IRI (Internationalized Resource Identifier): A universal resource identifier. It takes into account the characters used by the different languages of the world grouped in a universal directory.

Malware: Malicious program introduced into a computer or a connected object, for example by a Trojan horse.

Metadata: A set of information not apparent in the file at first sight, but present inside. For example, for an image, the date, time and GPS position where the photo was taken, but also, for a text, the name of the person who registered the software license or their keyboard dynamics. Metadata are often written implicitly in the files, without the user being explicitly informed. A technical manipulation must be done to read them, often simple (go to a special menu like "properties") but unknown or sometimes more complex.

Modified (augmented or diminished) reality: The ability of a video device to incorporate digital elements as "superimposed" on the reality that is being filmed and that the human eye naturally perceives. This has applications in medicine, for example during an operation, when the surgeon sees additional elements in their helmet. It can also be virtual elements placed in a landscape, such as the Pokémon in the game *Pokémon Go*.

NBIC: The progress of AI should be placed in a broader context, that is, of NBIC (for nanotechnology, biotechnology, information technology and cognitive science).

NFC (Near Field Communication): This is a global standard initially developed in 2004 by Nokia, Philips and Sony, a clear convergence between the worlds of telecommunications, consumer electronics and computing; NFC corresponds to one of the 12 RFID standards, a technology known since the 1940s. It came into its own with the integration of a tag reader in a cell phone which, by "touching" an object with a tag, makes it possible to obtain information, write information (in the tag or to a social site), emulate a payment card, open a door or make secure exchanges in P2P mode (peer-to-peer). The NFC standard (ISO/IEC 14443 and Felica) includes three basic operating modes: reading and writing a tag, card emulation (payment access cards) and peer-to-peer. An international organization dedicated to NFC (in addition to the recognition of the standard by ETSI and IEEE) has more than 130 members in 2011: the NFC Forum, which validates new versions of the standard[2].

Phishing: A way to get a user to click on a seemingly innocuous link that can actually allow a malicious program onto their computer.

QR code: The QR code is a (free) 2D code that can be read for free by a mobile phone; a URL or 4,296 characters (compared to about 10 for a traditional 1D barcode) can be easily associated with it. The QR code has been in open source since 1999, the flashcode is the proprietary version of French mobile operators.

Quantum: Quantum physics, along with optics and biology, is one of the avenues for building faster and faster computers. The specificity of quantum physics lies in the fact that if we manage to make it really work (which was still not the case in 2020), the "jump" in speed could be spectacular and "break" an encryption key. What takes, for example, 40 years now, without a flaw, could then take four minutes. The face of world security would be changed! It is important to stay informed on this issue.

Security protocol: A set of rule exchanges that allow for the exchange of messages that cannot be intercepted or corrupted, or any other function that requires computer security.

---

2 See: www.nfc-forum.org.

Serendipity: Meaning to make a discovery by chance when one is initially looking for something else. It finds its full meaning in the field of information and communication technologies. Browsing the Web is most often a matter of discovering, with the help of a mechanical click, information that we were not necessarily looking for but that will prove useful.

SOLID: A personal digital sovereignty project set up by Tim Berners-Lee, inventor of the Web, in the face of his disapproval of the mechanisms of dependence and manipulation of users set up by the hegemonic Web companies and those practicing the same business model.

Tags: Real-world labels that can be read by the end-user's mobile phone; these tags can be two-dimensional barcodes, radio frequency identification (RFID) tags such as NFC (Near Field Communication), sound tags or even invisible tags, pattern recognition tags such as Google or SNAPnSEE, Tokidev tags such as the *Spin Off* from MBDS.

Traces: Everything you do, say and write when you surf the Internet, including dates, times and places, as well as the characteristics of your equipment and your IP address, is recorded, often without your knowledge, by the servers.

Tracker: A program that allows a user's actions or location to be tracked covertly.

Trackers: Special categories of "bugs"; trackers are software embedded in mobile applications. They are part of the application's functioning, notably for sending notifications, and are also used to send advertising messages. Installed without the user's knowledge when an application is downloaded, trackers collect a large amount of personal data.

Trojan horse: A seemingly legitimate software that allows malicious features to be installed on a computer or connected object, for example a virus.

Unpatched vulnerability: A vulnerability identified by users of a computer system that has not yet been addressed. An unpatched vulnerability can be exploited for malicious purposes.

URI (Uniform Resource Identifier – short form: universal identifier): A name that conforms to an Internet standard, which allows an abstract or physical resource to be uniquely and permanently identified on the Internet, for example, a universal address is a type of universal resource identifier.

URL (Uniform Resource Locator): This is the address of a resource located on the Web. For example, the CNIL website has the URL: https://www.cnil.fr.

Virtual reality: In this case, it is no longer a question of filming reality and adding elements to it. Everything is calculated by the machine according to external parameters, such as the gyroscopic effect when you turn your head, a movement that is in fact totally calculated. The problem is that the computing power is such that it may soon be possible to fool a human brain.

Wikipedia: This system illustrates forms of collaboration, cooperation and online production based on new procedures inspired by democratic principles, such as the regulation of conflicts through discussion (Cardon, Levrel). It revisits the determinants of giving through the processes of voluntary contributions and the creation of a common good.

XML (eXtended Markup Language): This is a way of describing data that are flexible and easy to implement, both for writing and reading.

List of Authors

Sarah ABDELNOUR
IRISSO
Université Paris Dauphine – PSL
France

Serge ABITEBOUL
INRIA
Paris
France

Jean-Pierre ARCHAMBAULT
EPI
Villejuif
France

Jacques BAUDÉ
EPI
Paris
France

Corinne BAUJARD
CIREL
Université de Lille
France

Gérard BERRY
Collège de France
Aigaliers
France

Thierry BERTHIER
Hub France IA
CREC ESM
Saint-Cyr Coëtquidan
France

Laurent BLOCH
Institut de l'iconomie
Paris
France

Bruno BOIDIN
CLERSE
Université de Lille
Villeneuve d'Ascq
France

Gérard DE BOISBOISSEL
CREC ESM
Saint-Cyr Coëtquidan
France

Dominique BOULLIER
CEE
Sciences Po
Paris
France

Michel BRIAND
IMT Atlantique
Brest
France

Joana CASENAVE
GERiiCO
Université de Lille
France

Marie CAULI
Université d'Artois
Arras
France

Julien CEGARRA
SCoTE
Université de Toulouse
Albi
France

Aline CHEVALIER
CLLE
Université Toulouse – Jean Jaurès
France

Francis DANVERS
CIREL-PROFEOR
Université de Lille
France

Bruno DEFFAINS
Université Paris 2 Panthéon-Assas
France

Frédéric DEHAIS
ISAE-SUPAERO
Université de Toulouse
France

Jean-Paul DELAHAYE
Université de Lille
France

Laure DELRUE
University libraries
Université de Lille
France

Roberto DI COSMO
INRIA
Paris
France

Gilles DOWEK
INRIA
ENS Paris-Saclay
France

François ELIE
ADULLACT
Montpellier
France

Chantal ENGUEHARD
LS2N
Université de Nantes
France

Laurence FAVIER
GERiiCO
Université de Lille
France

Zhenfei FENG
GERiiCO
Université de Lille
France

Samia GHOZLANE
Grande École du Numérique
Paris
France

Manel GUECHTOULI
IPAG Business School
Nice
France

Jean-Yves JEANNAS
AFUL
Université de Lille
France

Philippe LE GUERN
CRAL-EHESS
Université Rennes 2
Angers
France

Fabien LOTTE
INRIA Bordeaux Sud-Ouest
Talence
France

Jean-Michel LOUBES
Institut de Mathématiques
ANITI
Université de Toulouse
France

Valèse MAPTO KENGNE
Université de Yaoundé I
Cameroon

Christophe MASUTTI
SAGE
Université de Strasbourg
France

Dominique MÉDA
IRISSO
Université Paris Dauphine – PSL
France

Ilham MEKRAMI-GUGGENHEIM
CYBERELLES
Paris
France

Vincent MEYER
Université Côte d'Azur
Nice
France

Serge MIRANDA
Université Côte d'Azur
Nice
France

Christophe MONDOU
ERDP-CRDP
Université de Lille
France

Widad MUSTAFA EL HADI
GERiiCO
Université de Lille
France

Jordan NAVARRO
EMC
Université Lumière Lyon 2
France

Fabrice PAPY
Université de Lorraine
Nancy
France

François PELLEGRINI
LaBRI
Université de Bordeaux
Talence
France

Nicolas PETTIAUX
Collège Saint-Hubert
Auderghem
Belgium

Edwige PIEROT
Aix-Marseille Université
Marseille
France

Nathalie PINÈDE
MICA
Université Bordeaux Montaigne
Pessac
France

Jean-Pierre PRUVO
CHU
Université de Lille
France

Julien ROCHE
University libraries
Université de Lille
France

Sophie SAKKA
LS2N
Centrale Nantes
France

Mylène SANCHIZ
CERCA
Université de Poitiers
France

Juliette SÉNÉCHAL
Université de Lille
France

Sébastien SHULZ
LISIS
Paris
France

Guy THUILLIER
LISST
Université Toulouse – Jean Jaurès
France

Ismaïl TIMIMI
GERiiCO
Université de Lille
France

Serge TISSERON
Université de Paris
France

Ornella ZAZA
Aix-Marseille Université
Aix-en-Provence
France

Éric ZUFFEREY
Consultant
Fribourg
Switzerland

Index

Other titles from

in

Science, Society and New Technologies

2022

AIT HADDOU Hassan, TOUBANOS Dimitri, VILLIEN Philippe
Ecological Transition in Education and Research

CARDON Alain
Information Organization of The Universe and Living Things: Generation of Space, Quantum and Molecular Elements, Coactive Generation of Living Organisms and Multiagent Model (Digital Science Set – Volume 3)

ELAMÉ Elosh
Sustainable Intercultural Urbanism at the Service of the African City of Tomorrow (Territory Development Set – Volume 1)

KAMPELIS Nikos, KOLOKOTSA Denia
Smart Zero-energy Buildings and Communities for Smart Grids (Engineering, Energy and Architecture Set – Volume 9)

2021

BARDIOT Clarisse
Performing Arts and Digital Humanities: From Traces to Data (Traces Set – Volume 5)

BENSRHAIR Abdelaziz, BAPIN Thierry
From AI to Autonomous and Connected Vehicles: Advanced Driver-Assistance Systems (ADAS)
(Digital Science Set – Volume 2)

DOUAY Nicolas, MINJA Michael
Urban Planning for Transitions

GALINON-MÉLÉNEC Béatrice
The Trace Odyssey 1: A Journey Beyond Appearances
(Traces Set – Volume 4)

HENRY Antoine
Platform and Collective Intelligence: Digital Ecosystem of Organizations

LE LAY Stéphane, SAVIGNAC Emmanuelle, LÉNEL Pierre, FRANCES Jean
The Gamification of Society
(Research, Innovative Theories and Methods in SSH Set – Volume 2)

RADI Bouchaïb, EL HAMI Abdelkhalak
Optimizations and Programming: Linear, Non-linear, Dynamic, Stochastic and Applications with Matlab
(Digital Science Set – Volume 1)

2020

BARNOUIN Jacques
The World's Construction Mechanism: Trajectories, Imbalances and the Future of Societies
(Interdisciplinarity between Biological Sciences and Social Sciences Set – Volume 4)

ÇAĞLAR Nur, CURULLI Irene G., SIPAHIOĞLU Işıl Ruhi, MAVROMATIDIS Lazaros
Thresholds in Architectural Education (Engineering, Energy and Architecture Set – Volume 7)

DUBOIS Michel J.F.
Humans in the Making: In the Beginning was Technique
(Social Interdisciplinarity Set – Volume 4)

ETCHEVERRIA Olivier
The Restaurant, A Geographical Approach: From Invention to Gourmet Tourist Destinations
(Tourism and Mobility Systems Set – Volume 3)

GREFE GWENAËLLE, PEYRAT-GUILLARD DOMINIQUE
Shapes of Tourism Employment: HRM in the Worlds of Hotels and Air Transport (Tourism and Mobility Systems Set – Volume 4)

JEANNERET Yves
The Trace Factory
(Traces Set – Volume 3)

KATSAFADOS Petros, MAVROMATIDIS Elias, SPYROU Christos
Numerical Weather Prediction and Data Assimilation (Engineering, Energy and Architecture Set – Volume 6)

KOLOKOTSA Denia, KAMPELIS Nikos
Smart Buildings, Smart Communities and Demand Response (Engineering, Energy and Architecture Set – Volume 8)

MARTI Caroline
Cultural Mediations of Brands: Unadvertization and Quest for Authority (Communication Approaches to Commercial Mediation Set – Volume 1)

MAVROMATIDIS Lazaros E.
Climatic Heterotopias as Spaces of Inclusion: Sew Up the Urban Fabric (Research in Architectural Education Set – Volume 1)

MOURATIDOU Eleni
Re-presentation Policies of the Fashion Industry: Discourse, Apparatus and Power (Communication Approaches to Commercial Mediation Set – Volume 2)

SCHMITT Daniel, THÉBAULT Marine, BURCZYKOWSKI Ludovic
Image Beyond the Screen: Projection Mapping

VIOLIER Philippe, with the collaboration of TAUNAY Benjamin
The Tourist Places of the World
(Tourism and Mobility Systems Set – Volume 2)

2019

BRIANÇON Muriel
The Meaning of Otherness in Education: Stakes, Forms, Process, Thoughts and Transfers
(Education Set – Volume 3)

DESCHAMPS Jacqueline
Mediation: A Concept for Information and Communication Sciences
(Concepts to Conceive 21st Century Society Set – Volume 1)

DOUSSET Laurent, PARK Sejin, GUILLE-ESCURET Georges
Kinship, Ecology and History: Renewal of Conjunctures
(Interdisciplinarity between Biological Sciences and Social Sciences Set – Volume 3)

DUPONT Olivier
Power
(Concepts to Conceive 21st Century Society Set – Volume 2)

FERRARATO Coline
Prospective Philosophy of Software: A Simondonian Study

GUAAYBESS Tourya
The Media in Arab Countries: From Development Theories to Cooperation Policies

HAGÈGE Hélène
Education for Responsibility
(Education Set – Volume 4)

LARDELLIER Pascal
The Ritual Institution of Society
(Traces Set – Volume 2)

LARROCHE Valérie
The Dispositif
(Concepts to Conceive 21st Century Society Set – Volume 3)

2018

Trestini Marc
Modeling of Next Generation Digital Learning Environments: Complex Systems Theory

2017

Anichini Giulia, Carraro Flavia, Geslin Philippe, Guille-Escuret Georges
Technicity vs Scientificity – Complementarities and Rivalries
(Interdisciplinarity between Biological Sciences and Social Sciences Set – Volume 2)

Dugué Bernard
Information and the World Stage – From Philosophy to Science, the World of Forms and Communications
(Engineering, Energy and Architecture Set – Volume 1)

Geslin Philippe
Inside Anthropotechnology – User and Culture Centered Experience
(Social Interdisciplinarity Set – Volume 1)

Goria Stéphane
Methods and Tools for Creative Competitive Intelligence

Kembellec Gérald, Broudous Evelyne
Reading and Writing Knowledge in Scientific Communities: Digital Humanities and Knowledge Construction

Maesschalck Marc
Reflexive Governance for Research and Innovative Knowledge
(Responsible Research and Innovation Set - Volume 6)

Park Sejin, Guille-Escuret Georges
Sociobiology vs Socioecology: Consequences of an Unraveling Debate
(Interdisciplinarity between Biological Sciences and Social Sciences Set – Volume 1)

Pellé Sophie
Business, Innovation and Responsibility
(Responsible Research and Innovation Set – Volume 7)

2016

Printed and bound by CPI Group (UK) Ltd, Croydon, CR0 4YY

27/10/2024

14580248-0004